Bettering Our Condition

American University Studies

Series V
Philosophy

Vol. 137

PETER LANG
New York • San Francisco • Bern • Baltimore
Frankfurt am Main • Berlin • Wien • Paris

Philip J. Chmielewski, S.J.

Bettering Our Condition

Work, Workers and Ethics in British and German Economic Thought

PETER LANG
New York • San Francisco • Bern • Baltimore
Frankfurt am Main • Berlin • Wien • Paris

Library of Congress Cataloging-in-Publication Data

Chmielewski, Philip J.
Bettering our condition : work, workers and ethics in British and German economic thought / Philip J. Chmielewski.
p. cm. — (American university studies. Series V, Philosophy ; vol. 137)
Includes bibliographical references and index.
1. Work. 2. Working class. 3. Work ethic. 4. Labor. 5. Labor economics. 6. Labor movement. 7. Machinery in industry. 8. Quality of work life. I. Title. II. Series. 92-13304
ISBN 0-8204-1851-X CIP
ISSN 0739-6392

Die Deutsche Bibliothek-CIP-Einheitsaufnahme

Chmielewski, Philip J.:
Bettering our condition : work, workers and ethics in British and German economic thought / Philip J. Chmielewski.—New York; Berlin; Bern; Frankfurt/M.; Paris; Wien; Lang, 1992
(American university studies : Ser. 5, Philosophy ; Vol. 137)
ISBN 0-8204-1851-X
NE: American university studies/05

The paper in this book meets the guidelines for permanence and durability of the Committee on Production Guidelines for Book Longevity of the Council on Library Resources.

Printed in the United States of America.

To my Mother and the memory of my Father

Acknowledgements

In the course of this project, many generous individuals have taught me much concerning sources, methods and learning. Louis Dupré, Margaret Farley, Branko Horvat, Timothy Jackson and Gene Outka, members of the faculty of Yale, examined and guided the early stages of this undertaking. Also at Yale, George Hogenson, David Montgomery, Maurice Natanson, Cornel West and Sidney Winter; Philip Land and Joseph Holland of the Center of Concern; Peter Kilby and William Barber of Wesleyan; Jaroslav Vanek of Cornell; Georg Elwert of Bielefeld; and Friedhelm Hengsbach and Philipp Schmitz of Sankt Georgen—each offered advice concerning the unfolding of the research and the development of the concept. Ian Shapiro of Yale provided both early guidance concerning the research and valuable, systematic criticism of the manuscript. At a subsequent stage, Franz H. Mueller of the University of St. Thomas and James Walter of Loyola University Chicago each extended encouragement and direction. I thank Mr. Dupré especially. His keen interest in this venture at its inception, sustaining encouragement, detailed suggestions, sure direction and enduring support contributed greatly to its structure and progress.

Among the fellow graduate students and many friends who, each with his or her own distinction, provided repeated and ready attention, questions, ideas, leads and care, I single out Christina Büchmann, Peter Casarella, Karen Kilby, James Murphy, Marisa Moolick-Gutierrez, R. Marisol Ravicz and William O'Neill. The reference staff of Sterling Memorial library continually offered full and shrewd assistance. My fellow Jesuits of the Midwest, especially Michael Brophy, Robert Harvanek, James McCann, George Traub and the members of Arrupe House freely both gave immediate and long-term support.

My family, especially my mother, provided sustenance in the long course of the writing and, indeed, much of the impetus for this project.

Grateful acknowledgement is made to those mentioned here, for their approval to make use of excerpts:

Reprinted by permission of the publisher from Emmanuel Levinas, *Totality and Infinity: An Essay in Exteriority*, translated by Alphonso Lingis, Duquesne Philosophical Series 24. Pittsburgh: Duquesne University Press, 1969.

Reprinted by permission of the publisher from Alfred Marshall, *Principles of Economics*, Vol. 1. Ninth Variorum Edition with annotations by C.W. Guillebaud. London: Macmillan, 1961.

Reprinted by permission of the publisher from Karl Marx, *Capital*, Vol. 1, translated by Samuel Moore and Edward Aveling, edited by Friedrick Engels. New York: International Publishers, 1967.

Reprinted by permission of the publisher from Oswald v. Nell-Breuning, *Wirtschaft und Gesellschaft Heute*, 3 Bände. Freiburg: Herder, 1956-1960.

Reprinted by permission of the publisher from David Ricardo. Volumes 1 (*Principles*) and 4 of *The Works and Correspondence of David Ricardo*, 11 vols., ed. by P. Sraffa with the collaboration of M.H. Dobb. Cambridge: Cambridge University Press, 1951.

Reprinted by permission of the publisher from Adam Smith, *An Inquiry into the Nature and Causes of the Wealth of Nations*, ed. by W. B. Todd, vols. 2:1 and 2:2 of *The Glasgow Edition of the Works and Correspondence of Adam Smith*, ed. by R.H. Campbell and A.S. Skinner. Oxford: The Clarendon Press, 1976.

Reprinted by permission of the publisher from Georg Wilhelm Friedrich Hegel, *Philosophy of Right*, trans. and ed. by T.M. Knox. Oxford: Oxford University Press, 1952.

Translations are my own, unless otherwise indicated.

Table of Contents

Introduction

> . . . The servant feels especially responsible for the sweeping; one can see how he, the moment the young man approaches him, tries to grasp the broom more firmly with his trembling hands; he even stands still and stops sweeping so as to direct his full attention to the ownership of the broom. The assistant [who wants to sweep] doesn't actually plead in words, for he is afraid of Blumfeld, who is ostensibly doing his accounts; moreover ordinary speech is useless, since the servant can be made to hear only by excessive shouting. (Franz Kafka, "Blumfeld, An Elderly Bachelor.")[1]

Kafka's scene presents a vivid, compendious and neutral view of the work world: neutral because it is set more than a century later than the onset of the industrial world, the start of the rise of the worker to prominence in society and the origin of economic science, and because it is situated at the turn of this century, far removed from the conflicts, problems and questions which arise in the contemporary world of work.

Contemporary discussions of the practical, socio-political and theological issues which surround work have prompted this research, which presents a normative theory in response to these current, socio-economic concerns. Pope John Paul II's encyclical letter *Laborem Exercens* (*On Human Work*) provided the impulse for some of the trajectories of the initial inquiry. The economic thought-world out of which the document emerged was, to be sure, related to those which I had studied but was, ultimately, distinct and unusual. It would be necessary to sketch the syntax and list the vocabulary of this theoretical position in continuity with those of other traditions of normative reflection on economics, if actual workers and managers were to be able to make use of it in articulating the difficulties and opportunities afforded by the situations and systems within which they labor. The present study in social ethics analyzes the writings of the major British and German forebears of contemporary economics and of two prominent figures in Roman Catholic social teaching who explored the relation of theology and economics. In an effort to discover a framework for discourse among the traditions, this book examines the grammar of the ideas presented in each writer. A conceptual anatomy of work, which is derived from continuities manifest in the literature itself, provides the foundation for the normative theory advanced in these pages.

The dependence of the sciences of their age on the model of the laws of nature as found in physics influenced the classical British economists, with the result that they assigned rationality primarily to nature. Consequently, in the broad, economic, natural processes which they depicted, no room remained for ethics insofar as the freedom of persons was constricted or even elided within systems of impersonal forces. On the Continent, the proponents of the German school integrated historical development with their understanding of economics. However, the tendency toward relativism and the proclivity to accept the current customs of the collective as ethical norms often vitiated the promised, comprehensive and systematic treatment of ethics and economics.

In order to avoid the impasse caused by both the British and German defects in economics and in order to make possible a fruitful exchange between ethical and economic norms, this presentation, in the examination of these earlier theories, concentrates on the human person who works.[2] In addition, this analysis proceeds on the basis of a broad understanding of the science of economics. Economics is not merely a technical, impersonal business enterprise; it is a sociocultural activity.[3] The basis for economic activity is human work. The understanding of work utilized here implies that workers are rational agents involved in a historically situated and socially specific system. This conception focuses on working persons along four lines: as they confront the limits impinging on them, as they collaborate in society, as they utilize technology and as they come to the service of each other.

In "Blumfeld, An Elderly Bachelor," F. Kafka presents a hierarchical scene in a linen factory. This vignette epitomizes the patterns and tensions of the work world presented in the story. Blumfeld directs a small department in charge of specialty articles fashioned at home as piecework by dozens of women. Beneath him, the two young assistants are the bane of his workday. One of these young employees threatens the elderly maintenance man, who wields his broom both as a tool and as an emblem. The concern for the limits to work capacity (expressed both in the ages of Blumfeld and the servant and in Blumfeld's worry about the amount of work his department is capable of completing), the intensely social life in the workplace (involving here, for example, relations of authority, camaraderie and competition), the figurative presentation of technology (in the broom, not only a device which can represent all efficient implements but also a device which betokens the worker's invested personal action and his position relative to other workers) and an array of service activities which surround and undergird the production which takes place in the plant—all four of these features characterize this world of work.[4]

Varying and developing emphases on these same four features of the activity of working persons mark the evolution of the understanding of work in the modern age. This study begins at the end of the eighteenth century not only because A. Smith (*An Inquiry into the Nature and Causes of the Wealth of Nations*, 1776) is the frequently acknowledged initiator and fountainhead of economics as a systematic discipline[5] but also because a significant shift took place at this time, one in which a re-conceptualization of work figured prominently. In the mercantilist literature of the preceding age, the predominant interest in the nation considered solely as a whole tended to seek the reduction of individuals' wages for the sake of creating a favorable balance of trade.[6] National economic planning took precedence over the attention to individual needs—even those of the individuals upon whom the economy itself depended. The system amassed national wealth; the same system effectively excluded the majority from partaking in the benefits of this wealth.[7] Another strand of mercantilist thought, however, took more interest in the condition of the workers and their wages as these contributed to economic development.[8]

The thinkers of the dawning era understood human work activity as the basis of the economic and social system devised for securing the social benefit not exclusively of the whole, as the earlier mercantilists had designed their system, but of its individual participants. With the rise of economics, the Physiocrats[9] and the early classical economists conceived work activity in terms of a eudaemonism and an egalitarianism. Workers were now included in society. Furthermore, as economic science became theoretical, principled and systematic, working persons figured prominently in the analysis of society. Whereas demand earlier comprised the requirements and caprices of the rich, now the satisfaction of natural and common drives constituted demand. Legal, social and economic changes began to free work from the rigid structures of social divisions and guild controls.[10] At this time of transition, the concern for providing the material requirements for the sustenance of persons—as they face ceaseless attacks by time and nature on their capacities—represented one dimension of the new concept of work. Whereas the mercantilist policy of protection engendered a "fear of goods" which required that a concern for the workers' needs ranked lower in priority than the aggrandizement of national wealth, with the end of the eighteenth century a "hunger for goods" developed. The "policy of provision" required that the workers of a nation be sustained and strengthened.[11] Theorists and politicians began to regard workers as society's prime source of value and wealth.

In the popular view, work is an activity in which one engages alone, to which machines are extrinsic, which benefits exclusively the worker and those whom he decides to include in this benefit. In contrast, the theory presented here maintains that, as a particularly human activity, work is intrinsically social. Whereas work became a significant concept for the study of modern society when people realized that their own free efforts could alter the shape of their common world, later, in the early part of the present century, work verged on disappearance as a significant, conceptual tool when people realized that their efforts were puny compared with the effective control of corporate agents. Yet work can still remain a critical category when it is clear that the concept incorporates responsible participation by each not only in the delimited workplace but also for the planning and patterning of the work world.

The dramatic shift which occurred during the late eighteenth century has usually been named the Industrial Revolution, after the extensive technical developments which either were introduced or became widely used. It is preferable to conceive this shift as one in which workers, through tools, machines and new organizations, began to reshape their own work activity. Persons develop machines for the purpose of altering the shape of their activity, an activity designed to contribute a service to other members of the community. Such an understanding of work may be found in the writings of major economic writers of the modern age.

With the shift of focus to the workers and their needs, the service component of work activity started to emerge. This concept appears in inchoate form even in the early writers whose writings I examine. How the sources of economic science treat work, in part, as a service to others and how the concept of service develops subsequently will be a key elements in the normative intent of this research. The service constituent represents a central development both in the conceptualizing of work reality and in its actuality—a development for which the concern of contemporary business for the provision of services, alongside the production of goods, offers a significant indication.

Service activity, which emerges both in theory and in the actual work world as a predominant element in work activity, supports a conceptual approach which aids in the re-establishment of work as a category central to social change and development. The service element of work entails both the freedom of personal efforts and the direction of these for the sake of one's fellows. Rather than the common conception of labor as servile, the concept of work activity which I shall present will show it to be personal and free activity.

The first three chapters will treat the early British and German economists. Chapter one will be devoted to the British writers Adam Smith and David Ricardo, chapter two to the Germans Georg Wilhelm Friedrich Hegel and Karl Marx. The third chapter will examine the work of Alfred Marshall, Gustav Schmoller and Heinrich Pesch at the end of the nineteenth century. Chapter four will examine the thought of Oswald v. Nell-Breuning concerning work, with particular attention to his arguments in support of *Mitbestimmung* (codetermination). The treatment of codetermination, an innovative, modern structure of participation in the workplace, will indicate how the theory of work activity can specify practical arrangements and must develop in response to social and structural developments. The concluding chapter will present a theory of work activity which benefits from the preceding analysis of the historical figures in economic thought and makes use of contemporary, philosophical reflection about work, especially that of Paul Schrecker and Emmanuel Levinas. This chapter will stress the ethical significance of work as recognitive and productive activity. In work persons actualize their freedom through the acknowledgement and configuring of their limits, through their participation in social institutions, through their exercise of liberating discovery and through their reflective, progressively more inclusive service of one another.

1. Franz Kafka, *The Complete Stories*, ed. by Nahum N. Glatzer (New York: Schocken Books, 1971), p. 204.

2. This anthropological approach does not fall into the double danger to which Francis Schüssler Fiorenza calls attention, namely of viewing working persons as free spirits or as merely wage workers. The four central characteristics of work activity attest not only to the free and spirited character of persons but also to their embodiment within society. Further, the theory developed here is also careful to examine work as it takes place apart from the institution of wage labor and attempts to discover how it might develop into social forms more advanced than the class-linked role of "worker." See "Work and Critical Theology," in *A Matter of Dignity: Inquiries into the Humanization of Work*, ed. by W.J. Heisler and John W. Houck (Notre Dame: University of Notre Dame Press, 1977), p. 40.

3. Louis Dupré, *Marx's Social Critique of Culture* (New Haven: Yale University Press, 1983), pp. 165-166.

4. Kafka knew whereof he wrote. As a lawyer employed in two insurance firms, he composed three professional pieces concerning workers' insurance: "On Mandatory Insurance in the Construction Industry"; "Measures to Prevent Accidents [in Factories and on Farms]; and "Workers' Accident Insurance and Management." Indeed, from 1908-1922 he held a post in the semi-governmental Workers' Accident Insurance Institute. See *Complete Stories*, p. 475-476.

5. Phyllis Deane, *The Evolution of Economic Ideas* (Cambridge: Cambridge University Press, 1978), pp. 4-7, 11-13.

6. E.S. Furniss, *The Position of the Laborer in a System of Nationalism: A Study in the Labor Theories of the Later English Mercantilists* (Boston: Houghton Mifflin Co., 1920), pp. 153-154.

7. Eli F. Heckscher, *Mercantilism*, rev. 2nd. ed., Vol. 2 (New York: Barnes and Noble, 1955), p. 166.

8. Richard C. Wiles, "The Theory of Wages in Later English Mercantilism," The Economic History Review Series 2, No. 21 (1968).

9. G. Myrdal, *The Political Element in the Development of Economic Theory* (New York: Simon and Schuster, 1954), pp. 29-30.

10. W. Conze with M. Riedel, "Arbeit" in *Geschichtliche Grundbegriffe: Historisches Lexikon zur politisch-sozialen Sprache in Deutschland* (Stuttgart: Ernst Klett Verlag, 1972), p. 171.

11. These terms are Heckscher's: for the "hunger for goods," see *Mercantilism*, Vol. 2, pp. 57, 103, 104, 108, and 110; for the "fear of goods," see Vol. 2, pp. 104, 108, 111, 114, and 118; and for the "policy of provision," see Vol. 2, pp. 80-111.

Chapter 1
The Classical Economists: Adam Smith and David Ricardo

The Free Development of Skills and Virtues

In the concluding decades of the eighteenth century, because of the development of new, non-agricultural sources of value and profit and after a long preparation in the studies and policy recommendations by earlier economic writers, the focus of economic analysis shifted decisively from the productivity of the land to the work effort of persons.[1] Smith, of course, figures centrally in this shift. While many interpretations of Smith's achievement give attention to the innovation achieved in his economic analysis, another important feature[2] of Smith's approach receives considerably less notice. There is a marked stress in Smith on compassion for those who suffer as a result of their work.[3] Smith realizes that the industrial systems of the large manufacturing plants and cities neglect individual workers, endanger their character and thereby corrode the bases of social cohesion.

Smith concerns himself with workers not only as they suffer from poor conditions or the narrowness of their tasks but also insofar as they weaken as a result of even the good treatment they receive in the economic system. For example, he writes:

> Workmen . . . when they are liberally paid by the piece, are very apt to over-work themselves, and to ruin their health and constitution in a few years. . . . Almost every class of artificers is subject to some particular infirmity occasioned by excessive application to their peculiar species of work. . . .[4]

Furthermore, whether in Bk. V, ch. 1 or here, Smith focuses less on the conditions of employment than on the worker's activity amidst these conditions.

His anthropology presents each person in his or her entirety as oriented toward productive action.[5] This natural human inclination toward action, indeed the desire for it, contributes to the welfare of society. The entirety of each person in his or her particular strengths is thrown into the action so that it might "produce" not commodities but those natural ends which secure the advance of persons. In addition, the progress receives social approbation.

In Smith's conception of the person, given the pivotal role assigned to activity, one particular human activity, work, carries a still more salient personal and social signficance. From a presupposition in favor of a minimal striving to maintain oneself in existence, there follows the desire for "bettering"; from this desire proceeds "personal industry."

> . . . The principle which prompts to save, is the desire of bettering our condition, a desire which, though generally calm and dispassionate, comes with us from the womb, and never leaves us till we go into the grave. . . . An augmentation of fortune is the means by which the greater part of men prosper and wish to better their condition.[6]

The industrious virtue which Smith praises, however, requires strenuous effort which exceeds the minimum and which corresponds but little to the grudging labor provided by actual workers whose reward rarely promised to meet even that minimum. Nevertheless, the desire to better one's condition figures decisively in Smith's conception of personal action.

Realizing the changes and demands visited by nature upon human limits, Smith sought a secure basis for the maintenance of human life. He favored that motivation which promises not a periodic impulse but a constant and lasting force over the entire life of the person. Although the desire to better one's condition includes a factor of dissatisfaction, more importantly, it includes a longing for "bettering," for improvement and for prosperity. In Smith's concept of the person, the desire for improvement encompasses the recognition of finitude. The desire for "bettering" does not entail a constant accumulation of material necessities. Given a choice, once past a certain degree of security, not all individuals would choose such an amassing as the sole means of improving their lives. For Smith, persons realize their freedom in their activity with others: either directly for their benefit or indirectly, that is, with some degree of attention paid to their interests. In cooperation with others, free activity is powerful.

Individual efforts toward self-improvement both benefit and advance society despite the obstacles government might interpose:

> The natural effort of every individual to better his own condition, when suffered to exert itself with freedom and security, is so powerful a principle, that it is alone, and without any assistance, not only capable of carrying on the society to wealth and prosperity, but of surmounting a hundred impertinent abstractions with which the folly of human laws too often incumbers its operations. . . .[7]

In calling this effort of improvement "the natural," Smith alludes to its inherence in the person. He also implies that the social framework must permit the exercise of freedom. In opposition to Quesnay's social construct, Smith declares the power of this individual, "bettering" activity to overcome the governmental, legislative contrivances which threaten to block the advancing commonweal.

> He [Quesnay] seems not to have considered that in the political body, the natural effort which every man is continually making to better his own condition, is a principle of preservation capable of preventing and correcting in many repects, the bad effects of a political economy. . . .[8]

Both the social quality inherent in work and the embeddedness of one type of social action—exchange—secure work a place within the larger, political matrix. Furthermore, this matrix need not be so extremely regimented as Quesnay proposed, so as somehow to arrange paradoxically for attaining liberty. Rather, the intrinsic, natural inclination of persons to improve themselves conduces to social benefit, especially provided the political framework maintains a field for the free play of individual liberty.

That Smith focuses his analysis on the workers indicates a continuity with his thinking in ethics concerning the critical function of individual free activity: the interior time located in and governed by the organic embodiment of persons constrains work; yet through work activity, the person may achieve happiness. Within the sphere of economics, free individuals may better themselves through their cooperation, indeed, within a society which itself moves forward.[9]

An examination of Smith's treatment of wage differentials reveals still other significant elements of his conception of the working person. In Bk. I, ch. 10, Smith examines five factors which account for these differentials.[10]

The first factor, the "agreeableness" of the employment, indicates that the sensitivity of working persons keys itself in part to the social interaction centered about honor and its opposite. Even in the tasks they undertake as individuals, the sensitivity of the workers attunes itself to the social context of these efforts. The second and third factors concern the temporal limits affecting a lifetime of work. The learning of skills indicates that a concern for their own transience determines the preparation and exercise of intelligence. Because of the constancy or inconstancy of employment, working persons must understand the probability of a work engagement over time in order to relate this probable activity to the necessity of maintaining themselves and their dependents.

Wages, furthermore, vary directly with regard to the trust invested in the worker. That workers accomplish their tasks most often outside the sphere of the choice and direction of the owner, manager or recipient of the work or service leads to the conclusion that trust bridges the gap between work's sociality and the exercise of individual liberty. The sociality of work activity emerges once again as an element in the activity of the workers—this time in conjunction with their own views of the future as they hope to earn the respect of their collaborators and fellow citizens through their work. Also immanent in work activity lie the optimistic conceptions which people hold concerning what they themselves may achieve by way of their individual actions. This is another element within Smith's theme of the person's free activity toward self-improvement.

Smith's attention to wages reflects his concern with the plight of the workers. It springs from an ethical center. Established upon the natural right to wages, Smith establishes his argument for higher wages on the promise of improved productivity and also on universal claims of equity.[11]

Yet it must be noted that Bk. I, ch. 10 elucidates not only wages but also wage differentials. This particular aspect of the analysis also indicates Smith's concern for the liberty of individuals. In the differentiation of wages, the employer must take account of the degrees of hardship as well as the ingenuity and other factors which depend on the preferences and choices of irreplaceable persons and which shape their work.[12] Smith confesses the difficulty of this commensurating, a difficulty which results from his own emphasis on the freedom and equality of persons. He does not attend so much to the differences which proceed from nature, because he wishes, rather, to emphasize the differences which proceed from free choice. He emphasizes the free activity of equal persons.[13]

Just as in Smith's discussion the differences in wages do not result solely from impersonal market forces but depend on the free and perspicuous choice of persons, so too Smith highlights the significance of the worker by his analysis of the development of persons which results from their work activity. The *Wealth of Nations* presents the skill and dexterity which are required in labor activity and asserts that work strengthens these skills and the person's virtues. The accomplishments of skill and dexterity are inseparable from and intimate to the person of the worker:

> The property which every man has in his own labour, as it is the original foundation of all other property, so it is the most sacred and inviolable. The patrimony of the poor man lies in the strength and dexterity of his hands. . . .[14]

The sacredness and inviolability of work in relation to the person indicate that the worker's skills and dexterity are not peripheral to him and are not commodities merely, but dimensions of the activity which inheres in him. Smith's criticism of the narrow employments created by the division of labor (Bk. V, ch. 1) may mislead the reader into believing that Smith criticizes all labor. He, of course, decries the results of a particular set of modern employments. He also believes, however, that work may foster personal growth. Farm labor—in connection with which drudgery is not unknown—leads to a development of a wider range of skills in the person than that found in manufacture.[15] The variety of materials upon which farmers work and the variation within some of the tools with which they work (for example, oxen) leads to the development not just of particular skills but of intelligence itself.[16] This flexibility in responding to a wider range of objects resembles, at the level of a high skill, which is one type of intelligence, the liberty which marks Smith's conception of the person.

Work efforts either develop or diminish not just skills but also personal virtues. F. Hutcheson, Smith's teacher, underlines the importance of the virtue "industry,"[17] which persons exercise and develop in the course of their work activity. Here again the social quality of work appears, in that these personal efforts affect and shape the individual at the same time as they produce value for and create structures in society. For example, these efforts create industry as a particular sphere of social production.[18]

In this light, we see that Smith is not concerned for the wages of the laborers as a stimulating incentive to which a response is necessarily conjoined. Rather, the equitable reward to labor represents an appeal to the free choice of a worker to develop his own virtues:

> A plentiful subsistence increases the bodily strength of the laborer, and the comfortable hope of bettering his conditions and of ending his

> days perhaps in ease and plenty, animates him to exert his strength to the utmost.[19]

Thus, higher wages encourage industry but do not cause it or compel it. Upon the basis of this industry, which Smith regards precisely as a human quality, economic processes are founded. In this passage, Smith refers first to the embodiment of the worker whose strength is requisite for any industry whatever, and, secondly, to the vision of a future over the range of which the workers hope to better themselves. He thereby emphasizes the immersion of workers in a time measured by their histories.

Some writers criticize Smith for a supposed separation of work from the person,[20] but Smith's theoretical foundations for work he had already presented in his *Theory of Moral Sentiments*. He repeated this approach—in terms of the desire for improvement—in the *Wealth of Nations*. Smith bases the prudence[21] of workers upon their self-interest, which includes an attention to the needs of others.[22] The activity which proceeds from this broadly conceived self-interest does not depend on the consequences of mechanistic processes but, rather, on a forward-looking optimism. This activity contributes to the social whole. The worker is not merely an exchangeable element in some aggregate but an integral member of a community.[23] Through attention toward and work for the needs of others, a person can both benefit society and earn the esteem of others.[24] Work is integral to the person, as is the person to society.

The person who works thereby engages in free activity for the attainment of personal and social betterment. One field in which workers may improve themselves and their society is the market, whose equilibrating processes depend upon perfect competition. The prerequisite for the latter—apart from the institutional requirements—is the liberty of individuals.[25] Work activity includes some sacrifice of liberty. The disutility of work reflects the experience of persons, that work's trouble is costly to them:

> The real price of every thing, what every thing really costs to the man who wants to acquire it, is the toil and trouble of acquiring it. What every thing is really worth to the man who has acquired it, and who wants to dispose of it or exchange it for something else, is the toil and trouble which it can save to himself, and which it can impose upon other people.[26]

The disutility of work is the costliness of effort and the difficulty endured in the course of work activity. This toil and trouble must not be understood simply in terms of subjective pain or weariness. The fatigues of labor entail the laying down of a portion of one's ease, but the disutility includes the more significant sacrifice of a portion of one's liberty.[27] This basis in the liberty of the person is central to Smith's pursuit of an invariant measure.

In his pivotal discussion of the measure which labor provides, Smith emphasizes the role of liberty.[28] At all times, the value of equal quantities of work are the same because of the disutility concomitant to such effort. That is, such effort costs the individual. Smith reckons that this disutility, beyond the obvious sacrifice of ease and happiness in labor, includes the sacrifice of a certain degree of liberty. Work, then, constitutes a free activity, the enjoyment of which constrains one to particular tasks and which, in turn, bring one against the limits of health, strength, skill and spirit, thereby constraining the person. The market estimate of the value of labor may differ from that of the worker. Nonetheless Smith argues for the equivalence of equal quantities of labor, not solely in an intrapersonal comparison but also interpersonally, antecedent to the market. His postulates make this possible: action takes place through individuals who are equal in their liberty.

Personal, free choices make possible within society the free movements of the market. Virtue maintains its role in Smith's system as the modality through which free action may both shape and be shaped by society. For equal persons within a social network, choosing and self-expression actualize their freedom;[29] the person's work constitutes the pre-eminent instance of this free activity.

The Value of the Constrained Efforts to Meet Needs

In viewing economic processes which are centered on productive labor and which are thereby subject to the variations induced by the limits of this activity and by the threats against it, Smith searches for an invariant measure of value. Opposed to mercantilist presuppositions, Smith uses a labor measure of value, "to bring into relief changes in the value of specie, and this in turn helps to weaken the claims of money to the homage formerly paid it."[30] A system which claims invariance as a characteristic of its measure may then assert that measure as the dominating good of the system. Smith derives the measure of invariance by focusing on the working person.

The embodied person, subject to time and pressed by external necessity, must work in order to satisfy the needs requisite for life. The necessities required to maintain life at or above its minimum determine the degree of this effort. Smith describes the worth of the person's labor: "Its real price may be said to consist in the quantity of the necessaries and conveniencies of life which are given for it. . . ."[31] Yet Smith conceives this maintenance of life not solely in terms of the individual:

> A man must always live by his work, and his wages must at least be sufficient to maintain him. They must even upon most occasions be somewhat more; otherwise it would be impossible for him to bring up a family, and the race of workmen could not last beyond the first generation.[32]

Labor sustains life not only for the workers but also for those who share their lives.

Smith describes some larger dimensions of the social need which work meets. He cites farmers and workmen as groups which require both the provisions for present needs and "advances" to maintain and sustain them.[33] Their obvious requirements for living now as well as their orientation toward further living, together constitute their neediness. Smith expands the perspective yet again: "The most decisive mark of the prosperity of any country is the increase of the number of its inhabitants."[34] Beyond the nurture of individuals, their families and groups of workers, work must support—in terms of its results in 'prosperity'—the lives of persons as expressed in the size of the population. Work fosters the whole of life in its social reality.

Still another limit confronts the workers. Those elements in society which have administration of stocks and thereby stand insulated from some of the threats upon the maintenance of life may turn their own relative security into a threat to the work activity of laborers. In discussing the dispute between masters and workers with regard to the reward of labor, Smith writes:

> Many workmen could not subsist a week, few could subsist a month, and scarce any a year without employment. In the long-run the workman may be necessary to his master as his master is to him; but the necessity is not so immediate.[35]

This indicates one reason for Smith's opposition to corporations: they threaten workers. Those social forces which control the greatest ability to last because of their control of stocks and because of their form, which extends beyond the limits of an individual's power and time, can deploy themselves to great advantage in the market when seeking access to productive activity, that is, work. The advantage of these social forces evolves to the disadvantage of the workers. Smith presents work as contending with and constrained by finitude. His opposition to corporate forces derives in part from the exemption of these agencies from the usual circumstances of persons; this exemption then results in a constraining of the freedom of persons.

Human work activity struggles against the finitude of the human condition. For Smith, personal activity must combat nature, which everywhere presses in upon persons. Indeed, the consumption needs of the person may be regarded as the hunger of nature which demands satiation lest it consume the person. Smith emphasizes human activity, which can control the mechanism of nature to the advantage of persons, indeed must do so, so that the same encompassing and mechanical nature does not overcome the free activity of frail human persons.

The treatment of risk-aversion in career choice sheds even more light on the conflict between work and nature. The bricklayers and masons represent occupations whose engagement in their proper tasks the vagaries of the weather strictly govern—that is, the workers must submit to nature's power. Beyond this irregular, transient constraint in particular periods, young workers consider wages over lifelong careers. However, some of these careers—for example, life at sea—must submit to nature in its rawest power. In facing such natural risks, the confidence of each in his or her own good fortune and abilities[36] finds particular strength among the young. This difference in risk-aversion and risk-attraction results from the perception of different ratios between external risks and those chances which are dependent upon personal capacities.[37] Those who are least acquainted with the despotism of nature—especially within their own bodies—are most apt to assert themselves in the struggle with it. Indeed, this confidence in their own capacities provides a particularly illuminating instance of the human propensity to engage in free activity against nature's limiting force.

Upon the field of natural limits, human effort produces value:

> The annual produce of the land and labor of any nation can be increased in its value by no other means, but by increasing either the number or its productive labourers, or the productive powers of those labourers who had before been employed.[38]

According to this view of Smith, for the purpose of increasing production, either the productive powers must be developed (this matter of technology will be handled below) or the population involved in production must be increased. Whereas the Physiocrats had seen the land as bountiful, the classical economists shifted the focus of productivity to the activity of laborers. Any expansion of production was finally dependent upon the working person. Furthermore, in contrast to the goal of earlier mercantilist thought and in an extension of a significant strand among the later mercantilists, the classical economists desired the increase of the population not to keep wages down but to raise productivity for the whole of society, to which they considered workers integral.

Labor produces value and also measures it:

> Labour, therefore, it appears evidently, is the only universal, as well as the only accurate measure of value, or the only standard by which we can compare the values of different commodities at all times and at all places.[39]

The measurement central to the project of economic analysis is, for Smith, anthropocentric. In a world suffused with change, Smith sought a stable measure within the human activity of work. Smith intends to establish this measure on the basis of personal equality, yet the means selected for the apprehension of this equality is the "toil and trouble" of effort. In understanding the basic equality in this manner, Smith betrays his own project in two ways: first, that such "toil and trouble" is equal among persons remains uncertain; second, the prominence given to this disutility betrays the significance of the other dimensions of work activity. For example, the calculation of individual costs in pain elides the social dimension implicit in the activity.

At least since the time of John Locke,[40] acquisition from nature is the model for production. In a world of propertied and legal relations, that which can be newly acquired must be wrested not from others but from nature. A shift takes place, however, in Smith's own conception of productivity; the emphasis shifts from human activity in the face of nature to the new object produced.

At the same time, Smith manifests a double understanding of the product. In his earlier conception the product is the result of an activity—productive labor—which draws income to itself in order to maintain and develop the physical and social life from which it emerges. In the second interpretation, the set of objects produced are material and durable but disjoined from the activities of those who produced them.[41] Influenced both by the tradition of British empiricism and in his Physiocratic heritage, Smith placed emphasis on the material means requisite for social existence. This material orientation concentrated Smith's analysis on the production of independent objects, thus leading to a shift away from the working persons toward a focus on the exchange of objects in the market. How Smith's theory focuses on objects, one sees both in the earlier conception, the "labor-embodied" approach, and in the second interpretation, the "labor-commanded" theory.

Both these approaches concern one particular basis of value in the Smithian framework: human work produces value. Smith distinguishes two types of value produced:

> The word value, it is to be observed, has two different meanings, and sometimes expresses the utility of some particular object, and sometimes the power of purchasing other goods which the possession of that object conveys. The one may be called "value in use," the other, "value in exchange."[42]

One must not equate this "value in use" with the perceived utility of a commodity at the level of exchange.[43] Smith's construction disjoins the labor which stands behind the products offered from the relations of the market.

The theories of Smith and David Ricardo give prominence to working persons. Their attempts to formulate theories of real value proceed from their intentions to uphold the significance of persons within the economic processes. Yet Smith offers two theories for the creation of value which do not cohere with this, his own intention. First, in the "labor-embodied" theory, Smith attempts to mediate between the activity of individual working persons and the social exchange process:

> But the labour of the manufacturer fixes and realizes itself in some particular subject or vendible commodity, which lasts for some time at least after that labour is past. It is, as it were, a certain quantity of labour stocked and stored up to be employed, if necessary, upon some other occasion. That subject, or what is the same thing, the price of that subject, can afterwards, if necessary, put into motion a quantity of labour equal to that which originally produced it.[44]

Smith undertakes to locate the value in the durability of the product. The durable product circumvents transience; thus needs may be met at a later time. The capacity to put into motion another quantity of the activity of labor at a later date provides the sole explanation offered for the fixedness or realization of labor activity in the object. This activity, however, re-emerges only at one remove, behind the price of the product. The use-value of the product would meet the biological and social needs of working persons; Smith's understanding of durability means that a product lasts so that it can be exchanged. At the level of market exchange, labor has taken the form of object; no longer does it remain personal activity.

In addition to the "labor-embodied" approach, Smith also defines value in terms of its "labor-commanded" quality:

> The value of any commodity, therefore, to the person who possesses it, and who means not to use or consume it himself, but to exchange it for other commodities, is equal to the quantity of labor which it enables him to purchase or command.[45]

The phrase "to the person" highlights an important aspect of value. The concern here lies not with value in use; however, the citation does not describe value in exchange strictly speaking but, rather, the value of the work obtained through exchange as a person may determine that value subsequent to the exchange and prior to another.[46]

The labor-commanded model involves power over the labor of others. In the use of the labor-commanded theory as a basis for guiding the growth of social welfare, human reproduction takes on a mechanical quality because the interests and choices of the majority of those affected are left out of account. Although the labor-commanded theory requires that somehow products "embody labor" and requires that this be done in such a fashion that the processes of exchange never diminish this embodied amount of value, nevertheless the personal character of work activity, upon which perduring value depends, remains peripheral to the theory. The processes of exchange take the wishes of individuals into account simply as those of producers and consumers, and not as the wishes of persons qua persons. Within the labor-commanded theory, the synchronization of production and consumption results from a mechanical equilibrating achieved at the cost of human activity and of the existence of a number of individuals. In such a system, only adventitiously does employment become the locus for the exercise of free human activity. Such activity is possible only for those who happen to possess commodities or to have power over others.

As I have related, Smith puts forth two flawed approaches toward the understanding of the production of value by labor. Smith's theory, although capable of attaining a balance by way of comprehending the centrality of free, human activity, remains an unstable amalgam.[47] Personal activity ranks second either to material goods in the market or to the power of some individuals in society. Desirous of establishing the science of economics in analogy with the laws of nature, he reduced his agents to functional members of a system modelled on such laws. Yet this makes him unable to account for the freedom within work, although this liberty is integral to his philosophy and economics.

The Market Process in a Society of Equals

Equality is a primary characteristic of the social context in which persons expend their efforts. Into this context persons enter as equals. The differences which do exist in persons result from the complex organization of their work activity.[48] For the sake of studying the social mechanism, Adam Smith concentrates his attention on these equal individuals. Unlike earlier writers on economics, who examined the processes of society as a whole, Smith shifts the focus of his attention to the individual actors while continuing to examine the social relation in which these individuals work.

Smith firmly rejected the mercantilist position which held that the masses of laborers were inferior to the members of certain privileged groups. Favoring the equality of working persons, Smith's position also opposed more modern policies which would force these workers into an inferiority of status and power vis-à-vis the emerging capitalist employers.[49] Smith heeds the plight of the laborers:

> No society can surely be flourishing and happy, of which the far greater part of the members are poor and miserable. It is but equity, besides, that they who feed, cloath, and lodge the whole body of the people, should have such a share of the produce of their own labour as to be themselves tolerably well fed, cloathed [sic] and lodged.[50]

These persons "of the lower ranks" are regarded not only as numerical members of society but also as integral members. In fact, their welfare provides a critical determinant of the state of society. Moreover, in conjunction with the production of goods for and the provision of services to these people, Smith appeals to "equity" to justify society's maintenance of them.[51] They who provide necessities should be secured necessities. All members of society ought to have their needs met. An equitable reward to these workers, previously excluded from or held inferior in society, benefits the whole.[52] The sustenance of natural life for all secures a sphere of free action on the part of each of these equal members of society[53] and also benefits society as a whole.

The primary social interaction which Smith as economist examines is, of course, the market. The basis for the market lies in each person. Smith writes of:

> . . . A certain propensity of human nature . . . the propensity to truck, barter, and exchange one thing for another. . . .

> Whether this propensity be one of those original principles in human nature . . . or whether, as seems more probable, it be the necessary consequence of the faculties of reason and speech, it belongs not to our present subject to enquire.[54]

Smith presents this propensity as quintessentially human (possibly based on "human nature" or on "reason" and "speech"). In still other passages, he indicates that the embodied limits on the human capacity to consume lead to circumstances where exchange appears rational.[55] Human nature, systems of communication, and the constraints and possibilities of embodiment create a special, complex social situation antecedent to the market. The limits of the human body as they are formed within a context both of product surplus and of the social recognition of mutual benefit are the foundation for the particularly human interaction of exchange.[56]

The equal participants in society follow their natural inclination and engage in the exchange of objects which they have produced. Those objects themselves represent the social structure. That is, persons in their work activity do not create objects which then subsequently become social; rather, the object produced is social because it is an object for further production and use within a particular social context. This object is thus already social because the pursuit of self-interest on the part of the worker guides his efforts toward the service of the whole. Further, the object is directly social because the worker takes up a work-role which the social division of labor has assigned.

As had the Physiocrats, Smith recognized the interaction of persons, groups and sectors both on the basis of their work activity as well as on the basis of the exchange of the products which result from this activity. But now Smith focuses on economic growth and capital accumulation. The socially integrated process of work benefits the entire society.[57] Through the integration of the division of labor and exchange, the society as a whole is progressively improved through the increase of goods, through the incorporation of more persons (with their free activity) into the system and through the discovery of new knowledge. These processes weave the very fabric of society more firmly as persons benefit one another through the increase in productivity.[58]

Smith recognized that his presentation of economic realities covered only a narrow range of human social activities and that the framework of these economic activities requires an antecedent social bond. Smith held that sympathy cemented the bonds of the members of society among one another.[59] Upon this basis, justice could be enforced among the members.[60] The proper functioning of the processes of production and exchange so that they benefit the whole depends upon the existence of these two pre-economic forces, sympathy and justice. Within the functioning system, free individuals, in their decisions concerning how they will act, find their consciences guided and shaped by social institutions and by their particular experiences in society.[61] Still further, "It is . . . in social life and in the creation of good customs that the individual, even when he does not fully intend this, finds his own 'propriety' and happiness."[62] The person works not only to produce goods, not only to produce the system of labor organization and product exchange but also to produce a social fabric of virtue and custom which makes other productive activity and organization possible.

Granted that Smith fought the subordination of the individual to the state which was found in mercantilist doctrine, this does not necessarily imply that his laissez faire position foresaw the subordination of the state to its individual members—to individuals freely exchanging in a market. Rather, both the state and individuals were in the service of the community.[63] The work activity of producers forms the basis for this shared service of the community which occurs in those areas of life together that are free and economic. In the service of the community, their activity thus entails a reciprocal benefit. The natural liberty of each person, of each worker, lays claim to justice.[64] The complex social relation of the division of labor and market exchange, which is guided by government,[65] secures this justice as the framework for free activity. Smith pursues the creation of lasting structures so that, through the processes of production and exchange, the community of persons may secure the social well-being.[66]

Machines and Work Activity

Smith grasps the distinct benefit of machines to production:

> In manufacturing the same number of hands, assisted with the best machinery, will work up a much greater quantity of goods than with the more imperfect instruments of trade.[67]

Yet technology provides not only a benefit to production processes but also, and more importantly, an advantage for the good of the whole society. The division of labor represents Smith's perception of technology at the service of society. The division of labor depends on the social process of exchange:

> As it is the power of exchanging that gives occasion to the division of labour, so the extent of this division must always be limited by the

> extent of that power, or, in other words, by the extent of the market.[68]

In this view, the division of labor itself makes up a machine which the power of exchange impels. Smith's broad view of the division of labor as a composite tool accounts for the frequent yet unexamined shifting in Smith's use of the difference between a social and an industrial (plant) arrangement of subordinated and production efforts.[69]

Smith stood within a tradition that saw the social activities of persons as integrated on the model of a machine.[70] For J. Millar, A. Ferguson and W. Robertson (all of the Scottish Historical School) the free actions of persons were the basic components of the social machine.[71] Whereas Smith's teacher F. Hutcheson had lectured and written on the advantages of the division of labor in a section of his *System of Moral Philosophy* entitled, "The Necessity of the Social Life,"[72] Smith's view progressed beyond this in that he linked the advantages of the division of labor to the technology of contemporary industrial organization.[73]

This system of technology stood at the service of society. The improvement of labor's productive powers brings about, by the increase of commodities in order to meet human needs and desires, an enlarged sphere for free activity. Smith presents a significant element of this free activity when he indicates that the technological system fosters personal capacities ("skill, dexterity, and judgment").[74]

Smith adduces three reasons for the increased efficiency resulting from the division of labor.[75] He notes once again that the system causes the development of the worker's dexterity. The division alters two additional dimensions of the person's experience. First, time is saved; that is, the machine constituted by the division enables persons to economize their losses to transience and thereby to enlarge the sphere of their free activity. Secondly, the technological system spurs invention; that is, the division of labor stimulates this particular exercise of intelligence. The advance of personal intelligence and discovery benefits social life because the invention of machines contributes to the development of the arts and sciences, which ennoble human life.[76]

Although Smith explores the benefits attributable to the division of labor, he also examines its debilitating effects:

> The man whose whole life is spent in performing a few simple operations of which the effects too are, perhaps, always the same, or very nearly the same, has no occasion to exert his understanding, or to exercise his invention in finding out expedients for removing

> difficulties which never occur. . . . His dexterity at his own particular trade seems, in this manner, to be acquired at the expense of his intellectual, social and martial virtues.[77]

When persons are consigned to activity as subassemblies in the larger machine, this machine, constituted by the division of labor, blocks the development of their skills, intelligence and judgment. Furthermore, the division of labor, which functions in a way thus harmful to persons, also undermines the development both of their virtue and of their action in relation to society.[78] It is the obligation of government—at the service of the larger social processes—to prevent the technology of the division of labor from rendering persons "elements" of a composite mechanism.

Sustenance as Service and the Neglect of Unproductive Work

Up until now the focus of the service component of work as found in Smith has been the understanding that production hinges on the efforts of workers, who provide sustenance for all persons in society. Such service does not depend on good will alone:

> . . . Man has almost constant occasion for the help of his brethren, and it is in vain for him to expect it from their benevolence only. He will be more likely to prevail if he can interest their self-love in his favour, and shew them that it is for their own advantage to do for him what he requires of them.[79]

Because the neediness of persons is continual, Smith requires a more constant motivation than benevolence. Self-love is selected as this principle and then situated in a system of production and exchange so that human needs might be met with greater certainty.

Smith's move to a humanitarian treatment of workers represents an initial step toward a still more complete comprehension of what the service dimension of work involves. Smith's project of protecting individual liberty required the dissolution of particular social bonds (servants in the employ of a master) structured around one type of service. The dependence on the forces of exchange then supplanted the subordination to particular masters; the forces of the market promised to form a field for independent activity upon which persons could foster their "intellectual, social, and martial" virtues. The search for liberty within the interaction of the market replaced the dependence on particular persons. At the same time as the actual, non-market service relationships were being abolished, this shift resulted in the concomitant neglect of the concept of service activity.

Smith's treatment of the distinction between productive and unproductive labor leads to a still clearer explanation for his weak handling of the service component of work activity:

> There is one sort of labour which adds to the value of the subject upon which it is bestowed; there is another which has no such effect. The former, as it produces a value, may be called productive; the latter, unproductive labour.[80]

Unproductive labor—which encompasses many different service activities—Smith holds to be valueless. On the contrary, farmers, manufacturers, merchants and retailers he regards as productive workers.[81]

Analogous to his double analysis of the creation of value (which I have discussed in an earlier section), Smith understood labor as productive in two ways: labor is productive when it replaces its own expenses and also yields a surplus; alternatively, labor is productive when it creates an object fit for purchase.[82] Both these conceptions indicate that Smith has weakened, if not eliminated, the service component of production. In the first version, although it manifests an orientation toward the future through the creation of a surplus, the durable, nonconsumed character of the commodities which embody labor is the more pronounced characteristic. The objects come to matter apart from the human needs they might meet. In the second version, the focus of reproduction once again lies upon the objects; the predominance of their reproduction conceptually subordinates the work of persons. In both cases, the activity of production strives to augment exchange value, not to secure the provision for necessary consumption. Smith equates useful labor with work for production.[83] He does not conceive how work whose usefulness lies in service might be included within the category of "productive."

In Bk. 2, ch. 3, Smith distinguishes between the labor of a manufacturer which adds to value and is, therefore, productive, and that of a menial servant who does not add to value by his "unproductive" efforts.[84] Yet this distinction holds true only in particular instances. Indeed, an innkeeper may provide much value to the society and in the economy by his provision of "productive services."[85] Contrary to the assumptions, activities which are nonenduring may yield value. These activities provide those goods which result from the activity of service.

The following citation reveals the core of Smith's failure to conceptualize the work of service within his analysis:

> The labour of some of the most respectable orders in the society is, like that of menial servants, unproductive of any value, and does not fix or realize itself in any permanent subject, or vendible commodity, which endures after that labour is past, and for which an equal quantity of labour could afterwards be procured.[86]

He once again betrays his materialist bias toward the object and away from the activity: the activity of service produces no commodity which can be grasped, the action of service work cannot be made lasting, cannot be held, and is therefore overlooked. Although he clearly understands the multiple tasks requisite for the maintenance of the social fabric, Smith ignores how the "respectability" of some of these orders—for example, officials of government—necessarily devolves from a recognition that their tasks are valuable precisely because they orient the economic processes so that the latter achieve the common welfare.

Smith valorizes the act of work only in terms of a particular category of effects:

> The protection, security, and defence of the commonwealth, the effect of their labour this year, will not purchase its protection, security, and defence, for the year to come. . . . Like the declamation of the actor, the harangue of the orator, or the tune of the musician, the work of all of them perishes in the very instant of its production.[87]

Performance both in the arts and the military provides services to the community. Smith, however, counts only those effects which are physical, in the sense of perduring apart from persons. The actions leading to the defense of the commonwealth this year or to an artistic performance all generate effects. Indeed, the results—say, the protection of a particular coastline or the impact on a particular spectator—can last; in this sense they are durable. Nevertheless, because the effects occur within persons (those who avoid attacking a particular coastline or the playgoers whose views have been altered), Smith does not calculate the effects as lasting and, consequently, does not regard these service activities as productive.

Smith does not adequately grasp the impermanent and dynamic character of service activity. In his view, work "produces" if society can utilize the object of this activity. The works of service offered in the two preceding citations constitute activities whose transience appears to entail that the object disappears too quickly for possible use; yet society "uses" the service activities in and through their transience. Whereas society can use the lasting objects produced to sustain the passing life of the commonwealth, so too it can make use of the varying service actions provided in order to maintain the lasting structure of society. Faced with human limits, Smith sought the perduring; consequently, he regarded the production of objects as central to confronting these limits.

In focusing on these objects, however, he overlooked the significance of service activities. Smith failed to realize that, through their alteration, the structures of society abide. Exactly the suppleness of transient activities serves these structures as they unfold. Smith, the liberal, fought the stasis of ossified governmental and commercial tradition; however, in so doing, he ignored the ongoing and varied service activities. The services promote the maintenance of life and the development of structures within a social tradition; indeed, service activities constitute the mainstay of such an ongoing tradition.[88]

Drawing upon earlier Continental and domestic reflection on economics and productive effort, Adam Smith synthesized major developments both in the science and in the conception of work activity. Owing to Smith, work became a key term in economics, both as a study and as a world view. The emphasis on work embodied an egalitarian dimension; on the basis of an order directed toward performance, no room remained for the non-achieving, privileged estates. Furthermore, Smith viewed work not as a sum of individual contributions but, rather, as an activity on the part of individuals who were already socialized.[89]

Ricardo's Concern for the Working Person

Smith stiffly criticizes mercantilist structures; Ricardo aims at a different target:[90] the failure to provide sufficient and adequate employment for work activity. This omission results in poverty for persons which may be measured not solely in material losses but also in the loss of other, proper concomitants of work. The exercise of power, the pursuit of objects beyond subsistence, the development of intellectual distinction and the broadening of the areas upon which these intellectual capacities may be deployed, all characterize work. The Poor Laws block these natural attainments which result from work activity. The impoverishment which results is both personal and social, individual and universal.

Labor is slow to respond to changes in wages. This stickiness, that is, a delay and inadequacy of response, indicates one basis for the unique position of the working person in Ricardo's thought.

> . . . Labour is a commodity which cannot be increased and diminished at pleasure. . . . You cannot increase their number [that is, of persons] in one or two years when there is an increase of capital, nor can you rapidly diminish their number when capital is in a retrograde state; and, therefore, . . . there must be a considerable interval before the price of labour is exactly regulated by the price of corn and necessaries. . . .[91]

Labor differentiates itself from other commodities. Although an agent endowed with rational and free choice provides this particular commodity, nonetheless labor resists rapid response to significant change even if it be in those funds provided for its own maintenance. The increase of human life exceeds rational planning; the quick decrease of the number of persons is prohibited. Human life—in the midst of Ricardo's deliberately dry analysis—remains the focus of his interest. Because this life, the basis of work activity, belongs to persons whose lives are necessarily colored by threats and transience, their intentions and desires deserve care and attention. Were Ricardo's concern here solely physiological—that is, merely the number of years before a child matures to a given work capacity—he would not have concerned himself with the disparity which results from the different rates of change. Within this gap stands the fragile condition of the laborers. His examination of this interval reveals Ricardo's concern for the condition of working persons as they are affected by macroeconomic processes.

Chapter five of the *Principles* explores some of the characteristics of work activity more explicitly.[92] The Poor Laws ruin those human characteristics needed in work. The remedy lies in this: to reemphasize that independence, exertion, careful judgment and evaluation provide the means for the advancement of individuals and society. Ricardo considers these traits the central features of the work activity of the person.

Persons, freely expending efforts for the benefit of themselves and their society, are the focus of Ricardo's system. He offers a *labor* theory of value to set neither capital nor land but, rather, persons as the linchpin of his theory. The role of labor in the process of production is both unique and fundamental. One could also say, as did the Physiocrats, that the role of fertile land is unique, though in a different sense. That human labor is fundamental, however, is the central point for Ricardo.[93] Human labor is fundamental because the system operates both on the basis of the efforts of working persons and on their behalf. While Ricardo did not argue for the assigning of the entirety of net product to labor,[94] he is nonetheless committed to the centrality of labor in his normative evaluation of economic processes.[95] Ricardo himself recognized the inadequacies of his labor theory as a measure of value. Despite this recognition, until the end of his life he continued the pursuit of an adequate articulation of the absolute value located in the work activity of persons. The arguments introduced in the subsequent section on value will highlight exactly this: Ricardo was committed to a labor theory as the basis for his normative prescriptions because of his interest in working persons and attention to their needs.[96]

As did Smith, so Ricardo also examines the relationship between different types of employment. His analysis here also indicates his conception of the working person:

> In speaking, however, of labour, as being the foundation of all value, and the relative quantity of labour as almost exclusively determining the relative value of commodities, I must not be supposed to be inattentive to the different qualities of labour, and the difficulty of comparing an hour's or a day's labour, in one employment, with the same duration of labour in another. The estimation in which different qualities of labour are held, comes soon to be adjusted in the market with sufficient precision for all practical purposes, and depends much on the comparative skill of the labourer, and intensity of the labour performed. The scale, when once formed, is liable to little variation.[97]

That the market leads to adjustments in the estimation does not imply that the market determines the variations in the quality of work. This estimation depends upon two parameters: the skills of the worker and the intensity of effort required. Skill is not a type of incorporated tool which a particular individual has acquired in the market at the cost of an apprenticeship, for this skill would then be subordinate to the work activity and thus capable of accounting for its productivity but not for a difference in labor. In order to account for the differing quality of labor resulting from variations in skill, the acquisition of a particular degree of skill must be regarded as a development instrinsic to the person.

The assumption that wage differentials remain constant is puzzling because Ricardo concerns himself, after all, with changes in the structure of relative prices.[98] Why are these prices exempted? For the sake of analytic simplicity? To hold stable this element in particular would appear to be problematic because these differences in the quality of work lie so near the core of his system.

"The scale once formed, is liable to little variation." Ricardo's actual formulation suggests that some time elapses before the ratio of qualities has been established. This would allow for the introduction of new employments.

Furthermore, once the scale has been established, little variation ensues because work activity as the critical factor remains stable. In other words, once society, through the market evaluation, comes to the comparison of this particular set of skills with others, then the scale fluctuates but little because of the unchanging quality of human work activity.

Work activity articulated in specific and personal skills provides the focus of Ricardo's analysis. Ricardo portrays labor not as an abstract process but as a result of our physical and moral capacities. Ricardo cites Destutt de Tracy with approval concerning the measure of value, even though he disagrees with him concerning the definitions of value, riches and utility:

> . . . As it is certain that our physical and moral faculties are alone our original riches, the employment of those faculties, labour of some kind, is our only original treasure and that it is always from this employment, that all those things are created which we call riches. . . .[99]

His references to the skill and industry involved in work[100] imply that such capacities are capable of development. The failure to develop such capacities brings with it deleterious social consequences, which can nevertheless be remedied. Better governance and the education of persons provide solutions which enable them to regain and enhance their capacities to work.[101] Knowledge, discipline and civilization characterize the fullness of these work capacities.

Such little treatment as Ricardo does provide concerning the skills and virtues of workers appears so much the thinner when one recalls Smith's handling of these. When Ricardo does talk of improvement, it stands remarkably distant from a concern for developing skills and virtues.[102] The inviolate person, the origin of work activity, appears in Ricardo's theory as a provider of labor. While Ricardo emphasizes and re-emphasizes the centrality of personal work effort, the main characteristics which color and shape this capacity are left nearly unexamined. I shall explore in a subsequent section a similar and similarly costly omission, namely, the thinness of Ricardo's treatment of the service component in work activity.

Unlike Smith, Ricardo does not present improvement in terms of the individual's desire to better himself. Rather, such progress as may take place results from social processes which better the lot of the class of workers.[103] Although one component of welfare consists in the reduction of human effort expended per unit of output, Ricardo did not have in mind the elimination of the need for the worker (see the chapter "On Machinery"). The minimizing of the effort per unit of output would benefit the workers as consumers taken together:

> . . . The invention of machinery, . . . the establishment of roads and canals, or . . . any means of abridging labour either in the manufacture or in the conveyance of goods. These are causes which operate on price, and never fail to be highly beneficial to consumers; since they enable them with the same labour, or with the value of the produce of the same labour, to obtain in exchange a greater quantity of the commodity to which the improvement is applied . . . .[104]

All consumers, but especially those who face the constant threat to life, will benefit from such innovations. As the "machinery" chapter cautions, however, it cannot be welfare when the worker can no longer exercise his proper activity.

In his treatment of capital rent and taxation, Ricardo maintains his focus on working persons. Directed to the service of persons, capital improves the social situation. "Capital is that part of the wealth of a country which is employed in production, and consists of food, clothing, tools, raw materials, machinery, &c. necessary to give effect to labour."[105] Capital—in all its individual components—is subordinate to labor because it provides the means to bring about the achievement unique to working persons: production.

In contrast to capital, rent results from population pressure, or the pressure of increased demand in general, upon land of ever-diminishing fertility. Rent does not figure into production itself. This is key to Ricardo's intense interest in ameliorating the lot of the workers. Since labor figures into the product whereas land-ownership does not, Ricardo argues for a redistribution from rents to wages.[106]

Ricardo also evaluates taxation in terms of its bettering or worsening the situation of working persons. Taxes threaten to worsen, not better, the situation of the worker through waste, through decreased provisions, through the diminishment of capital (which provides the means of work) and through the decreased opportunities for employing persons.[107] The government must levy taxes, insofar as they are necessary, in such a way that each sector of society bears its fair share of the burden of this necessary evil and so that the effects of the taxes do not hold back reproduction.[108] Reproduction offers a generic term for the replication of sustenance, capital and persons themselves. Reproduction makes it possible that work activity, that is, uniquely personal activity, may continue for the benefit of all.

Some critics indicate that Ricardo laid too great an emphasis upon productivity. Yet the distributive aspects of the social problem matter greatly to him because he focuses on the needs of laboring persons. Ricardo sought a rising income secured to the working class as a key resource for achieving their improvement.[109] Since the productive activities of persons create the goods which are distributed, the purpose of this distribution ought to guarantee progress in the workers' circumstances. Productive activity ought to lead to progress in the lives of workers.

The Maintenance of Labor and the Single Source of Value

Ricardo recognizes that the improvement of persons presupposes the requisites for life:

> The natural price of labour is that price which is necessary to enable the labourers, one with another, to subsist and to perpetuate their race, without either increase or diminution.[110]

Subsistence matters not only for the laborer but also for this group as a whole: their needs must be met so that the group can last through time. Ricardo knows well the dire circumstances which face those whose very choices of means to subsistence are sorely circumscribed:

> The friends of humanity cannot but wish that in all countries the labouring classes should have a taste for comforts and enjoyments, and that they should be stimulated by all legal means in their exertions to procure them. . . . In those countries, where the labouring classes have the fewest wants, and are contented with the cheapest food, the people are exposed to the greatest vicissitudes and miseries. . . . On any deficiency of the chief article of their subsistence, there are few substitutes of which they can avail themselves, and dearth to them is attended with almost all the evils of famine.[111]

Gains in the desires and tastes of a people can augment their needs and prevent such a marginal state on the verge of catastrophe from coming to pass. More importantly, society and government must encourage exertions directed toward the procurement of these "comforts and enjoyments." Ricardo views work activity as confronting human limits in two ways: in the obviously required efforts for sustenance, and also in further exertions for multiplied wants, which stimulate the entire system to greater productivity.

Opposed to the efforts of persons to sustain themselves is the range of forces which causes their means of sustenance to diminish or deteriorate. The production of essential necessities may be arduous or the worker himself may be in a "deteriorated condition."[112] More broadly still, Ricardo relates the perishable character of goods at the service of human efforts to the varying degrees of fixedness of capital.[113] Capital—both fixed and circulating—comprises a range of objects used to sustain life and augment human efforts in the sustenance of life. Human need provides the impetus to measure the perishability of these objects utilized in the maintenance of persons.

With regard to the opposition between persons and nature, Ricardo recognizes the "indestructible powers of the soil."[114] Even outside agriculture, nature benefits persons in society: "There is not a manufacture mentioned, in which nature does not give her assistance to man, and give it too, generously and gratuitously."[115] As is well known, Ricardo differentiates his own position from that of the theory which he had received concerning productive services.[116] Rent is that effect of prices which is a "compensation, which is paid to the owner of land for the use of *its original and indestructible powers*."[117] This fertile, personified nature, substantial and lasting, stands in stark contrast to the limited persons who are obliged to struggle for their lives. With an increase in need, for example, by way of a population increase, a worse quality of land must be put into use in order to provide food necessities. In these cases, "Man would do more by the sweat of his brow, and nature perform less; the land would be no longer pre-eminent for its unlimited powers."[118] The power of nature, in the case of agricultural products, remains indestructible; but its potency depends on the exertions of persons. Nature in abundance, as represented by land, cooperates with persons. The disappearance of abundance results from the extension of human need; at the onset of scarcity, human efforts create value out of recalcitrant nature.

The land is eminently serviceable to persons until their needs exceed its fecundity. Ricardo opposes human effort to the passivity of the land, human freedom to the immobility of these natural powers. Further still, he regards the land as representing the vested interests of landowners, against which he sets the free power of individuals who carry out production.

Time constrains the very efforts of individuals in opposition to nature. The foundation of Ricardo's analysis lies in the specification of value in terms of time expended in labor.[119] However, sundry exceptions to the first approximation of value in terms of labor time interfere with precise valuation and prediction. Different proportions of fixed and circulating capital, different durabilities of fixed capital and differences in the time required to bring a product to market complicate the ascertaining of value in terms of labor time. Yet both fixed and circulating capitals result from labor. That is, they both derive from an activity taking place over varying periods of time. In addition, the distinction between these two categories of capital and the variation within the category of fixed capital devolve from their durability with regard to human use. Market time likewise depends on the time which workers require in bringing a particular item to market. Human activity and need are the gauges of the exceptions to the measure of value.

When Ricardo suggested that the labor embodied in commodities could be conceived in terms of the differences of time, he decided that "a commodity produced by labour employed for one year" represents the mean for all the variances in the production processes.[120] One year, of course, represents the natural production period in agriculture. The average Ricardo selects derives not from observation and statistical calculation; rather, he settles on this average because the selection of the annual period emphasizes the production of subsistence goods.[121] The choice of the average cycle indicates once again Ricardo's focus on persons whose production, especially for life maintenance, is caught in time.

Workers live and put forth their efforts through time; they produce over time. The activity of production consumes time and deteriorates the workers. The sundry goods of which they make use to sustain their individual lives and facilitate their particular works, themselves last a longer or shorter period in use. Some variable quality independent of human standard is not the measure of the durability of the various capital goods. Rather, in his measurement of durability Ricardo calls attention to the various human needs marked by the varying natural processes to which persons must submit.

In their analysis of the relation between work activity and economic wealth, both Smith and Ricardo proposed labor theories of value, which make possible the evaluation of goods on the basis of labor alone, independent of any other economic factor. Although Ricardo first began to concentrate on economic analysis as a result of reading the *Wealth of Nations*, he took exception to one element of Smith's theory of value.

> Sometimes he speaks of corn, at other times of labour, as a standard measure; not the quantity of labour bestowed on the production of any object, but the quantity which it can command in the market: as if these were two equivalent expressions. . . .[122]

Because the market can cause the amount of labor commanded by a fixed quantity of a commodity to vary from period to period, they cannot be equivalent expressions. Ricardo rejects the labor-commanded theory, while he recognizes that Smith does present, in Bk. I, ch. 5, the correct position.[123] This emphasis lies upon the goods produced, for in these inheres the labor of their production.

In a letter to James Mill from December, 1818, he asserts the centrality of labor as a determinant of value:

> In opposition to him [Smith], I maintain that it is not because of this division into profits and wage, it is not because capital accumulates, that exchangeable value varies, but it is in all stages of society, owing only to two causes: one the more or less quantity of labour required, the other the greater or less durability of capital: that the former is never superseded by the latter, but it is only modified by it.[124]

As was presented in an earlier section, here again capital and its durability result from the efforts and the requirements of labor. Through capital, laboring persons save themselves; they spare themselves in the course of production. Thus, from the activity of production which incorporates value into commodities are derived both the origin and the variance in value.

While Smith emphasized grain, a central element in the provision for human sustenance, Ricardo moves a step further from nature to a position more intimate to the person—the measure now becomes precisely the human activity of working. So much does Ricardo wish to locate value within the laboring person that he excises from his understanding of value any notion of value in use:

> Dr. Smith's error throughout his whole work, lies in supposing that the value of corn is constant; that though the value of all other things can, the value or corn can never be raised. Corn, according to him, is always of the same value because it will always feed the same number of people. In the same manner it might be said, that cloth is always of the same value, because it will always make the same number of coats. What can value have to do with the power of feeding and clothing?[125]

In the rush to his rhetorical question, Ricardo overlooks the significant difference between the completed, non-repetitive action of feeding a single person a portion of grain and the production of a coat which may remain used or unused by one or several persons over periods of varying lengths. Smith's measure in grain represents a commodity, the use of which sustains persons. Ricardo wishes to hurry past even this measure, since it is exogenously determined in two ways: first, by the limitations imposed by nature within the person's body and second, by the variance created in the course of market processes. To his question, Ricardo desires the answer: value has nothing to do with the power of feeding and clothing; value concerns the power to produce; value concerns the power of persons to act against and apart from nature. Smith directs the focus to the laboring person, while Ricardo attempts to expunge any element of passivity in the face of nature by emphasizing the human activity at the source of value. Work produces value.

Ricardo presents a single source of value. As was noted in the preceding subsection, he wishes to correct Smith's error concerning value, which he interpreted as the command over commodities. In addition, he rejects the theories of contemporary economists: Say's utility theory, Malthus' commandable labor measure and Torrens' theory of value based on the quantity of required capital.[126] For Ricardo, the quantity of labor used to produce a commodity determines its value.

> The value of a commodity, or the quantity of any other commodity for which it will exchange, depends on the relative quantity of labour which is necessary for its production, and not on the greater or less compensation which is paid for that labour.[127]

Not wages, determined in the market, but the productive activity itself determines value.

Ricardo distinguishes, further, between value and those riches which labor produces:

> Value, then, essentially differs from riches, for value depends not on abundance, but on the difficulty or facility of production. The labour of a million men in manufactures, will always produce the same value, but will not always produce the same riches. By the invention of machinery, by improvements in skill, by a better division of labour, or by the discovery of new markets, where more advantageous exchanges may be made, a million of men may produce double, or treble the amount of riches, of "necessaries, conveniencies, and amusements," in one state of society, that they could produce in another, but they will not on that account add anything to value; for everything rises or falls in value, in proportion to the facility or difficulty of producing it, or, in other words, in proportion to the quantity of labour employed in its production.[128]

So the source of value Ricardo considers not in terms of produced objects, external to persons but, rather, in terms of the activity of persons. Given his empirical leanings, Ricardo expresses this in the physical terms of "difficulty or facility." Three sorts of difficulty must be distinguished, however: technical difficulty, pain cost and time measure. The first refers to processes, while the second reduces the third to a single range of human work investment. By "difficulty or facility of production" Ricardo refers to the entire range of human investment through activity. Time measures the quantity of labor. It is the condition which affects the entire, active living of limited persons. Hence it is the condition which represents the entirety of their lives in action. Ricardo draws attention to the basis of value:

> . . . It [the wealth of a country] may be increased by employing a greater portion of revenue in the maintenance of productive labour,—which will add not only to the quantity but to the value of the mass of commodities; or it may be increased, without employing any additional quantity of labour, by making the same quantity more productive,—which will add to the abundance, but not to the value of commodities.[129]

The population provides the basis of value. A society utilizes "a greater portion of revenue" in order to maintain more workers. The number of workers may be increased either by putting more of a given population to work or by raising the population so that more persons are available for work in the community. Not simply an increase in the number of persons matters; rather, only an increase in the persons who are active in work can augment value in society. All means which may make a given population more productive are peripheral to the source of value: employed, that is, actively working persons.

In the late essay, "Absolute Value and Exchange Value," Ricardo continues his pursuit of a firm measure of value over time: "It is a great desideratum in Polit. Econ. [sic] to have a perfect measure of absolute value in order to be able to ascertain what relation commodities bear to each other at distant periods."[130] Ricardo describes the perfect measure of value: ". . . Any commodity which continues always to require the same quantity of labour to produce it will be a perfect measure of value."[131] Yet in the later version of the essay, Ricardo realizes that he cannot devise such a standard: "If all commodities were produced by labour alone, without any advances, and were brought to market in one day, then indeed we should possess an uniform measure of value. . . ."[132] If the provision for human sustenance were to require no advances, if the admixture of capital (that is, past labor resulting from varying rates of interest and of varying degrees of durability) were not requisite to production and if all product cycles not only were uniform but also were completed in such a period so short that advances for sustenance were not needed, then labor would advance from the status of an imperfect measure to that of an absolute and unvarying standard. Although labor remains an imperfect measure of the value of commodities, one must recognize that in a critical sense Ricardo does not consider labor itself a commodity. Work activity antecedes the commodity. This activity produces value prior to the insertion of the object into the market, that is, before it becomes a commodity.

The human inquirer exists as a person antecedent to the market. The way to speak of the value of a commodity at two different times lies in the evaluation from the vantage of such a person. Both question and evaluation proceed from the person; the person's work activity determines the result of the evaluation. The historical dimension of the comparison provides the clue that the estimation of value arises first in terms of the working person and then seeks a resolution in those same terms.

In his project to discover the invariable mean of value, Ricardo sought the circumstances determinative of value, not necessarily an actual commodity.[133] The circumstances which determine the value of a commodity are the conditions in which a person works. The conditions which a commodity would have to satisfy for invariability are the contrary-to-fact conditionals examined just above. These conditions would require the elimination of temporality from human production. These conditions cannot be met; thus labor cannot stand as an invariable measure of value. It remains nonetheless a measure of value. Furthermore, in his emphasis on transience and change, Ricardo has repeatedly drawn attention to the condition of persons. The circumstances of finitude and transience are the conditions which must be taken into account in founding the laws of an economic system upon the axiom of human work. The upholding of personal work activity as the source of value requires that the formulation of the law of value include the factor of change, yet this very move renders impossible the use of labor as an invariant measure.

In addition to presenting a theory of value which proceeds from the uniquely human capacity of productive work, Ricardo often refers to disutility as a way of measuring value.[134] Does Ricardo fall into an error analogous to the one of which he accuses Smith? Does he double the sources of value? The stress in Ricardo does not fall upon this disutility interpretation. As Ricardo progressed in his thinking he shifted the emphasis increasingly to the work activity of the person.[135] It is significant that Ricardo inserted in the third edition of the *Principles* the statement by Destutt de Tracy (which I cited above) concerning human physical and moral faculties as the creative capacity for the production of riches.[136]

Ricardo thus shifts his analysis of value so that it gives prominence to the potentials and activity of the person.

In the later version of the "Absolute Value and Exchange Value" essay, Ricardo sets out some particular requirements for the measure of value:

> There can be no unerring measure either of length, of weight, of time or of value unless there be some object in nature to which the standard itself can be referred and by which we are enabled to ascertain whether it preserves its character of invariability, for it is evident on the slightest consideration that nothing can be a measure which is not itself invariable.[137]

The usual treatment accorded this passage would focus on the requirement of invariability. Also meriting attention, however, is Ricardo's specification of the locus for this invariability, namely, "some object in nature." Ricardo's concern to establish his analytical system on unchanging, basic elements and to establish his system as an explanation of lived reality draws upon the thinking of his advisor James Mill. Like J. Mill, Ricardo sought, as the foundation of his system, simple, concrete elements.[138] As we have seen, Ricardo rejected an understanding of value which held that products and riches commanded certain quantities of labor, because he wanted to avoid the variability of market relations and to draw attention to the paramount value of human activity. Ricardo was struggling to determine not only the invariability of the measure but also its locus. He sought "some object in nature." Given his materialist propensity, this would appear to indicate some physical object. When, in the final essay, he names "the labor of men" as this "object in nature" that meets the requirements for the measure of value, the reader is drawn up short. Since it cannot be conceived as some physical object, "the labor of men" can only be an object in some derived sense. Viewed as abstracted from the products it creates, work activity is the simple source of value.

The "labor of men" is not an object in nature like some physical substance; it is not inert and passive. The "labor of men" is "of nature" because it is suffused with finitude (for example, the recurrent need for bodily sustenance) and because it works in nature. Human work sets itself against nature; it is active and changing. Whereas the Physiocrats conceived nature as bountiful, the classical conception, especially under the influence of Ricardo, saw nature as niggardly—particularly so in response to waxing human needs—and saw human activity as productive of bounty. Whereas earlier laborers expended efforts upon nature and she rewarded them in abundance, now workers disciplined their own efforts; thus they themselves produced. In this way, the disutility to which Ricardo sometimes refers may be read as the efforts of human persons constrained by their own choices to continue a selected activity over time. The disutility, then, results from freedom exercised upon nature over a period of time for the purpose of production.

The charge often brought against Ricardo alleges that his understanding of value is "metaphysical" in the sense that it appeals to entities which are not susceptible of demonstration or whose influence on everyday realities is obscure. Ricardo's unchanging value may both be a metaphysical notion and at the same time still represent a concept of rational science.[139] According to this view, the ontological underpinnings may remain irresolvable; nevertheless, the analysis of social structures may fruitfully employ the concept and may make it possible to predict on the basis of laws. This offers a suggestive, sympathetic account of Ricardo's intention. The problem which remains is that in his move to locate value within the individual's activity antecedent to the market, even though the social structures shape that person's work in multiple ways, Ricardo leaves no indication how he conceives the connection between the internal, active value and the external, fully social, deposited value. Ricardo's move to universalize value within the action of persons has disjoined it from social realities. This theoretical separation does not prevent him from examining the social factors amidst which work takes place.

The Social Needs and Interchange Which Shape Work Activity

Especially in view of his close association with James Mill, some writers have been tempted to view Ricardo as a utilitarian and have thus reduced his views to a mechanistic interpretation of social interaction.[140] Both the earlier examination of Ricardo's understanding of the individual worker and his firm grasp of the obvious human and social (that is, non-mechanical) context of work activity argue for the rejection of such an interpretation of Ricardo's position. The social interaction which he describes encompasses an interdependence both of production and of exchange.[141]

Ricardo's explanation of the natural price of labor clarifies, in terms of this key element in his system, two aspects of the sociality of work activity:

> The power of the labourer to support himself, and the family which may be necessary to keep up the number of labourers, does not depend on the quantity of money which he may receive from wages,

> but on the quantity of food, necessaries, and conveniences become essential to him from habit, which that money will purchase.[142]

The exercise of the worker's power in his activity must achieve not only his own sustenance but also that of his family members. The familial bond offers the first dimension of the sociality of work activity. Moreover, a second dimension immediately enters in: the conveniences are held to be—along with food and other necessities—essential. The habit which determines these conveniences to be "essential" is a social habit; that is, the customs of the particular society in which the worker lives establish the conveniences not needed for strictly physiological survival. These needs, which social customs have defined, contribute to the delineation of what individual workers expect as a result of their work.

A theory of wages attentive to the subsistence minimum focuses on embodied, laboring entities. In his treatment of the relation between the market and the natural price of labor, Ricardo explicitly affirms that social customs determine some necessary comforts.[143] If the minimum for subsistence were determined so that it were independent of "habit and custom," the system would be unable to account for the activities of actual persons affected by historical time and social interaction. For this would be a theory whose standard treated the agents as detemporalized and individuated, and their interaction would not affect the standard. Ricardo, however, does not offer a theory in terms of such a timeless minimum. He demonstrates this in his insistence on the essential conveniences (the necessity of which custom creates) as a factor in the natural wage.

In delimiting the field of his concern, Ricardo indicates a still more comprehensive dimension of sociality implicit in work:

> . . . We mean always such commodities only as can be increased in quantity by the exertion of human industry, and on the production of which competition operates without restraint.[144]

Human labor produces those objects which concern Ricardo, and these objects enter into a situation of exchange wherein competition governs. Upon the exchange of these fruits of production between needy persons, competition exerts its force. Competition is one form of sociality.

The competitive market presents one dimension of social interchange, which, in turn, emerges from a still larger social matrix. The determination of the rate of profit does not proceed solely from intrasystemic factors. A key factor in this determination lies outside the economic system as such. This is the particular set of social and productive organizations which provide the framework and foundation for each market.[145]

The sociality of work activity shapes even the notion of nature with and against which humans labor. People comprise society; they must be provided with sustenance for life. If no persons existed, then the land would not be called fertile or infertile, niggardly or generous. The necessities and conveniences requisite for persons in society, the availability of capital and the efforts of workers are social vectors which shape, in large part, the determination of the fruitfulness of the land. The fertility of the soil is a synecdoche for the richness and generosity of nature. The social and productive ranges which structure work activity set the level of the soil's fertility. The structures which enframe human work determine the characteristics of nature, the opponent of human productive activity.

Ricardo introduces a critical dimension into the sociality of work activity, one which paves the way for Marx's studies. In the situation where social demands press against the natural limit, a differentiation, a fissure, occurs within society. Because of the difference between the rise in wages and the rise in rent, the situation of the workers deteriorates, while the return to the landowner increases[146] The optimism which suffused the system of individual free activity found in Smith here evaporates. Returns begin to diminish in the sphere of production most vital for the sustaining of life, namely, agriculture. The pessimism increases: the market processes which theorists opined would facilitate the commonweal both diminish the welfare of the workers, who activate the system, and weaken the infrastructure of social bonds because some but not all segments of society now enjoy an improved situation.[147]

The Benefit and Threat of Machines to Workers

Living amidst the rapidly changing technical and social circumstances brought about in the Industrial Revolution since the time of Smith, Ricardo attempted to take into account the role machines were playing in the process of production.

Even in the "Preface," Ricardo carefully notes (in addition to the fertility of the soil and human skill) the contribution of instruments to production.[148] He implies that the constructions of technology are intrinsically connected with work activity:

> Not only the labour applied immediately to commodities affect their value, but the labour also which is bestowed on the implements, tools, and buildings, with which such labour is assisted.[149]

These technical implements result from human work and, in turn, assist it.

Although the value of some products decreases, the increased productivity owing to the introduction of efficient machines redounds to the benefit of all because of the reduction in prices:

> The natural price of all commodities, excepting raw produce and labour, has a tendency to fall, in the progress of wealth and population; for though, on the one hand, they are enhanced in real value, from the rise in the natural price of the raw material of which they are made, this is more than counterbalanced by the improvements in machinery, by the better division and distribution of labour, and by the increasing skill, both in science and art, of the producers.[150]

The rise in the price of raw materials results from the employment of ever less fertile land; yet improvements in machine technology are one means which contains this tendency. In addition, an improvement in the "division and distribution of labour" also offsets the increasing difficulty in production. On one view, Ricardo's placing this in series with machine improvements suggests that these work relations are a type of technology. From another viewpoint, that he places machine improvements in the series with developments in work arrangements and in worker skill indicates once again that he allies developments in machince technology with human, both personal and social, elements in productive progress. In a passage from chapter six, Ricardo provides a still fuller description of his understanding of technology:

> . . . [The] gravitation as it were of profits, is happily checked at repeated intervals by the improvements in machinery, connected with the production of necessaries, as well as by discoveries in the science of agriculture which enable us to relinquish a portion of labour before required, and therefore to lower the price of the prime necessary of the labourer.[151]

Again he refers to the countervailing force which "improvements in machinery" bring about. Here, however, Ricardo clearly extends this type of improvement to the sphere of agriculture. Furthermore, in this area he speaks of technological improvements, not only in terms of machines but also in terms of the expansion of knowledge through discovery.

In the third edition (1821) of the *Principles*, Ricardo added chapter thirty-one, "On Machinery," which muted his earlier, optimistic tone concerning technical improvements. His two central conclusions are:

> 1) . . . The motives for employing machinery are always sufficient to insure its employment, if it will increase the net produce, although it may and frequently must, diminish both the quantity of the gross produce and its value.
>
> 2) The opinion entertained by the labouring class that the employment of machinery is frequently detrimental to their interests, is not founded on prejudice and error, but is conformable to the correct principles of political economy.[152]

The reduction of the value mentioned at the close of the first point implies that production will require less work. In this light, the workers (as indicated in the second point) are correct to see a restriction of their sustaining activity. The dictates of the economic process may benefit society *in toto*, but, as Ricardo emphasizes, the cost will be borne by the fundamental operative segment of society: the productive workers. Ricardo specifies the problem in terms of the survival and welfare of the persons who work:

> . . . As the power of supporting a population, and employing labour, depends always on the gross produce of a nation, and not on its net produce, there will necessarily be a diminution in the demand for labour, population will become redundant, and the situation of the labouring classes will be that of distress and poverty.[153]

One may object that when the demand for workers plummets in a certain production sector, those persons can move to locales or sectors where the demand for workers maintains itself or increases. Two arguments favor a reading supportive of Ricardo's concern. Since he was concerned with change, far-reaching in its effects, and since exactly this sort of change was occurring in contemporary England subsequent to the Napoleonic wars in broad sectors of the economy, Ricardo may have assumed that such mobility was a phantom because the decline in demand was systemic and widespread. In addition, the actual demands which such a theoretical mobility presented often exceeded the actual capacity on the part of many familied workers with geographic ties and limited means for easy movement.

Ricardo, attentive to the competitive context provided by international investment, realized, even in the chapter "On Machinery," that machines must be utilized for the sake of the laborers themselves.[154] The construction as well as the operation of machines require human work; these two requirements will necessitate the employment of some persons, even though the utilization of machines reduces the overall number of the employed.

The question concerning the net benefit of machines and other technological improvements remains unsettled for Ricardo. Workers, if their society may advance at all through increased productivity and leisure, must make use of machines for the most effective investment of their own efforts. More profoundly:

> The consequence of a rise of food will be a rise of wages, and every rise of wages will have a tendency to determine the saved capital in a greater proportion than before to the employment of machinery. Machinery and labour are in constant competition, and the former can frequently not be employed until labour rises.[155]

As the demands of society press against the limits to productivity, as wages then rise, a shift to machine production will take place. As the limits of simple nature are approached, this integument of machines and technology which settles upon social processes becomes a second nature. The use of machines increases productivity but reduces the demand for labor. Persons become "redundant." Without machines, persons stand over against nature to eke out, by dint of their limited efforts, the maintenance of life; the introduction of machines brings with it an opposition between these economizing devices and those workers whose efforts have been rendered superfluous. Opposed by machines which do not need labor for sustenance, unable to counter instruments which benefit other segments of society, these persons retain their capacities; but no field exists wherein their efforts may be expended for the maintenance of life. The creation of the second nature means that some persons lack a world for their potential work activity. As society approaches the limits of nature, its economic processes compel these workers beyond both natures, outside these limits, indeed beyond their own limits.

Ricardo's concern for personal work activity—and not subjective demand—provides the basis upon which he escapes from a purely technical understanding of disutility and production.[156] Even in his treatment of the technological, Ricardo's concern remains with the worker.[157]

In his parliamentary speech of May 30, 1823, Ricardo asserts that the poor would exert forethought and prudence with regard to the supply of labor, ". . . if they were not made such machines of by the poor laws."[158] Characteristic of Ricardo is his concern for the persons who face decisions concerning their own reproduction when they are considering the possibilities for work activity. His revulsion at the reduction of the person to a mechanical state implies, conversely, an emphasis on the service of persons by machines. His pursuit of the absolute value established upon work activity indicates his concern to derive a norm which both operates within an economic system and seeks the sustained development of human life.

Ricardo's Inchoate Treatment of Service

Ricardo's firm interest in the welfare of workers, as demonstrated by his insistence on basing his system upon the work activity of persons, leads nevertheless to a merely inchoate grasp of the service dimension of work. Ricardo conceptualizes the results of production as wealth and value. I have already pointed out the difficulty of mediating between value's location in work activity and wealth's position in commodities. Because wealth as a set of objects appears far removed from the service which wealth provides, Ricardo's understanding of wealth leads him to a deficient grasp of service. These services, however, as well as these objects are the results of production.[159]

Ricardo's closest approximation to an understanding of service occurs on those occasions when he sees wealth as sustenance or as something which facilitates the provision of sustenance. Despite this, the emphasis lies not upon the provision by persons of goods for each other but, rather, upon the entire system's output of wealth. From this mass, society distributes goods. The labor theory arose in a context where the sociality of the market established a possible foundation for service activity.[160] Smith's concept of society certainly presents the interdependence within the market. In Ricardo's society, however, the interdependence of persons, although implied, recedes in analytical significance in favor of the interaction of the abstract factors of production. With this move, it becomes far more difficult to consider the interpersonal provision of services. To be sure, Ricardo's high valuation of the work activity of the individual worker in the face of natural limits permits an understanding of sustenance as service within his analysis; nonetheless, the analytical abstraction of labor makes this more difficult. Further, the pre-eminence assigned to work activity has isolated it from external exchange realities. Thus, this precious, personal capacity of each individual for work remains conceptually sequestered from the needs and requirements of other persons.[161]

In the preceding I have offered an interpretation which describes a twin development in Ricardo's theory: an increasing emphasis on the centrality of work activity and a widening disjunction of this core activity both from market processes and from some dimensions of the working person himself. That the significance of work activity became theoretically disjoined from actual processes and persons undermined its normative force.[162] Working persons cannot—because of the great significance of their unique work activity—enter the sphere of commodities.[163] The question concerning how the value of work activity becomes related to the sphere of commodities remains unresolved.

Ricardo, disquieted by the contemporary circumstances for the workers and desirous of alleviating their plight, sought a solution to the distributional question by concentrating his analysis on production and producing agents. After Ricardo, however, the self-conceived task of economics shifted away from a concern for labor. Subsequent to Ricardo's death, with the growth in the strength and organization and activity of the workers, his theory came to be judged as too great a threat to the commercial and industrial leaders who became established in a position stronger than that of the landed interests.[164] In such a climate the voices of the Ricardian Socialists and, later, of Marxist economists, did much to force classical economic analysis of the mainstream away from a further wrestling with the concept of the work activity of finite persons who make use of technology to serve their mutual needs in society.

After 1830, by abstracting from productive activity, economic analysis moved away from Ricardianism. Ricardo himself had set the course for this abstraction. Although it was accomplished for the sake of the workers, his shift to the systemic analysis inevitably held those persons involved in producing and consuming subordinate to the macroanalysis, in part because of the empiricism and objectivism of the system which could not account for the uniquely human. Faced with this difficulty himself, Ricardo asserted the unique and pre-eminent value of work activity. Yet this very uniqueness and pre-eminence remained unintegrated into the system. Many subsequent writers would abstract still further from the actual conditions of producing persons. Georg W.F. Hegel, however, carefully studied Smith and other writers involved in the initial formulation of economics as a science, in order to investigate the role of work in a still more comprehensive and cohesive system of thought.[165]

1. Ronald L. Meek, *Smith, Marx and After: Ten Essays in the Development of Economic Thought* (London: Chapman and Hall, 1977), p. 7.

2. Some of Smith's expressions, for example, "the demand for men, like that for any other commodity," may distract from his broader and more significant concerns for the person. See *An Inquiry into the Nature and Causes of the Wealth of Nations, ed. by W.B. Todd, vols. 2:1 and 2:2 of The Glasgow Edition of the Works and Correspondence of Adam Smith*, ed. by R.H. Campbell and A.S. Skinner (Oxford: Clarendon Press, 1976), p. 98. This work will henceforth be abbreviated as *WN*.

3. Concerning Smith's favoring of an interventionist social policy which worked on behalf of the poor, see G.V. Rimlinger, "Smith and the Merits of the Poor," in *Adam Smith: Critical Assessments*, ed. by J.C. Wood (London: Croom Helm, 1984), Vol. IV, p. 231.

4. *WN*, I, viii, p. 100.

5. Adam Smith, *The Theory of Moral Sentiments*, ed. by D.D. Raphael and A.L. Macfie, vol. 1 of *The Glasgow Edition of the Works and Correspondence of Adam Smith*, ed. by R.H. Campbell and A.L. Macfie (Oxford: Clarendon Press, 1976), II, iii, 3, 3, p. 106.

6. *WN*, II, iii, p. 341.

7. *Ibid.*, IV, v, b, p. 540. Also see *Ibid.*, II, iii, p. 343.

8. *Ibid.*, IV, ix, p. 674.

9. "Happiness is found not in isolation but in the society of one's fellows and only in a dynamic, creative, and forward-looking environment wherein the individual is absorbed in the pursuit of new excellence, new knowledge, and new invention (or to use Smith's terms: Wonder, Surprise, and Admiration)." See E. West, "Adam Smith and Alienation," in *Essays on Adam Smith*, ed. by Andrew S. Skinner and Thomas Wilson (Oxford: Clarendon Press, 1975), p. 541.

10. *WN*, I, x, b, pp. 116-117.

11. Andrew Skinner, "Introduction," in *The Wealth of Nations*, Books 1-3 (Harmondsworth: Penguin, 1974), p. 61. Also see D.A. Reisman concerning the merit of better standards of living: *Adam Smith's Sociological Economics* (New York: Barnes and Noble Books, 1976), p. 146. Also see Wiles, "The Theory of Wages" for the long debate in the English economic tradition preceding Smith concerning the benefits of low or high wages.

12. See *WN*, I, v, p. 48.

13. T.J. Lewis presents Smith's view that the market for labor is the natural rights context in which freedom may be exercised and through which it is threatened. See "Adam Smith: The Labor Market as the Basis of Natural Right," in *Adam Smith: Critical Assessments*, ed. by J.C. Wood, Vol. I, pp. 672-673.

14. *WN*, I, x, c, p. 138.

15. *Ibid.*, I, x, c, p. 143.

16. *Ibid.*

17. Francis Hutcheson, *A System of Moral Philosophy*, 2 vols. in 1 (London: A. Millar and T. Longman, 1755; reprint ed., New York: Augustus M. Kelley, 1968), Vol. 2, pp. 318-319.

18. In the tradition of civic republicanism, independent action achieves virtue in a situation of interdependence. See E.R. Gill, "Civic Virtue, Adam Smith and the Division of Labor," Address at the Annual Meeting of the American Political Science Association, 1984, pp. 5-6.

19. *WN*, I, viii, p. 99.

20. For example, see O. Bordollo, *Der Begriff "Produktiv" bei Adam Smith und bei Friedrich List*, Inaugural-Dissertation zur Erlangung der Doktorwürde der hohen philosophischen Fakultät der Friedrich-Alexanders-Universität zu Erlangen (Borna-Leipzig: Buchdrückerei Robert Noske, 1909), p. 24.

21. In part VI, section one of Smith's *Theory of Moral Sentiments*, he weights his account of prudence so as to emphasize the economic: frugality and industry are characteristic of the prudent person. A. Macfie, *The Individual in Society: Papers on Adam Smith* (London: George Allen and Unwin, Ltd., 1967), p. 73.

22. Macfie, *The Individual in Society*, p. 75. On the desire for approbation see the *Theory of Moral Sentiments*, pp. 113-122, 126-127, and 170.

23. Macfie, *The Individual in Society* p. 34. J.M.A. Gee emphasizes the communitarian dimension in Smith's understanding of prudence. See "Adam Smith's Social Welfare Function," in *Adam Smith: Critical Assessments*, ed. by J.C. Wood, Vol. IV, p. 88.

24. ". . . The self-love of both books [*Moral Sentiments* and *Wealth*] consistently implies not a hostile anti-social trait but a simple, constructive, and common-sense prudence; not the vice of greed but the virtue of economy." See E. West, "Adam Smith and Alienation: 'Wealth Increases, Men Decay'?" in *Essays on Adam Smith*, ed. by Andrew S. Skinner and Thomas Wilson, p. 548.

25. For the theme of labor mobility as an expression of natural liberty, see *WN*, IV, ii, p. 470 and P.J. McNulty, "A Note on the Division of Labor in Plato and Smith" in *Adam Smith: Critical Assessments*, ed. by J.C. Wood, Vol. III, pp. 352.

26. *WN*, I, v, p. 47.

27. Skinner, "Introduction" in *WN* (1974), p. 48.

28. "Equal quantities of labour, at all times and places, may be said to be of equal value to the laborer. In his ordinary state of health, strength and spirits; in the ordinary degree of his skill and dexterity, he must always lay down the same portion of his ease, his liberty, and his happiness. . . ." *WN*, I, v, pp. 50.

29. L. Billet, "Justice, Liberty and Economy," in *Adam Smith and the "Wealth of Nations,"* ed. by F.R. Glahe (Boulder, Colorado: Colorado Associated University Press, 1978), p. 86.

30. J.M. Clark, "Adam Smith and the Currents of History" in *Adam Smith, 1776-1926: Lectures to Commemorate the Sesquicentennial of the Publication of the "Wealth of Nations,"* J.M. Clark et al. (Chicago: University of Chicago Press, 1928), p. 68.

31. *WN*, I, v, p. 51.

32. *Ibid.*, I, viii, p. 85.

33. *Ibid.*, I, viii, p. 83.

34. *Ibid.*, I, viii, pp. 87-88.

35. *Ibid.*, I, viii, p. 84.

36. *Ibid.*, I, x, b, p. 123.

37. Albert Rees, "Compensating Wage Differentials," in *Essays on Adam Smith*, ed. by Andrew S. Skinner and Thomas Wilson, p. 344.

38. *WN*, II, iii, p. 343.

39. *WN*, I, v, p. 54.

40. *Two Treatises of Government*, 2nd edition, ed. by Peter Laslett (Cambridge: Cambridge University Press, 1970). See II, especially, chapter five, sections 25-45.

41. Karl Marx also discovers two versions of productive work in Smith: one which produces profit and the other which leaves an object. See Jacques Nagels und Klaus Schäfer, "Adam Smith über produktive und unproduktive Arbeit—Denkanstösse für die politische Ökonomie des Sozialismus" in *Adam Smith gestern und heute: 200 Jahre "Reichtum der Nationen,"* hrsg. von Peter Thal et al. (Berlin: Akademie Verlag, 1976), pp. 206-207.

42. *WN*, I, iv, p. 44.

43. See Samuel Hollander, *The Economics of Adam Smith*, Studies in Classical Political Economy, I (Toronto: University of Toronto Press, 1973), p. 136.

44. *WN*, II, iii, p. 330.

45. *Ibid.*, I, v, p. 47.

46. *Ibid.*, I, v, p. 48 corroborates this.

47. See L. Dumont's assessment of this instability: *From Mandeville to Marx: The Genesis and Triumph of Economic Ideology* (Chicago, University of Chicago Press, 1977), p. 203.

48. *WN*, I, ii, p. 28.

49. Eli F. Heckscher, *Mercantilism*, 2 vols., translated by Mendel Shapiro; revised, second edition by E.F. Söderlund (London: George Allen and Unwin, 1962); here, vol. 2, p. 167.

50. *WN*, I, viii, p. 96.

51. J.M.A. Gee indicates a particular danger to the whole of society when the poor are despised: the moral sentiments become corrupted. See "Adam Smith's Social Welfare Function," in *Adam Smith: Critical Assessments*, ed. by J.C. Wood, Vol. IV, p. 89.

52. Attention must be called to the limits of this equality. In opposition to mercantilism, the humanitarian schools directed their concern toward the working poor. See Heckscher, *Mercantilism*, Vol. 2, p. 329. The laissez-faire approach, however remained indifferent to those who had nothing to offer. This one type of equality of treatment depended upon the work activity of a person as it contributed to society.

53. J. Cropsey, *Polity and Economy: An Interpretation of the Principles of Adam Smith* (The Hague: Martinus Nijhoff, 1957), p. 96.

54. *WN*, I, ii, p. 25.

55. Macfie, *The Individual in Society*, p. 77. In addition see *Theory of Moral Sentiments*, IV, i, 184-185 and *WN*, I, xi, c, pp. 180-181. J.S. Davis severely qualifies Smith's assertions concerning the limits based in embodiment. "Adam Smith and the Human Stomach" in *Adam Smith: Critical Assessments*, ed. by J.C. Wood, Vol. II, pp. 38-46.

56. It is of some theoretical interest that Jacob Viner agrees with R. Raikes in tracing the origin of commerce as theorized by Smith to a residual dynamic of providence. "Smith never explicitly formulates the idea that providence favors commerce, but it is implicit in his writings; and I am not at all sure that Raikes [Anglican clergyman Richard Raikes, who in 1806 offered an exposition of the providential interest in commerce] has not hit upon the most satisfactory explanation available of why Smith explained the initial resort of men to trade by a psychological propensity, instead of by the usual appeal to a rational or utilitarian search for gain. *The Role of Providence in the Social Order: An Essay in Intellectual History*, Jane Lectures for 1966, the American Philosophical Society (Princeton: Princeton University Press, 1972), p. 47. D.D. Raphael presents an unconvincing argument against the religious import of the language of "invisible hand." See D.D. Raphael, *Adam Smith* (Oxford: Oxford University Press, 1985), p. 72.

57. E.G. West notes the dynamism which lies in the division of labor. See "Scotland's Resurgent Economist: A Survey of the New Literature on Adam Smith" in *Adam Smith: Critical Assessments*, ed. by J.C. Wood, Vol. IV, p. 286.

58. West, "Adam Smith and Alienation," in *Essays on Adam Smith*, ed. by Andrews W. Skinner and Wilson, p. 543.

59. The virtue of each person—in his or her regard for society and in the sympathy which guides all affections—provides the basis for this social bond. See K.R. Ranadive, "The Wealth of Nations: The Vision and the Conceptualization," in *Adam Smith: Critical Assessments*, ed. by J.C. Wood, Vol II, p. 258. D.D. Raphael offers a distinction concerning sympathy which helps avoid much misinterpretation of Smith: "The basic error lying behind all these interpretations is the supposition that Adam Smith in the *Moral Sentiments* gives sympathy a central role as a motive of action. In both books the influence of sympathy (or 'benevolence' or 'humanity') as a motive is very limited. In the *Moral Sentiments*, the influence of sympathy as a binding social force of approbation is immense." See *Adam Smith*, pp. 89-90.

60. Macfie, *The Individual in Society*, pp. 62-63.

61. *Ibid.*, p. 67. See also N. Rosenberg's interesting article which presents Smith's concern for the institutional arrangements which would secure the development of personal capacities. "Some Institutional Aspects of the Wealth of Nations," in *Adam Smith: Critical Assessments*, ed. by J.C. Wood, Vol. II, p. 108.

62. Macfie, *The Individual in Society*, p. 71.

63. Heckscher, *Mercantilism*, Vol. II, p. 328.

64. Macfie, *The Individual in Society*, p. 77.

65. See *WN*, IV, ix concerning the necessity for the government to provide defense and justice.

66. J. Viner indicates (in his famous "Laissez Faire" lecture) the range of broader, governmental guidance of investment, commerce and employment to which the process of production and exchange must submit, not only in order to secure the minimal framework of defense and justice but also in order to aggrandize the common benefits of society. See "Adam Smith and Laissez Faire" in J.M. Clark et al. *Adam Smith 1776-1926* (Chicago: University of Chicago Press, 1928), pp. 116-155.

67. *WN*, II, ii, p. 287 and also I, i, 8; II, iii, p. 32; and IV, ix, p. 35.

68. *Ibid.*, III, i, p. 31.

69. Bordollo, *Der Begriff "Produktiv,"* p. 43 and Thal, *Adam Smith Gestern und Heute*, p. 49.

70. While he analyzes both the social benefit of work and the social context of market activity, Smith tends to see social interaction in terms of a vast workhouse. "What takes place among the laborers in a particular workhouse, takes place, for the same reason, among those of a great society. The greater their number, the more they naturally divide themselves into different classes and subdivisions of employment." See *WN*, I, ix, p. 108.

71. R.L. Meek, *Smith, Marx and After*, p. 177.

72. Francis Hutcheson, *System of Moral Philosophy*, see especially Vol. 1, pp. 288-289.

73. W.L. Taylor, *Francis Hutcheson and David Hume as Predecessors of Adam Smith* (Durham, North Carolina: Duke University Press, 1965), p. 59.

74. *WN*, I, ii, p. 13.

75. *Ibid.*, I, i, p. 17.

76. E.G. West, *Adam Smith* (New Rochelle, New York: Arlington House, 1969), p. 94.

77. *WN*, V, i, pp. 781-782.

78. N. Rosenberg discusses how the same processes in the division of labor, which at first benefit the worker, grow to a degree that they debilitate him. "Adam Smith on the Division of Labour: Two Views or One?" in *Adam Smith: Critical Assessments*, ed. by J.C. Wood, p. 176.

79. *WN*, I, ii, p. 26.

80. *Ibid.*, II, iii, p. 330.

81. *Ibid.*, II, v, p. 362.

82. Maurice Dobb, *Theories of Value and Distribution Since Adam Smith: Ideology and Economic Theory* (Cambridge: Cambridge University Press, 1973), p. 60.

83. Edwin Cannan, *A History of the Theories of Production and Distribution in English Political Economy from 1776-1848*, 3rd ed. (London: King and Son, 1917; revised ed., London and New York: Staples Press, 1953), pp. 30-31.

84. *WN*, II, iii, p. 330.

85. Adam Smith, *An Inquiry into the Nature and Causes of the Wealth of Nations*, ed. by E. Cannan (New York: The Modern Library, 1937), p. 314, n. 2.

86. *WN*, II, iii, p. 330.

87. *Ibid.*, II, iii, p. 331.

88. One might speculate concerning the cause of Smith's failure to articulate within his analysis the service work which takes place within the household. Michael Perelman notes that, "Classical political economy did recognize the value of the services rendered by the household; however, that literature also excluded the household from all theoretical discussions." See *Classical Political Economy* (London: Francis Pinter, 1983), p. 23. It may well be the case that "households" could be taken to mean the estates of the rich landowners against whose ethos the *Wealth of Nations* strenuously struggled. In this case, it would not strengthen his argument for Smith to demonstrate the worth of work occurring in such establishments. From another viewpoint, Smith's compassion for the poor could have led him to discern the great work of service offered by parents within the family even of the poorest. Perhaps he saw the lot of these needy persons as so passive before the vagaries of employment and social power that action, properly speaking, and thus work, were excluded. Then again, Smith may have seen the lives of these persons taken up in the round of natality, growth and mortality and living so close to the margin of existence that their lives, let alone the results of work, were themselves, passing, that is, hardly lasting. In either of these latter views, it would be impossible to comprehend either work or service in these situations. Household "work" performed so near the margin of subsistence would be a natural drive, rather than a result of choice. Without choice there could be no free activity and, hence, no work.

89. West, *Adam Smith*, p. 94.

90. "The principle of gravitation is not more certain than the tendency of such laws [the Poor Laws] to change wealth and power into misery and weakness; to call away the exertions of labour from every object, except that of providing mere subsistence; to confound all intellectual distinction; to busy the mind continually in supplying the body's wants; until at last all classes should be infected with the plague of universal poverty." See *PPE*, ch. V, p. 108. References to the *Principles of Political Economy and Taxation* will be to volume 1 of *The Works and Correspondence of David Ricardo*, 11 vols., ed. by P. Sraffa with the collaboration of M.H.

Dobb (Cambridge: University Press, 1951). The abbreviation *PPE* will be used, followed by chapter and page numbers.

91. *Ibid.*, ch. IX, p. 165.

92. "The nature of the evil points out the remedy. By gradually contracting the sphere of the poor laws; by impressing on the poor the value of independence, by teaching them that they must look not to sympathetic or casual charity, but to their own exertions for support, that prudence and forethought are neither unnecessary nor unprofitable virtues, we shall by degrees approach a sounder and more healthful state." See *Ibid.*, ch. V, p. 107.

93. Ronald L. Meek, *Studies in the Labor Theory of Value*, 2nd ed. (New York and London: Monthly Review Press, 1976), p. 112.

94. See Samuel Hollander, *The Economics of David Ricardo*, Studies in Classical Political Economy, 2 (Toronto: University of Toronto Press, 1979), pp. 263 and 265 concerning the significance of this for normative theory.

95. H. Brems asserts the importance for Ricardo of the question of distribution over price theory. See "Ricardo's Long-Run Equilibrium," in *David Ricardo: Critical Assessments*, ed. by J.C. Wood (London: Croom Helm, 1985), Vol. III., p. 307.

96. Eleonore Lipschitz comes to a similar point from another direction—Ricardo's commitment to justice: "Die dominierende ethische und philosophische Ideen der Zeit, in der Ricardo lebte, finden sich auch in seiner Theorie wieder, besonders die Idee der Gerechtigkeit, übernommen aus der puritanischen Ethik, soll die Norm sein, nach der sich eine Wirtschaft auszurichten hat. . . . Warum ist der Wertmassstab der Güter bei Ricardo nur die aufgewendete Arbeitsmenge, obwohl er die wachsende Bedeutung des fixen Kapitals als Kostenelement erkannt hat? Weil damit in der Theorie der Zusammenhang zwischen Arbeitsopfer und Entgelt gewahrt bleibt." While finding this insightful, I suggest that the evidence indicates that Ricardo's concern for the limited lives and powers of working persons pervades his systematic analysis. Ricardo is capable of arguing for more just arrangements because of the trenchancy of his arguments based on a keen attention to finitude which effort attempts to meet. In addition, to cast the question in terms of effort and recompense, although accurate, remains too narrow. The merit of Ricardo's analysis stands in his comprehension of the entire economic system upon which the relation of effort and recompense depend—a system centered upon finite, working persons. *Die theoretischen Grundlagen David Ricardos im Lichte des Briefwechsels* (Berlin: Duncker und Humblot, 1957), p. 120.

97. *PPE*, ch. I, p. 20.

98. Hollander, *Economics of David Ricardo*, p. 193.

99. Antoine Louis Claude, Comte de Destutt de Tracy, *Eléments d'Idéologie*, IVe et Ve parties. "Traité de la volonté et ses effets" (Paris, 1815). Cited in *PPE*, ch. XX, p. 285.

100. For example, *PPE*, ch. VII, p. 145.

101. "In those countries where there is an abundance of fertile land, but where from ignorance, indolence, and barbarism of the inhabitants, they are exposed to all the evils of war and famine . . . a very different remedy should be applied. . . . To be made happier they require only to

be better governed and instructed, as the augmentation of capital, beyond the augmentation of people, would be the inevitable result." See *Ibid.*, ch. V, p. 99.

102. For example: "But improvements in agriculture are of two kinds: those which increase the productive powers of the land, and those which enable us, by improving our machinery, to obtain its produce with less labour. . . . For it is the essential quality of an improvement to diminish the quantity of labour before required to produce a commodity. . . ." See *Ibid.*, ch. II, p. 80.

103. Ricardo's maiden speech in Parliament pleaded the cause of laborers who were plagued with inadequate wages. See *Collected Works*, ed. by Sraffa, Vol. 5., p. 1. Also see Hollander, *The Economics of David Ricardo*, p. 566.

104. *PPE*, ch. VII, p. 133.

105. See *Ibid.*, ch. V, p. 95.

106. Hollander, *The Economics of David Ricardo*, p. 547. See *PPE*, ch. XXXII, pp. 424-425. E. Lipschitz emphasizes Ricardo's concern with theoretical analysis for the sake of justice: "Arbeit soll allein die Voraussetzung für den Erfolg sein, und Eigentum erscheint ihm im Grunde nur dann gerechtfertigt, wenn es durch Arbeit erworben ist. Wo dieser direkte Zusammenhang gelöst und diese normative Idee durch die tatsächliche Entwicklung in der Wirklichkeit verletzt ist, wendet Ricardo allen Scharfsinn auf, um durch wirklichkeitsferne Konstruktionen in der Theorie diesen Zusammenhang wiederherzustellen, und wo ihm das nicht gelingt, den Störungsfaktor als gegen das Glück der Menschen gerichtet darzustellen." *Die theoretischen Grundlagen*, p. 120. I note that Lipschitz's view of Ricardo's linking of work and property does not take into account that Ricardo respected property and did not call for its redistribution. In any case, the happiness of persons toward which Ricardo applies his efforts is that of working persons. Ricardo saw the systemic increase in rent as the *Störungsfaktor* which blocked this happiness.

107. "It must, however, not be forgotten, that the produce of taxes is generally wastefully expended, that they are always obtained at the expence of the people's comforts and enjoyments, and commonly either diminish capital or retard its accumulation. By diminishing capital they tend to diminish the real fund destined for the maintenance of labour; and therefore to diminish the real demand for it." See *PPE*, ch. XVI, p. 221-222.

108. See *Ibid.*, ch. IX, p. 167.

109. Hollander, *Economics of David Ricardo*, p. 540.

110. *PPE*, ch. V, p. 93.

111. *Ibid.*, ch. V, pp. 100-101.

112. *Collected Works*, ed. by Sraffa, vol. 4, "Absolute Value and Exchangeable Value," later version, p. 405.

113. *PPE*, ch. I, p. 31.

114. See *Ibid.*, ch. II, p. 67; also see p. 68.

115. See *Ibid.*, ch. II, p. 76, Ricardo's note.

116. See *Ibid.*, ch. XX, p. 283.

117. *Ibid.*, ch. II, p. 69; emphases added.

118. *Ibid.*, ch. II, p. 75.

119. "By far the greatest number of commodities which are objects of exchange are produced by the union of capital and labour, that is to say, of labour employed for a certain time." *Collected Works*, ed. by Sraffa, Vol. 4, "Absolute Value and Exchangeable Value," later version, p. 405.

120. *Ibid.*

121. Even in reference to the use of money (gold) as a gauge of value, he writes: "The circumstances of this measure being produced in the same length of time as corn and most other vegetable food which forms by far the most valuable article of daily consumption would decide me in giving it preference." *Ibid.*, pp. 405-406.

122. *PPE*, ch. I, p. 14.

123. *Ibid.*, ch. I, p. 17.

124. *Collected Works*, ed. by Sraffa, Vol. 7, p. 377.

125. *PPE*, ch. XXVIII, p. 374.

126. Meek, *Studies in the Labor Theory of Value*, pp. 114-115.

127. *PPE*, ch. I, p. 11.

128. *Ibid.*, ch. XX, p. 273.

129. *Ibid.*, ch. XX, p. 278.

130. *Collected Works*, ed. by Sraffa, Vol. 4, "Absolute Value and Exchangeable Value," earlier version, p. 396.

131. *Ibid.*, p. 367.

132. *Ibid.*, later version, p. 403.

133. In a letter to McCulloch of 21. August, 1823, Ricardo queries, "Is it not clear then that as soon as we are in possession of the knowledge of circumstances which determine the value of commodities, we are enabled to say what is necessary to give us an invariable measure of value." See *Collected Works*, ed. by Sraffa, Vol. IX, p. 358. In commenting on the above passage, P. Sraffa describes Ricardo's project, ". . . The problem which mainly interested him was not that of finding an actual commodity which would accurately measure the value of corn or silver at different times and places; but rather that of finding the conditions which a commodity would have to satisfy in order to be invariable in value—and this came close to

identifying the problem of a measure with that of a law of value." See "Introduction" in *Collected Works*, Vol. 1, pp. xl-xli.

134. One "sacrifice" text may be adduced as an example of this strand of Ricardo's thought: "That commodity is alone invariable, which at all times requires the same sacrifice of toil and labour to produce it. Of such a commodity we have no knowledge, but we may hypothetically argue and speak about it, as if we had; and may improve our knowledge of the science. . . ." See *PPE*, ch. XX, p. 275. Sraffa's note indicates that even this phrasing results only from an oversight which neglects to take into account a development in Ricardo's thought: "These observations on an invariable standard of value should have been altered in ed. 3 to agree with the changes introduced in Chapter I. . . ." *Ibid.*, p. 275, n. 2.

135. See Meek, *Studies in the Labor Theory of Value*, p. 112.

136. *PPE*, ch. XX, p. 285, n. 1.

137. *Collected Works*, ed. by Sraffa, Vol. 4, later version, p. 401.

138. "He [Mill] wants what is primitive to be simple, failing this, there can, in his view, be no possible explanation. He also wants what is primitive to be concrete: failing this his philosophy would cease to be a philosophy of experience." E. Halévy, *The Growth of Philosophic Radicalism* (London: Faber and Faber, 1928), p. 448.

139. Lipschitz, *Die theoretischen Grundlagen*, p. 65.

140. For example, see Wilhelm Hasbach, *Die allgemeinen philosophischen Grundlagen der von Francois Quesnay und Adam Smith begründeten politischen Ökonomie*, Staats- und sozialwissenschaftlichen Forschungen, hrsg. von Gustav Schmoller, 10. Bd., 2. Heft, der ganzen Reihe 43. Heft (Leipzig: Duncker und Humblot, 1890), p. 172.

141. See, for example, *PPE*, ch. VII, p. 132. Also, K.R. Ranadive contrasts the Ricardian emphasis on productive interchange with the neoclassical focus on exchange: "The Relevance of the Ricardian System," in *David Ricardo: Critical Assessments*, ed. by J.C. Wood, Vol. III, pp. 568-569.

142. *PPE*, ch. V, p. 93.

143. "When the market price of labour is below its natural price, the condition of the labourers is most wretched: their poverty deprives them of those comforts which custom renders absolute necessaries." See *Ibid.*, ch. V, p. 94.

144. *Ibid.*, ch. I, p. 12.

145. See Hollander, *The Economics of David Ricardo*, p. 265 and M. Dobb, "Ricardo and Adam Smith," in *Essays on Adam Smith*, ed. by Andrew S. Skinner and Thomas Wilson, p. 330.

146. "The fate of the labourer will be less happy [than that of the landlord]; he will receive more money wages, it is true, but his corn wages will be reduced; and not only his command of corn, but his general condition will be deteriorated, by his finding it more difficult to maintain the market rate of wages above their natural rate. While the price of corn rises 10 per cent.,

wages will always rise less than 10 per cent., but rent will always rise more; the condition of the labourer will generally decline, and that of the landlord will always be improved." *PPE*, ch. V, pp. 102-103.

147. M. Blaug asserts that "the postulate of diminishing returns is insufficient to derive his general theorems about the pattern of income shares in a growing economy." M. Blaug admits, when he continues, that ". . . the theorem does hold true on his own assumptions for money wages and the rate of profit, but it does not hold true for the relative shares of labor and capital." *Economic Theory in Retrospect*, 3rd ed. (Cambridge: Cambridge University Press, 1978), pp. 109-110. On these terms society would still begin to split along the lines dividing the different functional groups within society. On the basis of Ricardo's premises, Milton Myers suggests alternative accounts for the break up of society. See *The Soul of Modern Economic Man* (Chicago: University of Chicago Press, 1983), p. 129.

148. *PPE*, p. 5.

149. *Ibid.*, ch. I, p. 22.

150. *Ibid.*, ch. V, pp. 93-94.

151. *Ibid.*, ch. VI, p. 120.

152. *Ibid.*, ch. XXXI, p. 392. These are the second and third of the four conclusions offered by Ricardo. The other two may be taken as implicit in these for the purposes of this subsection. With regard to the second point here cited, attention may also be called to this: ". . . I am convinced, that the substitution of machinery for human labour is often very injurious to the interests of the class of labourers." *Ibid.*, ch. XXXI, p. 388.

153. *Ibid.*, ch. XXXI, p. 390.

154. *Ibid.*, ch. XXXI, pp. 396-397.

155. *Ibid.*, ch. XXXI, p. 395.

156. Hla Myint, *Theories of Welfare Economics* (London: Longmans Green & Co., 1948), p. 26.

157. A closing note, of some interest, will corroborate the foregoing exposition. In the following quotation, consider, first, that weapons for hunting represent a type of tool or, by extension, a type of machine: "Or suppose that the same quantity of labour was necessary to make both weapons, but that they were of very unequal durability; of the durable implement only a small portion of its value would be transferred to the commodity, a much greater portion of the value of the less durable implement would be realized in the commodity which it contributed to produce." See *PPE*, ch. I, p. 23. I contend that Ricardo's prime analogue for the machine and its manner of production, is the human person who, limited in his time and energies (note the durability of the machine), transfers from himself to the product a value which is therein "realized."

158. *Collected Works*, ed. by Sraffa, Vol. 5, p. 303. Also cited in Hollander, *The Economics of David Ricardo*, p. 568.

159. F.H. Knight, *On the History and Method of Economics* (Chicago: University of Chicago Press, 1956), p. 39.

160. Ronald L. Meek, *Economics and Ideology and Other Essays* (London: Chapman and Hall, 1967), p. 204.

161. Two references in Ricardo offer an adumbration of a fuller treatment of service activity. A reference from chapter three of the *Principles* corroborates the preceding presentation of Ricardo's implicit concern: "The metals, like other things, are obtained by labour. Nature, indeed, produces them; but it is the labour of man which extracts them from the bowels of the earth, and prepares them for our service." See p. 85. As the second reference, an interesting parallel to the concern in Smith over the difference between productive and unproductive labor offers itself in the *Principles*. In chapter thirty-one, Ricardo distinguishes between two types of luxury spending on the part of the rich: on menial servants or for luxury goods. See *Ibid.*, chap. 31, p. 393. If luxuries are not produced entirely by labor, then the total expenditure would further the interests of laborers more if menial servants were hired. Contrary to the tradition stemming from B. de Mandeville, which would see such servants as engaged in unproductive, useless consumption, Ricardo here pays heed to two types of service: that which the servants would offer their employers and the service of providing for a particular and pre-eminent need of a person: the need to find a place of insertion for his work activity.

162. Conze, "Arbeit," p. 182.

163. An examination of the following quotation concerning the "Ricardian effect" indicates the isolation of work activity: "[A] All commodities which are produced by very valuable machinery, or in very valuable buildings, or which require a great length of time before they can be brought to market, would fall in relative value, [B] while all those which were chiefly produced by labor, or which would be speedily brought to market would rise in relative value." *PPE*, ch. I, p. 35. Which of the two sets of criteria apply to persons, the providers of labor? Both or neither? This ambiguity indicates the difficulty of conceiving the commodification of labor.

164. Meek, *Economics and Ideology*, pp. 15-16. See also M. Reich who demonstrates that the charge of empirical irrelevance brought against Ricardo is without foundation. Reich avers that the middle class, once established in power, wished to excise the conflictual elements found in Ricardo's theory. "Empirical and Ideological Elements in the Decline of Ricardian Economics," in *David Ricardo: Critical Assessments*, ed. by J.C. Wood, Vol. III, p. 667.

165. Georg Lukács, *Hegel's False and His Genuine Ontology*, Ch. 3 of Pt. 1 of *Toward the Ontology of Social Being*, translated by David Fernbach (London: Merlin Press, 1978), p. 54.

Chapter 2
German Economic Thought: Georg W.F. Hegel and Karl Marx

Having completed the examination of work activity in the writings of Smith and Ricardo, I now explore the idea of the working person, the limits faced by this worker, the individual worker's social nature, the role of technology in work and the service component of work, as aspects of G.W.F. Hegel's and K. Marx's concepts of work. In my treatment of Hegel's thought, I show how he presents work as that factor through which persons develop their capacities and become social beings.

Hegel's handling of the limits which constrain human activity shows that, as these limits are transformed by work, the person develops. In a more extensive section, I then indicate how Hegel sees sociality as intrinsic to work activity. In particular I examine those institutions which Hegel describes as structures for securing productivity and reciprocity through work activity. Then I focus on Hegel's treatment of the tool as an instantiation of the rationality and universality of work. Concluding the section on Hegel, I show how the service component of work activity is implicit in the class and corporate structures which secure participation in the creation of a common world.

The Sociality of Persons in Hegel's Thought

Hegel understands work as an activity through which persons actualize their capacities. Lest his position appear naive and in order to show that a number of his insights critical of the social system antedate Marx's better known, similar criticisms, I begin with Hegel's treatment of those effects of work which debilitate the workers and frustrate their shaping of a world.

Even as early as the 1805-1806 *System*, Hegel describes how workers partake in a social organization of work, the operations of which remain mysterious to them, so that even as the their activity drives this system, it diminishes the value of their efforts, reduces the range of their activities, narrows their own capacities and multiplies their needs. Indeed, the system itself subjects whole groups of workers to the chance that their own work may no longer be required. While exposing many laborers to the dangers of deprivation, for others the system amasses wealth. The inequality in distribution and the neediness of many rend the social fabric as passions rise. All this Hegel attributes to the operations of the social organization of work.[1]

Yet this need not be the upshot of the system of work, for it can be constituted so as to further the human and social development of persons. Hegel's theory moves toward an understanding of work which does not remain solely at the level of the economic.[2] Hegel's conception of work extends beyond the limits of labor conceived along narrowly economic lines, for him work has become the primary human activity. In the next pages, I show that Hegel holds that this primary activity shapes consciousness, develops the intellectual powers, provides the foundation for volition, brings about the expansion of free choice through social institutions and creates persons as social beings. I begin with the first of these results of work activity.

Which characteristics of persons are actualized in their work activity? Through work these agents achieve self-consciousness. Through work, in which agents at one point become externalized and strange to themselves, each individual achieves a recognition of him- or herself as person.[3] Similarly and only shortly after this description of self-externalization, Hegel wrote, in the *Phenomenology of the Spirit*, that the bondsman severs every attachment to natural existence and thereby becomes conscious of himself through his work.[4]

Work affects the person by developing the individual's intellectual powers. As presented in the *Phenomenology*, the bondsman, compelled to labor for another, understands that the lord can take his life. Having come to this knowledge and now guided by concepts, he controls his inclinations and further directs himself to a peculiarly human activity: work.[5]

This capacity to choose, when fully developed, becomes the aptitude of the free person. In the *Phenomenology*, Hegel presents the lord as one who is separated from handling the world of things. Disjoined from the opportunity to work upon these, he cannot become a free, independent person; he remains bound to the desires which direct his consumption. In and through his labor, the bondsman, however, realizes his own capacity for thought, by which he is able freely to choose and effect projects.[6] Beyond the lord and bondsman and beyond the unhappy consciousness, the further development of consciousness at the level of reason (*Vernunft*) reveals the transcendent freedom of the subject which negates any determination in the work of this subject and implies that in the work the subject actualizes freedom.[7]

Later, in the *Philosophy of Right*, Hegel describes both theoretical and practical education as he considers the type of work found in civil society with its division of labor. He implies that the "numerous ends" and "different processes" of this work activity provide the "multiplicity of objects and situations" upon which theoretical education develops.[8]

Reason establishes purposes which guide the person's actions toward their accomplishment. These purposes require the reshaping of the life of desires. Workers shape both their world and their desires in accord with their individual wills.[9] The projects chosen or adopted lead to a development of the person's will through this individual's repeated efforts to conform his or her own desires and inclinations to the requirements of the project and to bring external factors in accord with this plan.

By means of institutions, work activity brings about the expansion of this liberty. In the *Philosophy of Right* Hegel calls attention to the advancement of the human persons from potency to actualization by dint of work activity when he speaks of the "moment of liberation intrinsic in work."[10] Hegel situates the treatment of this liberation within his consideration of civil society.[11] Not for Hegel is the British notion of a freedom independent of all external factors. Rather, the framework—provided by both civil society and state—presents the field upon which human freedom may reach its fullness.[12] The exercise of personal freedom thereby comes to be further developed because persons realize their freedom at a far higher level than that which the isolated actions of the bare individual may reach. They achieve this by way of social institutions and practices—the market, the corporations, the police and the branches of government. The reconciliation of subject and object, when realized through these social institutions and practices, enables persons to articulate their actions within a far wider and more varied world than would be the case were they each operating only within the narrow sphere of individual achievement. In addition, the larger range for reconciling work conduces to the richer development of the persons themselves.

Work creates persons as social beings. As an object, the product of work alters the world, not just for the worker but for all persons. Because the product achieved in work now exists apart from the worker it may be given in exchange or as gift.[13] Workers recognize in their products not only an expression of themselves but also the determination of these objects for others. This universality and the possible orientation of objects toward others conduce, through reflection, to the development of sociality in persons. Similarly, the recognition that no one person can satisfy his or her needs by way of his or her own labor deepens this sociality. Conversely, individuals realize that the particular work of each provides for others. In that work activity develops the ability to abstract, in that the workers realize that their wills develop and in that they see the mutuality of provision through work, the working person can come to recognize the interaction of separate wills and can come to learn the social skills requisite for the shaping of this interaction.

The Transformation of Limits

Work confronts the limits imposed by desires and need, by nature and by death. Self-consciousness is achieved not by the removal but through the transformation of the phenomenon of life with all its limits, contingent and embodied.[14] Life presents a field of the unrealized, upon which field human effort exerts itself to actualize its potential.

This potential begins with the embodied persons themselves. A change in the subject, such as the movement to universal consciousness, entails a change in embodiment.[15] Hegel addresses this point in the *Encyclopaedia*:

> . . . In like measure I cannot be recognized as immediate, except so far as I overcome the mere immediacy on my own part, and thus give existence to my freedom. But this immediacy is at the same time the corporeity of self-consciousness, in which as in its sign and tool the latter has its own sense of self, and its being for others, and the means for entering into relations with them.[16]

The movement to universal consciousness requires the employment of one's own embodiment as expressive and useful in the attaining of a universality in conjunction with others. This results in an altered self-consciousness, with a consequent alteration of one's view of one's own corporeity.

These human efforts are, naturally, centered in particular persons, each of whom is necessarily embodied and consequently shares in the limitedness of contingent life. Desires, centered in the embodied person, represent the intersection of the unrealized world with human striving. In desire the subject first comes to know his or her difference from the object sought. Self-consciousness presupposes desire—a biological, animal want—as its necessary but insufficient condition. Human desire exceeds that of the animal in that the human longing does not terminate in the consumption of the desired.

The recurrent, yet transient, processes of biological and personal life precipitate in sundry desires. Work stands as the force which confronts the natural waxing and waning of the desires. Hegel describes this engagement: "Work . . . is desire held in check, fleetingness staved off; in other words, work forms and shapes the thing."[17] The product of work perdures. Work uses the passing elements of nature to create an object remaining in nature which outwits this transience.

The maturing of Hegel's thought marks a movement from the treatment of individual desires to a system of social needs. In the *Phenomenology* Hegel directs his attention to the action of work upon desire, while in the later consideration as found in the *Philosophy of Right* and the *Encyclopedia*, he treats work activity within the system of *needs*. At the advanced, more complex level of civil society, Hegel clearly presents the specifically human character of these needs:

> An animal's needs and its ways and means of satisfying them are both alike restricted in scope. Though man is subject to this restriction too, yet at the same time he evinces his transcendence of it and his universality, first by the multiplication of needs and means of satisfying them, and secondly by the differentiation and division of concrete need into single parts and aspects which in turn become different needs, particularized and so more abstract.[18]
>
> Since in social needs, as the conjunction of immediate or natural needs with mental needs arising from ideas, it is needs of the latter type which because of their universality make themselves preponderant, this social moment has in it the aspect of liberation, i.e. the strict natural necessity of need is observed and man is concerned with his own opinion, indeed with an opinion which is universal, and with a necessity of his own making alone, instead of with an external necessity, an inner contingency, and mere caprice.[19]

Needs are particularly human because human efforts multiply both them and the means to satisfying them, because the needs themselves are subdivided and because their social character results from the combination of the natural with the mental needs. Human activity does not remove the limits inherent in the natural life: natural necessity is observed. Indeed, persons meet yet another limit, constituted by the complex set of needs which they themselves have fashioned.[20] Hegel writes that, "The means of acquiring and preparing the particular means appropriate to our similarly particularized needs is work."[21] This work activity pursues the satisfaction of needs, yet Hegel also maintains that this work operates within a fabric of complex requirements, the complexity of which continually grows as a result of work activity. The needs which spring from the potential of persons progressively exceed the offerings of the ever more elaborate world. In seeking to actualize themselves, persons continually face limits which spring from their desires and needs.

Not only do persons face the limits imposed by their own desires and needs, but also nature contours the field of human work through the limits which she imposes. The human activity which externalizes itself proceeds in a dialectic with the processes of nature. Neither side is eradicated; at each stage the novel developments have reshaped the interaction, but both sides—externalizing activity and natural forces—remain.[22] A significant Hegelian example: slaves have been freed from nature, master, and their own nature; the liberation, however, has been achieved from the previous status of their ties: from their specific ties to nature, from their relationship to the master and from their relationship to their own nature. Although these persons have now been altered, limits—which nature, the master and their own nature impose—remain.

In his later work, once again, Hegel makes clear this dialectical interaction with the limits imposed by nature:

> Mind attains its actuality only by creating a dualism within itself, by submitting itself to physical needs and the claim of these external necessities, and so imposing on itself this barrier and this finitude, and finally by maturing (*bildet*) itself inwardly even when under this barrier until it overcomes it and obtains its objective reality in the finite.[23]

Mind, instantiated in the acting person, accepts the limits imposed by nature, develops itself by means of activity within these limits, and achieves objective reality—the overcoming of the past does not eliminate the finite altogether. Nature poses a set of limits to which persons must not submit and to which they must oppose themselves. It is in this way that persons achieve their freedom.[24] This freedom is, however, not a means of transcending all bounds.

Work has been presented as a dialectic activity within limits. Death is the individual, yet common, limit of all human activity. In the struggle for recognition as found in the *Phenomenology*, the threat of death emerges as the result of a dual action: each contender threatens the other with death. The struggle for liberty maneuvers against the ultimate limit to life in order to move beyond the merely given. As earnest of one's own desire and effort, each stakes one's own life, so that freedom may be won. By staking one's own life one declares that what matters finally is not the given expanse of life but how activity stands apart from and determines that life. Hegel describes the bondsman, whose "consciousness has been fearful, not of this or that particular thing or just at odd moments, but its whole being has been seized with dread; for it has experienced the fear of *death, the absolute Lord.*"[25] In the struggle for recognition, the bondsman has chosen life instead of liberty because the proclivity toward activity cannot be fulfilled without life, constricted as this may be beneath a lord. Just as limits are the natural, constraining factors which the self accepts and works within in order to remake nature; so here: the bondsman faces death and thereupon "trembles in every fibre of his being, and everything solid and stable has been shaken to its foundations." Thus, everything fixed has been made fluid; out of the given emerges the possible. The resolute acknowledgement of the ultimate limit reveals the potential for activity which can alter any given. As the bondsman realizes the universal moment to which this final limit has delivered him and as this becomes explicit in the consciousness of his own being, the work in servitude remakes the constraints of death.

Later, in the *Philosophy of Mind*, Hegel reasserts that the significance of the life-and-death struggle for recognition is critical to the achievement of freedom. While he takes care to declare that this struggle which leads to lordship and servitude occurs only in the natural condition of separate individuals antecedent to civil society and the state,[26] he also notes that this first relation marks "the emergence of social life and the commencement of political union."[27] Since the struggle which leads to this double beginning depended on the threat of death and since it led to the dialectic of work and death in the person of the bondsman, the activity intrinsic to the lord-bondsman relation indicates in germinal form the significance of work activity for the shaping of subsequent social and political groupings.[28]

The Sociality of Production and Recognition

Contemporary analyses frequently overlook the sociality intrinsic to work.[29] Yet in actuality, work both creates society and reacts to the constantly changing society which it creates. In this section, I will examine Hegel's treatment of this sociality, which commences in his early writings, with an examination of the alteration in workers as their individual activities become universal and social. In his later work, Hegel presents sociality in terms of the "abstraction" which leads to the subdivision of needs and means. Of great significance is Hegel's presentation of those social institutions which are created on the basis of work activity so that persons can actualize reciprocity and productivity—two key aspects of work. Work establishes both civil society and also the classes and corporations which constitute much of this economic sphere. Work is also the basis for the pinnacle of society, the state, which secures the framework so that persons can accomplish production and recognition. I now turn to the first element in this complex approach of Hegel to sociality.

Even long before the *Philosophy of Right*, in the first of his Jena systems, Hegel asserted the intrinsic social dimension of work activity, when he declared that even the work of the single person entails the universal.[30] The desire to be recognized, the learning of skills and the implicit rationality of these accomplishments establish the universality of a group of persons.[31] Beyond the Jena position in which the person's activity leads to the social world on the basis of universal skills, in the *Phenomenology*, the personal activity requires the response of the other. The first individual can not make use of the other unless the independent and self-contained other agrees.[32] Just as in the examination of the universality emergent from work activity, so too here in his indication of the requisite response from the other self-consciousness, the link between the individual's activity and that of a group of people still requires a more firmly forged connection. This Hegel does in his later work.

As presented in the *Philosophy of Right*, in the system of needs at the stage of civil society, subjective particularity attains objectivity in two ways: both through external things and through the effort of work. First, in order to work upon some thing, one must come to terms with the others to whom this may belong as property[33] or, in a particular case, as their product. Second, in exercising effort, "the universal asserts itself in the bearing which this satisfaction [of subjective particularity] has on the needs of others and their free arbitrary wills."[34] That Hegel unfolds his understanding of the sociality of work in much the same terms as he did years earlier at Jena is revealed in this passage from the *Philosophy of Right*:

> When needs and means become abstract in quality, abstraction is also a character of the reciprocal relation of individuals to one another. This abstract character, universality, is the character of being recognized and is the moment which makes concrete, i.e. social, the isolated and abstract needs and their ways and means of satisfaction.[35]

Whereas in the earlier Jena citation the movement had been from the individual to the universal; here the development proceeds as a movement of reason from the individual through the universal to the concrete or social.[36]

The universalizing, objectifying vector in work activity brings about several specific, significant changes in the social world. The abstracting character of work leads to a subdivision of needs and of the means to them: ". . . The means to particularized needs and all the various ways of satisfying these are themselves divided and multiplied and so in turn become proximate ends and abstract needs."[37] The attaining of these subdivided means to particularized ends requires the delimitation of work activity by individuals to tasks removed from the satisfaction of primary needs, to tasks which are only fragments of an entire process which conduces to the meeting of a need; thus there results the division of labor. Furthermore, the abstracting of personal skills used in the means of production necessitates the reciprocal dependence of persons.[38] Through labor persons become related to others in society in terms of the satisfaction of their needs; they also become related to one another, as it were, technically. The reciprocity of relationship is not, in the first place, economic. These social developments have proceeded, on the basis neither of subsistence needs nor of power constraints but, rather, on the basis of the rationality inherent in the work process, a rationality through which human freedom actualizes itself.

Attention must be drawn to Hegel's particular concept of the division of labor. The work activity of a person, precisely in its dependence on others for its actualization in objectivity, that is, precisely because of its social character, must be partial. On this basis one must distinguish between a division of labor which is intrinsic to work activity and one which arises as a consequence of particular socio-economic circumstances.

Since labor is social, a discontinuity emerges between the efforts of individuals and the satisfaction of their particular, specific needs. One may speak still more fundamentally of the necessity of work activity to proceed both against and in conjunction with the other. The task which constantly molds the process of work lies not, then, in the disutility of effort but, rather, in the resistance to and the negation of the given, which includes the other. Working requires opposition as intrinsic to the process of the actualization of freedom. At the same time, work activity proceeds in collaboration with others.

In advanced forms of social production, work both brings about the differentiation of tasks and needs which distances persons from each other, and also achieves patterns of integration.[39] In applying their efforts to overcome the nature both outside of and within themselves, workers shape a world not only of products but also of institutions. The achievement of potential requires a properly ordered set of social institutions.[40]

In examining the social institutions in terms of their adequacy to the nature of work activity, two criteria must be applied: reciprocity and productivity. A non-reciprocal, structured social interaction undermines sociality, because this framework subverts the free and responsible work activity of some persons as it reduces them to mere providers of prescribed activities. Should social institutions inhibit productivity, then the practical orientation of work activity will be frustrated.[41] Neither case will actualize human potential. In order to permit an understanding of human activity within these institutions, Hegel presents, instead of a liberal division of social institutions in which the sundry powers check and compete with each other,[42] an organic model of interaction, wherein each person may contribute responsibly and effectively to the maintenance and reshaping the whole. In this model, the vital, organic substance lives through the incorporation of particular circles into the larger whole.[43]

I now turn to the specific structures which Hegel presents in the *Philosophy of Right*. Although work appears as a moment of abstraction at the level of the family,[44] the first level of "ethical life" and the proper field of its deployment is civil society. At this level, within the "system of needs," work activity accomplishes the division of labor and mutual dependence:

> The universal and objective element in work . . . lies in the abstracting process which effects the subdivision of needs and means and thereby *eo ipso* subdivides production and brings about the division of labour. . . . At the same time, this abstraction of one man's skill and means of production from another's completes and makes necessary everywhere the dependence of men on one another and their reciprocal relation in the satisfaction of other needs.[45]

Hegel realizes that work activity creates civil society and thereby contributes to the establishment, structures and vitality of the state, which caps the total articulation of society.[46] Hegel recognizes that although self-interest and self-assertion motivate activity in civil society, the individual can realize the ends sought only in established interaction with others. Moreover, at the level of the state, this same individual activity leads to the establishment of still further structures which integrate all persons into a whole, wherein they may achieve true recognition.

The class system emerges as one stage on the path to this full integration and recognition. The complex patterns and movements of need, production and dependence found in the system of needs crystallize "owing to the universality inherent in their content" into "particular systems of needs, means and types of work," that is, into class divisions.[47] The three classes Hegel designates as the substantial (that is, agricultural), the formal (that is, business), and the universal (that is, civil service).[48] Working persons achieve an integration into society by way of their particular classes:

> In this class-system, the ethical frame of mind . . . is rectitude and esprit de corps, i.e. the disposition to make oneself a member of one of the moments of civil society *by one's own act, through one's own energy, industry and skill,* to maintain oneself in this position, and to fend for oneself only through this process of mediating oneself with the universal, while in this way gaining recognition both in one's own eyes and in the eyes of others.[49]

The status of individuals in the class results not from some passive allotment. Rather, they themselves by their work energetically maintain this position. Indeed, their activities enable them to surpass their own interests and goals by integrating them into a larger whole; by means of this integration they attain recognition in society. Not the varied relationships to the means of production but, rather, types of work distinguish the classes. Each type requires a set of general skills and a group "ethos or consciousness" which results among those who undertake this work.[50] The classes are channels which integrate particular working persons into the entirety of society. These characteristics of the Hegelian understanding of class recur in the theory of Schmoller and Pesch concerning corporative work groups.

Corporations are means both for integrating persons into the larger group that provides the structure for their work and for articulating them into society. The corporation emerges from the needs of the person:

> Thus we have the corporation, in which the particular citizen in his private capacity finds the securing of his stock, whilst at the same time he in it emerges from his single private interest, and has a conscious activity for a comparatively universal end, just as in his legal and professional duties he has his social morality.[51]

The corporation proceeds from the activity of the working person toward its relatively universal end. In opposition to the centralizing tendencies of modern nation-states, Hegel revived interest in subordinate bodies such as the corporation.[52] One goal Hegel sought to achieve in this was an amelioration of the situation of the person caught in the individualized competition of the market system.[53] Given the relatively backward industrial economy of Germany at the time he was writing, Hegel may have attempted to prevent the emergence of the striking ills which he knew plagued Great Britain as a result of competition. Through action toward an attainable universal, through the opportunity for solidarity, and through the protective agencies offered by the corporation, individuals would be capable of maintaining the importance of their central life activity in the midst of otherwise destructive social forces.[54] The corporations enable working persons both to produce and to be recognized.[55]

Hegel saw the political articulation of the working person into society as providing another range of benefits which derive from the corporate structure. The varied work of persons bind them to the corporation; these persons contribute to the internal life of the corporation; the corporation contributes to the political life of the state. Thus their work draws them into the life of the state. The interaction of civil society and state requires the political action of the economically based corporations:

> It is a prime concern of the state that a middle class should be developed, but this can be done only if the state is an organic unity like the one described here, i.e. it can be done only by giving authority to spheres of particular interests, which are relatively independent, and by appointing an army of officials whose personal arbitrariness is broken against such authorized bodies. Action in accordance with everyone's rights, and the habit of such action, is a consequence of the counterpoise to officialdom which independent and self-subsistent bodies create.[56]

The interaction of the corporations and officialdom secures the field upon which action, including that of the individual person, may be assured as habitual activity leading, given the overarching purpose of the organic and total conception of society, to the actualization of human potential. The corporations offer persons, then, beyond the guarantees provided to each by the fiction of a mere, legal *persona*, a social and political basis for their own, individual development.[57] Neither connection due to birth nor the subjection to the arbitrary decisions of a ruler matters for the life opportunities of individuals. Rather, what carries weight is the freedom entailed in the exercise of their skills and abilities. The corporations secure this exercise by sheltering the person's capacity to work.

The corporations, although economically grounded, provide a necessary basis for the political power of the state. Without the organization offered by these associations, the masses of people would be a powerless "heap, an aggregate of atomic units."[58] The state grants a more formal and comprehensive recognition to these bodies and in so doing confirms that recognition already attained by the members of the corporation through the exercise of their work. This particularity supports the universal; persons discover union through diversity.[59]

The activity of persons achieved within the corporations develops in such a manner that the state completes the universality of their lives:

> . . . The corporation mind, engendered when the particular spheres gain their title to rights, is now inwardly converted into the mind of the state, since it finds in the state the means of maintaining its particular ends. This is the secret of the patriotism of the citizens in the sense that they know the state as their substance, because it is the state that maintains their particular spheres of interest together with the title, authority, and welfare of these. In the corporation mind the rooting of the particular in the universal is directly entailed, and for this reason it is in that mind that the depth and strength which the state possesses in sentiment is seated.[60]

The life within the corporation leads the particular individuals to link their own welfare with that of the corporation; consequently, when they see the state as necessary for the support of the corporation, they include the state within the ambit of their allegiance. Indeed, the articulation of the person within the corporation also provides the means by which individuals learn a similar but broader comprehension of their integration within the universality of the state.

The accomplishment of recognition which persons obtain at the level of the state, and not solely in civil society, derives from their particular work activity, which their individual corporations preserve and foster.[61] The state, in the complexity of its subordinate organs, provides the structures which enable persons to recognize their own work in addition to that of others as instrumental to private and common goods as well as socially integrative.

The fulfilling of personal freedom also results from the rationality which the state introduces into work activity. Persons must rationally grasp their integration into society; yet this would not be possible so long as the movements of production and consumption in civil society exceed the comprehension even of society as a whole. The state makes possible a rational conception of the economic activity which shapes the work the individual undertakes.[62]

Insofar as the state incorporates the individual's participation by way of the corporation, Hegel has achieved his purpose of obviating the danger of emptiness and narrowness posed at the level of the system of needs. The danger could take on one of two forms of alienation: that in which the central authority, at a distance from the person, dictates all important plans and decisions, and that in which no authority but only the brute mechanism of the market governs. In both cases freedom would remain unachieved. Work situated in the corporation establishes the middle ground between the twin dangers which threaten should either civil society or state dominate.[63] These benefits to the working persons—recognition, rationality and participation—become available to the person through the mediation between civil society and state which takes place in the unified group of working persons: the corporation.

The Rationality of the Tool: From Nature to Culture

Persons in their work activity confront natural limits and create a social world. In the interchange with nature and the social structures, both the tool and its successors, the machine and technology, present still another element of work activity. My intention here is to examine Hegel's analysis of the several meanings of the tool for the working person. Hegel describes how the tool reshapes needs and how rationality and universality endure in the tool. These tools do not make tasks merely easier, they remake the world. The development of the tool contributes to the reshaping of society so that machines take their place in work activity. In response to the work action of machines, nature takes revenge on working persons. This revenge signals that tools and machines cannot be utilized to subdue but, rather, only to reconstitute nature.

The analysis of Hegel's thought about technology begins with an understanding of the rational mediation of the tool. Hegel regards the tool as a mediation between work activity and nature: the tool intervenes between the continual demands of nature and the transient efforts of persons. In the tool, work finds its permanence, for beyond the contingencies of the worker and the worked material, the tool remains.[64] Work interrupts the round of insistent desires; the permanence of the tool results from work. The tool, itself unconsumed, is a product which remains so that the renewed arising of desires might be checked still more easily. In one sense, the tool narrows the range of recurring needs in that its employment transfers from the worker to the object tool much of the attrition entailed by work upon objects. The tool's action, its "work," causes the tool, not the worker, to deteriorate. The possession of the durable instrument reworks, then, the pattern of passing desires.

The tool is the deposit, the external and perduring mark of the rationality intrinsic to work activity. The lastingness of the tool makes available to others the discovery resulting from personal experience. One particular element of work activity, the process of discovery, highlights this rationality and universality intrinsic in work. Although even the simplest use of the tool places specific demands on the reason of its users, requiring of them particular decisions about the application of the tool (for example, the force to be used and the direction of the utensil's action) one special work activity singles itself out in manifesting work's rationality: the discovery and development of a tool yet more suitable for a task. This particular skill, inventiveness, is especially universal,[65] since the result of invention may be used by any individual, may be used in the fashioning of many similar objects, and may satisfy a similar need any number of times for any number of persons.[66]

In the inventive element of work activity, the particularly human capacity of rationality, in which each member of society is capable of partaking, concentrates on the way in which work itself focuses on nature.[67] The creation of tools determines the boundary between nature and culture, while the tools themselves are the boundary. Furthermore, they provide the mediation between nature and the world of human activity. The discovery of tools is a universal activity because its results reshape the world in which all persons act.

The same rationality which leads to the discovery of tools leads to the development of machines. As workers discovered and improved an object-tool to intervene between nature and themselves, so now they discover some process in nature which can provide the force that they will direct against nature itself.[68] The discovery, creation and utilization of machines emerges from the world of tools and at a particular moment in the social use of tools. The abstraction of the work of individuals, both from their own needs and from the engagement of the full range of their capabilities, means that their efforts become mechanical. As a result they can at last install machines in their own places.[69] This installation of machines concretizes the abstraction of the work process. Individuals, then, undertake their tasks using devices which abstract them further from their own work activity; even though these same machines are themselves external only as embodiments of human discoveries, skills and efforts.

Insofar as persons have cunningly subdued nature, it takes its revenge upon them. The greater their distance from a lively encounter with nature, the more mechanical does their work activity become. Hegel states that the use of machines, although it reduces the work required for the social whole, increases the work required of the individual, and thereby diminishes its value.[70] At the same time, owing to the intimate connection between persons and their work activity, their own worth is correspondingly diminished: skills are reduced, potentials unchallenged, capacities deadened.[71] Still more, the lifeless system of needs and production takes on a life of its own. Upon this artificial system, the lives of real persons blindly depend. Paradoxically, rational beings come to require a system whose irrationality subjugates persons to its own requirements.[72]

When persons pretend to subdue nature they mistake the goal of work activity. Rather than subjugation, a reconstitution of this common activity must be the project, lest the necessary claim of nature in all work activity be misrepresented by persons to themselves. The object of human work activity is thus a struggle for reconstitution, not a struggle for subjugation. When persons attempt the latter, especially through the equivocal means of technology, they perforce disjoin themselves from nature, and they themselves become unnatural agents, that is, machines.[73] They thereby give rise to that contradiction in which rational beings are dependent on an irrational system.

The Service of Maintaining Society

To what end does this work activity, which takes place in society and is assisted by technology, move? Hegel conceptualizes along several lines the service of persons which work activity engenders. In Hegel's early writings, personal recognition and productive provision are the bases upon which a concept of service may be founded. In the mature works, the several structures of classes, civil service and corporations extend the ideas presented in the earlier treatment by developing the concepts of the provision and maintenance of a world for the commonweal.

The constitution of Hegel's understanding of the service component immanent in work activity must begin with two elements of the lord and bondsman interaction: productive provision and the recognition of relative position. The placing of one of the contenders in a position of bondage requires that his own desires be checked as he executes his tasks for the lord. In this servitude can be seen an initial stage of the other-direction characteristic of service. This represents the formal aspect of service. As for its content, the bondsman acts in order to satisfy the desired consumption of the lord; this represents that element of service which the secure provision for the needs of the other constitutes. The formal aspect requires the execution of work on behalf of the other. The substantive aspect entails productive activity. Whereas the lord consumes and thus destroys, the bondsman produces, thus creates. I stress here not the productive objectification achieved through the servant's efforts; I wish rather to draw attention to the forming and shaping of a world which is initiated in the work of the bondsman. The bondsman both makes and forms. The formative shaping of a world as a component of work emerges more clearly in Hegel's later work.

A further, implicit basis for an understanding of service emerges more concretely from Hegel's later system when the three classes are examined. As was discussed above, a particular type of work is the basis for each of the classes.[74] One can easily add, however, that each of the classes maintains its position in civil society because its work provides either goods or services to that society: the agricultural class offering nourishment and raw materials, the business class providing a wide variety of manufactured goods and commercial services,[75] and the civil servants contributing not by the provision of any articles but by regulating through their sundry offices the interchange among all members of society.[76]

The service activity of the third of these classes evokes particular interest because Hegel unhesitatingly integrates their work into his scheme, although this work is "unproductive" in the sense employed by the classical economists—whose thinking had guided Hegel's own initial understanding of economic processes.[77] The type of service offered by the universal class is the maintenance of the fabric which makes the whole of society possible. The workers of this class stand apart from all sectional interests, both local and corporate. Through its administration of the mechanisms of society, the work activity of this class brings about the conditions which meet needs and produce goods. Their work then represents in an intensified manner two aspects of the service work which remains hidden in the labors of members of the other classes. To serve implies the maintenance of a world, of a set of structures and processes, for the commonweal. In addition, this service for the commonweal implies an inclusiveness: all those in the commonwealth are entitled to the service offered. Beyond the particularity of provision observed as the results of the efforts of the other classes, the universality of service found among the civil servants shows the extent of the service intrinsic to work activity.

The life of the corporations offer an insight into two other dimensions of service activity. With regard to its own members, a particular corporation provides protection against the risks of the market which harm the working interests of a particular member.[78] Whereas earlier the primary understanding of supplying for needs was the socially undertaken provision of those items which made life, activity and production possible, this internal, corporate provision expressly shelters the members from the deleterious effects of market activity.

The second dimension of service activity indicated in corporate life is connected with the experience in decentralized governance which the corporation exercises. The corporation provides one link between civil society and the state.

This link is not merely structural, that is, it is not a connection which mediates individuals into the whole while they remain merely passive. The possiblity of universalization at the level of the state on the basis of the activity of particular individuals necessitates that in their own mediating activity these persons themselves change. Specifically:

> Participation in the management of the corporation, in the administration of its affairs, in the decisions of its membership, and in the election of its representatives, all contribute to the preparation for politics.[79]

Activity develops capacities. The skills learned through this participation extend the capacity for universalization developed within the corporation to the larger universality of the state. In participating along these lines, the members of the corporation, furthermore, engage in a work of service. Just as the civil servants are "unproductively" active in order to secure the universal system of work, so here one range of the efforts of the corporation members is an "unproductive" work directed to the maintainance of their particular arena of activity within the community. Notice that this sphere is not solely economic, for the corporation both protects its own from the vagaries of the market and engages them in internal political action.

Human work activity requires the actualization both of its recognitive and its productive dimensions. The work of service, as exemplified by the efforts of civil servants and the participatory governance of the corporations, contributes to the establishment and maintenance of that community whose political interaction secures the frame within which, in turn, work activity may continue and develop.

A brief review can present the major elements of Hegel's thought in relation to work activity.

Through work, in Hegel's understanding, persons actualize their capacities, especially the capacity for freedom as it can be attained through institutions within society. Work transforms human limits whether these are expressed in individual desires or systems of needs. Requirements are re-shaped; new limits emerge. Both the natural, thwarting forces and work as externalizing activity remain. Work re-makes even the constraint of death.

Work actualizes rational freedom; this is the basis for work's intrinsic sociality. Productivity and reciprocity are the elements of work which are the basis for corporations, classes and civil society. As Hegel describes extensive, interpersonal interaction, the corporation secures the articulation of workers into political society. In this way, the society reaches an encompassing and mutual recognition and makes possible the fulfilling of personal freedom.

Tools and technology are both results and expressions of work activity. The permanent instrument—tool or machine—reworks the pattern of transient, insistent, recurring and incessant desires. The tool is the deposit of the rationality and universality of work. Human inventiveness and social orientation are borne within each tool. The continuing effort toward the discovery of new tools, uses and processes constantly reshapes the boundary between nature and culture.

For Hegel, personal recognition and productive provision establish the core of the service component of work activity. Upon this groundwork, in his later work concerning the civil servants, he derives the concepts of the maintenance of a world and inclusive participation. Similarly, in the analysis of corporative life he investigates both protection and participation—two additional elements of the service activity immanent in work.

Just as the reading of Steuart and Smith guided Hegel's early reflections on social and economic processes, so, in addition to the teachings of the earlier and later classical British economists, the works of Hegel himself directed the inquiries of Karl Marx. Based on his own critical understanding of Hegel's thought, Marx developed a theory of the significance of human work activity in society. Later, upon his Hegelian foundation, after his study of both Smith and Ricardo, Marx erected his own economic system. I now turn to the examination of the role of work activity in Marx's thought.

The Integral Development of Personal Freedom

Marx can both direct his charges against capitalist society and indicate how the better, future state of affairs might appear, on the basis of his understanding of the person. Marx's anthropological view of work activity may be briefly summarized: Marx asserts that the person (and particularly his or her consciousness) develops through the teleological positing characteristic of work activity. Through the individual's choice of ends and of the means for reaching them, this same positing develops the person's exercise of will. Similarly, work activity exercises one's freedom in the course of that person's self-realization. The personal capacities for freedom subsume labor power; Marx's consequent emphasis on the bond between persons and their work capacity attests the centrality of the person in his economic thought. Marx holds that when persons are integrally developed and when they are integrally related in community, they attain freedom through work activity. I begin the fuller presentation of Marx's view of the working person by examining some central characteristics of the person which work causes to evolve.[80]

Work shapes the person because it brings about the development of consciousness in the person. In their individual work, persons acquire particular relationships to an object and in those relationships establish their own purpose which alters the object. Work activity engages consciousness in a teleological positing as its principal activity. This positing through work alters both the object and the person:

> At the end of every labour-process, we get a result that already existed in the imagination of the labourer at its commencement. He not only effects a change of form in the material on which he works, but he also realizes a purpose of his own that gives the law to his modus operandi, and to which he must subordinate his will.[81]

The teleological positing enables persons to achieve a change in the material upon which they operate and also a change in themselves as they discipline their own wills to execute their envisioned purposes. The exercise of freedom within work activity depends not only on the selection of a particular end but also on the choice among means toward that end. As an important instance of this, one which will be examined below, the discovery and development of technology represents an expression of liberty in work activity.

Persons who work posit goals, examine alternative means and also attain the realization of these goals. In this actualization, the individual wills and activities of these persons make use of resources and tools to establish objectively a result of their own life activity.[82] Humans are characterized as beings whose actions transform both nature and themselves into a world of objects and relations respectively. This transformative action I call productive activity. I call this activity "productive" in order to emphasize that it leads to actual results. Production is a species of the genus productive activity. The productive activity central to Marx's analysis includes the creative freedom of persons in shaping their own self. "By . . . acting on the external world and changing it, he [the working person] at the same time changes his own nature. He develops his slumbering powers and compels them to act in obedience to his sway."[83] Work activity achieves the development of the person's capacities.

Work activity develops the person in two further ways. Since work requires the combined application of the several capacities, it conduces to the integral development of the person. Further, because persons bring nature itself to new possibilities through their work and since they realize themselves in interaction with nature, their work activity (through which the individuals interact with these new potentials in nature) actualizes still more and novel capabilities in themselves.[84]

Marx's critique of all socio-economic systems depends on a conception of the person, who in some key aspects remains unchanged:

> Work is the eternal natural condition of human existence. The process of labor is nothing but work itself, viewed at the moment of its creative activity. Hence the universal features of the labour process are independent of every specific social development.[85]

Marx's analysis of economic structures in terms of work, labor and labor power presuppose this valuation of persons and their freedom as well as this estimation of the work activity through which persons achieve themselves in actualizing their freedom. The emphasis on the unfolding of human capacities in their range and depth is pivotal for Marx because his critique of social arrangements proceeds from an examination of how these facilitate or hinder the development of personal capacities. These capacities are critical to persons as channels through which their own individual freedom works and furthers itself; thus, those structures which hinder human development are criticized.

Freedom is central to Marx's understanding of human work activity.[86] This freedom depends on the integral life of persons, which is ordinarily marked by the self-appropriation of their work activity. Marx purposefully departs from the tradition found in classical economics that regarded work as a limitation upon freedom:

> . . . Neither does it occur to A. Smith that the overcoming of . . . obstacles may itself constitute an exercise in liberty, and that these external purposes lose their character of merely natural necessities and are established as purposes which the individual himself fixes. The result is the self-realisation and objectification of the subject, therefore real freedom, whose activity is precisely labour.[87]

Work activity, in that it overcomes natural necessities by fixing them in terms of human purposes, imbues the world with freedom. Marx states more clearly than even Hegel that the work of production itself constitutes an action of freedom.[88] Freedom actualizes itself within the world by acting upon it in productive activity.

Marx's conception of freedom is richer than that found in the liberal tradition. Rather than requiring a negative liberty, it prescribes the positive development of capacities. Each person is to undertake this on the strength of a particular activity which encompasses the entire person: work activity. Secondly, although Marx shares with the liberals a loathing for constraints on the individual's self-chosen activity, he emphasizes the importance of the formation of self in the course of one's own projects and understands the accomplishment of this not by way of stoic withdrawal and reflection but by engagement with the world. Thirdly, Marx recognizes that freedom attains its fullness within a community, into which the person is integrated through the exercise of some aspects of one's work activity, namely, through participation in the community life. Finally, Marx presents a concept of free activity upon which impinge several social structures, often in such a way as to threaten self-determination. Yet through their work, persons come to realize that these structures are the creation of past work activity—either their own or that of others like them. Through their work, persons can alter these structures in order to shape a world adequate to their further, free, individual development. Indeed, for Marx persons are free in a rich sense, because no aspect of living presents itself as impervious to work activity.

Human freedom takes place in the world as productive activity or praxis. By means of his view of praxis as productive activity integral to persons, Marx criticizes in particular the economic system of his age. This activity of freedom ought to instantiate itself in the economic sphere.[89] As matters stand in the actual world of economics, the work activity of persons becomes commodified as labor power. Yet for Marx, free, productive activity, still present and vigorous within the distortions of the system of capitalist production, will emerge fully to supplant the processes of labor commodification and the extraction of surplus value.

Marx distinguishes human production from that of animals. Whereas animals are one-sided in their production, persons produce universally. While animals produce only out of necessity, persons produce even when free from physical need; indeed, it is particularly in this state that they can be said to produce in freedom. An animal produces only itself, while human activity reproduces the whole of nature. Persons can produce, in addition, according to the standards of each species. Furthermore, humans can confront the objects of their production and can bring forth objects of beauty.[90]

In this delineation of the universality characteristic of productive activity, Marx emphasizes the centrality of free choice in the work undertaken by persons. They are free to produce, not solely for particular ends but for the universal. Since human activity is the determinant of the entire world, as it reproduces the entirety of nature, free activity shapes even its own milieu. Not merely the norms of their own species but the norms of production found in other species may be freely chosen and executed by working persons. Both in their reflective awareness of the object produced and in the ability to shape objects according to norms which surpass the useful, namely the aesthetic, workers demonstrate the freedom involved in human production. In highlighting the universality of the productive activity of persons, Marx affirms a key dimension in his anthropology: the capacity for free creation.[91]

Marx, to be sure, did not rest content with a philosophical concept of productive activity but, rather, examined its subversion or actualization in real economic processes. Within the economic system, Marx analyzes the fulfillment or constriction of free activity in terms of the operation of labor. As Engels carefully comments in a well-known note, the English language can conveniently distinguish between "work" which produces use-value and "labour" which produces "(exchange-)value."[92] "Work" corresponds to productive activity, that is, praxis which is undertaken in the economic sphere. Marx himself makes a distinction between two types of "labor" which correlates with Engels's difference:

> As useful activity directed to the appropriation of natural factors in one form or another, labor is a natural condition of human existence, a condition of material exchange between man and nature, quite independent of the form of society. On the other hand, the labor which posits exchange value is a specific social form of labor.[93]

The "specific social form" of work activity which governs labor as productive of exchange value is, of course, capitalism. Praxis is characteristic of the human person, as work is a condition of all human life.[94] Under the regime of capitalism, this essentially human activity is incorporated as labor for livelihood; the person, rather than being regarded as a free agent, appears to be a "working animal."[95] With the growth, accumulation and concentration of capital, human potentials are lessened, dispersed and divided so that the person becomes a force in production, rather than an agent of productive activity.[96]

Within the social relations which lead to alienation, persons become separated from their own work activity. The continuance of those structures and the consequent alienation result from the activity of the persons themselves.[97] In the *Grundrisse* Marx expresses this disjunction as the alienation of actual labor from the capacity for labor.[98] That is, the acts of persons become separated from their own, individual potential. The activity itself becomes alien to persons so that they cannot affirm themselves, cannot actualize their freedom. They regard their own particular work with the same indifference as does capital, which sees only abstract value.[99]

The contemporary system of exchange and the division of labor also damages persons because this abstract society prevents the personal development integral to work activity from realizing its fulfillment. The division of labor fashions the market with a vitality that is self-governing apart from working persons; they become dependent on the fluctuations of the market. The market reduces them to providing labor and to consuming what is required for labor; and each of them is "depressed spiritually and physically to the condition of a machine."[100] While their society is distorted, when the products of their own efforts oppose them and since their own activity is turned against them, they themselves are deformed. They sacrifice their creative capacity for the opportunity to labor in heteronomously stipulated quantities.[101] Whereas social productivity increases through their efforts, the workers find that their own productive capacity is narrowed and even diminished.[102] The productive mechanism of society reduces the person to a functional part in this system, with consequent dissipation of capacities for fuller activity.[103] When malformed into labor, work activity which ought to develop human potential becomes the means for its dissolution.

Under this particular, capitalist social form, the division between self-activity and labor has become so pronounced that the latter manifests itself as the opposite of self-activity. At best, labor could become a means for the sake of self-activity.[104] This is to say that labor is accomplished over a certain period so that self-activity may be undertaken in another period. Alternatively, labor done by some persons could enable others to engage in self-activity. In this regime, however, the stunting quality which accompanies labor, that is, work effort in the capitalist economy, appears as a necessary adjunct of all work. This misconstrual leads to the further deception that there can be no liberation from this supposedly compulsory dulling of capacities.[105]

The free praxis which was discussed above represents what Marx in the *German Ideology* calls self-activity. Implicit in his treatment of the disjunction between self-activity and labor stands the expectation that, although such a hope is just now instantiated in meager form, present effort will ultimately bring about a new situation. There work for the maintenance and development of life will exist as free activity, after the contradictions between the means and the relations of production have become more pronounced. This hope will be realized amidst the actual, complex and interworking forces of the economic system.

An examination of the concept of value-created-by-work reveals Marx's continued emphasis on the centrality of the person in the economic sphere. Marx began to accept the labor theory of value only subsequent to the composition of the

1844 Manuscripts and perfected this theory in conjunction with the theory of surplus value only fourteen years after those manuscripts.[106] Without treating the controversy concerning the transformation of labor values into price, I present an interpretation of Marx's labor theory of value which accounts for his initiating the analysis with labor rather than with capital or some other factor.[107] In part because of his reading in the classical economists, Marx came to reconceive his idea of self-activity so that he might explore its application in the economic sphere. Although he criticizes the ambiguity introduced by Smith in the double method (the labor-embodied and labor-commanded approaches) for determining value, Marx nonetheless acknowledges that Smith's theory offers some sense of a distinction which is also central to his own later comprehension of work as a characteristically personal activity:

> . . . He [Smith] has a presentiment, that labour, so far as it manifests itself in the value of commodities, counts only as the expenditure of labour-power, but he treats of this expenditure as the mere sacrifice of rest, freedom, and happiness, not as at the same time the normal activity of living beings. But then he has the modern wage-labourer in mind.[108]

Smith comes close to seeing both the difference between and the relation of labor and labor power. He errs, however, in viewing the outlay in labor power solely as deleterious to the person. Because Smith does not realize that the capitalist economy utilizes a specific form of the expenditure of labor power, namely labor, he fails to recognize that this outlay characterizes normal human activity; indeed, it is an activity which develops freedom.[109]

Marx provides a definition of the key term critical for his understanding of work activity:

> By labour-power or capacity for labour is to be understood the aggregate of those mental and physical capabilities existing in a human being, which he exercises whenever he produces a use-value of any description.[110]

Work activity, outside the impositions of the capitalist economy, is intrinsically related to personal capacities which make possible interaction with the world.[111] Marx emphasizes, in terms of the activity of the laboring process, the generality of this identification:

> The labour-process . . . is human action with a view to the production of use-values, appropriation of natural substances to human requirements; it is the necessary condition for effecting exchange of matter between man and Nature; it is the everlasting Nature-imposed condition of human existence, and therefore is independent of every social phase of that existence, or rather, is common to every such phase.[112]

Each person possesses, as integral to himself, the capacity of labor power which actualizes itself in the labor process in whichever social construct this person operates.[113]

Within the capitalist system, labor power, the capacity to labor, is purchased so that the worker may then provide labor itself.[114] Labor itself constitutes the use value of labor power;[115] capital purchases the capacity for work and activates it as labor so as to produce exchange value. In so doing, capital controls the person to whom this capacity for work is inseparably linked. Capital regards the human solely as a provider of a key commodity, not as a person. Notwithstanding the restriction of consideration to exchange value, the worth of labor power as a commodity depends on factors which are "produced" only in an extended and comprehensive sense; these factors are strongly associated with the persons qua persons: the expenses of developing the capacities of labor power and the natural diversity that exists among individuals.[116]

Within the capitalist system, the actualization of labor power results in "abstract labor," in which all the particular forms of work activity are left out of account with the result that only a single factor remains that is common to all these tasks productive of use values.[117] This abstract, common factor is the production of exchange value, which can appear only as borne by particular, useful "concrete" labors.[118] Abstract labor results from an abstraction which can occur only under a system in which the production of exchange value is a governing principle.[119] Abstract labor produces exchange value, while concrete labor produces use value.

The labor theory of value is founded upon a factor which is itself valueless:

> That which comes directly face to face with the possessor of money on the market, is in fact not labour, but the labourer. What the latter sells is his labour-power. As soon as his labour actually begins, it has already ceased to belong to him. Labour is the substance and immanent measure of value, but *has itself no value*.[120]

Value as such is value in the market, that is, exchange value. Labor has no value, first, because the market transaction between the possessor of money and the laborer has already occurred. Labor, in this sense, is not a market item, for it no longer remains in the market. When labor commences, it is no longer a commodity open to the free sale and purchase of the market. It is now bound in a particular engagement of contract and task.

Smith and Ricardo spoke of the value of labor, but this is an erroneous expression for ". . . the value of labour-power, as it exists in the *personality of the labourer*."[121] Hence labor lacks value in a second sense, since the prime indicator of worth lies within the person. As he did in his discussion concerning the relation of freedom and work activity, Marx once again connects the capacity for work activity with the person of the worker. Because labor power is intimately bound up with the person, it surpasses any attempted quantification. The labor process, when viewed as productive of exchange value, is quantifiable in terms of the time expended.[122] The same process, when viewed as a particular set of acts resulting in a specific, useful product, can be regarded under its qualitative aspect alone.[123] Marx holds that there can be no equivalence in the expressions "value of labor" and "quantity of labor" when the exchange occurs between materialised and living labor.[124] Living labor means the capacity of living persons to engage in work activity; their activity within a particular economic system may be quantified in time units; they themselves cannot be.[125]

G.A. Cohen's interpretation of the labor theory of value further promotes the understanding of this theory as a particular way of heeding the human person in his activity. Cohen argues that the labor theory of value, as it is usually interpreted, cannot be true. He supplies a corrective interpretation. Speaking of the laborers he writes:

> They create the product. They do not create value, but they create what has value. The small difference of phrasing covers an enormous difference of conception. What raises a charge of exploitation is not that the capitalist gets some of the value the worker produces, but that he gets some of the value of what the worker produces.[126]

Exchange value is a result of a contingent, historical constellation of forces, which inheres in objects fabricated by workers. Both the distinction between value and its vehicle and the consequent shift of attention to the objects make possible an emphasis on the work activity which shapes these objects. This distinction and this emphasis lead in turn to a concentration on the persons who activate their capacities and externalize themselves in this shaping work.[127] Behind the abstracted and anonymous forces and relations, persons in action are the conceptual focus of value. Persons, through their present activity, are the living and effective basis for the transformation of the conditions of their work and participation.

Unlike Hegel, Marx maintained a vision of how persons and their activity would come to a completion in the future. Indeed, he regards the revolution as aimed not solely at a redistibution of results but at a transformation of activity itself.[128] The form which activity takes in the capitalist system will itself bring about this transformation:

> Capital's ceaseless striving towards the general form of wealth drives labour beyond the limits of its natural paltriness, and thus creates the material elements for the development of the rich individuality which is as all-sided in its production as in its consumption, and whose labour also therefore appears no longer as labour, but as the full development of activity itself, in which natural necessity in its direct form has disappeared; because a historically created need has taken the place of the natural one.[129]

The work activity of the future accomplishes the complete unfolding of human activity. In this fullness of activity, persons themselves reach a multidimensional richness. Both present production and the natural set of needs become a collection of needs and a creative activity directed entirely by persons. In this final development of the activity of persons as concretely universal, they appropriate their own talents specifically in coming to a scientific understanding of nature and in the attainment of their own, individual freedom in community.[130] In sum, beyond work as the natural activity of the person, beyond labor under the capitalist system, human work activity will become fully praxis.

Limits Denied and Reconstituted

Marx's examination of the limits upon work treats both the capitalist reality and the future ideal. The manner in which work activity confronts the world in the modern economy inhibits the integral development of the person. In his examination of the limits upon work activity, Marx makes clear how the capitalist system arrogates the freedom which properly belongs to persons. Indeed, so far does this proceed that the limited persons who provide labor power are drawn into an interminable pursuit. The system even alters the conceptualization of time so that it no longer is a reckoning of the lives of persons but, rather, a standard of clocking the expenditure of labor-time. Nevertheless, another understanding of work effort is possible, one in which work activity, instead of striving interminably and futilely to eradicate limits, acknowledges and reshapes the limits of the human. In the ideal scheme, persons through their particular work activities will achieve their individual liberty both in the realm of necessity and in the realm of freedom. If its limits be recognized, human life itself will remain as the motive force of the economic system. Such is the future ideal. My presentation of the Marxian examination of the limits upon work starts with the capitalist attempt to deny limits.

Faced by nature, capitalist production strives to control its processes, dominate its energies and overcome those limits which nature imposes. With deleterious consequences for workers, capitalist production strives to expand itself beyond any limit imposed by nature. The concept of exchange value, which is the focus of capitalist production, entails an abstraction which results in the negation of the limits presented by the physicality of useful labor and use value.[131] In its pursuit of exchange value and ever more exchange value, production becomes its own goal.[132]

Although the capitalist mode begins with an apparent recognition of the person and an impulse to liberation, it swiftly moves beyond any limit:

> This political economy . . . displays a cosmopolitan, universal energy which overthrows every restriction and bond so as to establish itself instead as the sole politics, the sole universality, the sole limit, and sole bond.[133]

The rebellious energy of the system overcomes both social and natural limits. Furthermore, not content to make full use of the available labor power, capitalist production ". . . requires for *its free play* an industrial reserve army independent of these *natural limits*."[134] The system arrogates to itself a free activity which then reshapes the society of persons who ought properly to be those who are able to exercise their own freedom.

Rather than free movement on the part of persons, there is instead the free circulation of capital as money. This circulation, in the endless struggle to realize profit, takes on a life of its own. Money, the agent and substance of this circulation, has no natural limit, for it is only quantitatively determined. Not met by limits, money as capital hastens to increase its quantity.[135] Now since persons, whether as capitalist or worker, also strain to acquire money and since the struggle is shaped by the goal, their own efforts become limitless, measureless.[136] This very struggle dispels security and fixity in the lives of the workers.[137] Capital's movement may deprive them of subsistence; capital's movement also draws their own activity into its own unrestricted pursuit.

Production pursues the maximization of surplus value, not the satisfaction of human needs. Consequently, as Marx writes, ". . . production for the sake of production must be confronted by consumption for the sake of consumption."[138] Unless this be so, mounting overproduction would block the realization of increases in capital. Pursuing its own aggrandizement, capital dominates both the productive activity and the consumption on the part of persons, without regard to their own needs and limits.

Capitalist production seizes and transmutes the significant measure of human limits, time, into a standard of human activity, but this is one which provides a scale far removed from actual human limits and activity. Products are denominated in crystallized labor-time; they are the deposit of a certain number of hours and days of "social labor."[139] The time measure of political economy differs sharply, however, from that of a possible moral economy. For example, the pursuit of exchange value brings about a division of labor in the plant, which represents an attempt to utilize this scarce resource (labor power) more efficiently. The field of evaluation is not the activities and limits of the working individual but the processes and constraints of production; the time of the former matters only in terms of the timing of the latter. That time which is specified through the duration of the efforts of embodied persons denotes the pre-eminent value of the system; yet the objectification and quantification which dominate the system pervert this subjective index.[140] The system appears to commend labor time as the sole measure and origin of wealth, yet the system tries to overcome its own limits by reducing this time expenditure to a minimum.[141] The calculation of time as labor time converts the working persons into "workers," that is, they are perceived only in their function within the abstract system of production.[142]

In terms both of the individual and of society, every form of economics can be reduced to an economics of time.[143] The standard of measurement determines in which sense an economy may be judged moral or otherwise. Marx locates this standard in the activity of working:

> Labour is the living, form-giving fire; it is the transitoriness of things, their temporality, as their formation by living time. In the simple production process—leaving aside the realization process—the transitoriness of the forms of things is used to posit their usefulness.[144]

Living time, namely, human activity, shapes the limited forms of products as goods useful to persons. While freedom can characterize human activity within the limits of time, when such limits are regarded solely in terms of an abstracted temporality, the system redirects them to impinge upon the sphere of freedom. An attention to these limits from the vantage of personal work activity, however, maintains the priority of human activity over the productive processes and helps secure the potential for freedom in work activity.

Some characteristics of Marx's own understanding of freedom lead to an inattention to the life time and limits of persons. A criticism may be raised against Marx's idea of freedom, since the ideal he intends in the fulfilled freedom of the individual offers no content save what may be drawn from a ceaseless, renewed activity. The lack of content brings a disquiet that individuals might undertake such activity solely for its own sake and not for the fulfillment of particular, personal and social potentials. The emphasis on activity of the person might lead to a failure to recognize the individual who acts. Marx here approaches the dangerous error to which the early economists themselves succumbed: an emphasis on production too readily conduces to a failure to attend to the individual producer.

The capitalist system of production attempts to overcome all limits, including those of persons themselves. Marx does, in fact, often realize the critical significance of these limits in the concept of work activity. The needs which constitute these limits—the requirements of food, rest, assistance, care—operate centrally in Marx's understanding of the distinction of persons from animals at the beginning of history.

> The first historical act is thus the production of the means to satisfying these needs [eating, drinking, housing, clothing], the production of material life itself. . . .
>
> The second point is that the satisfaction of the first need, the action of satisfying and the instrument of satisfaction which has been acquired, leads to new needs; and this creation of new needs is the first historical act.[145]

Work activity's original orientation lay in the satisfaction of needs. In satisfying these needs, work transforms them from natural and biological requirements into the historical and social, thus into the human. Work transforms them; it does not eradicate needs. In fact, although cooperation and knowledge may reduce the difficulty of satisfying a particular need, these same dimensions of activity create still more and more varied needs. Work reshapes the limitedness of the human.

Marx envisions that limits will continue to condition human work activity in the fully communist society. In that age, science and technology will have contributed to the conquest of nature; nevertheless, the "realm of necessity" will persist:

> . . . [There] always remains a realm of necessity. The true realm of freedom, the development of human powers as an end in itself, begins beyond it, though it can only flourish with this realm of necessity as its basis. The reduction of the working day is the basic prerequisite.[146]

Society's development will have reduced but not eliminated the working day—this, even though work activity itself will have been liberated so as to be the chief means for personal and social realization. The two realms will dialectically interact so that each presents a type of necessity and each presents a way of freedom. Two surprises result from this dialectic. First, in the realm of necessity, persons, now nondominated, will achieve their self-realization in working toward the satisfaction of their physical and social requirements.[147] In this manner, they accomplish one significant goal of their freedom. Second, in the realm of freedom, persons strive to complete their freedom. The necessity here lies in the disposition to actualize potentials.[148] On this view of work activity, in the society of the future, they will acknowledge their (albeit altered) limits, while they will understand freedom not as the conquest of nature, not as the cessation of work but as the development of their capacities.[149]

In contrast to the skewed and dangerous stance of capitalist production toward limited, personal life, Marx underlines three ways in which the life of the person is significant in the productive process. First, the totality of the limits that face individuals frame their lives. In order to maintain their lives, they work. In the capitalist system, they sell parts of their individual lives, their particular activities, in order to meet the needs which limits impose on them.[150] Marx clearly associates the potential for work activity, labor power, with the person, in that it proceeds from the life of the individual.[151]

Secondly, in addition to its providing the locus of labor power, the life of the person, according to Marx, has a further role in and through this individual's work activity. In putting machines and tools to use, the worker brings them to life.[152] Without the interaction with living labor, the machines fall prey to the limits which impinge upon them. They fall into disuse and disrepair; they decay.

Living labor communicates life in the negative sense: it rescues from death. Living labor communicates life positively, for it brings the machines from potential to actual value in use.

Thirdly and finally, the life of the person figures significantly in the productive process as the communicator of life to other persons. Work activity reproduces itself. Faced with their own limits, persons struggle to provide the means that others might live afterwards.[153] Workers also make it possible that others may themselves develop through work activity. In this way, the final limit, mortality, is not overcome but, rather, transformed within society and history.

I have emphasized this three-fold manner in which the living work activity of the person, though fettered about with limits, provides the basis of the productive system. My earlier criticism analyzed a productive system which, in eradicating limits, misshapes and destroys the form of the life activity on which the system itself depends. The functioning of the system which observes human limits only to eliminate them is countered even now by the ongoing, living work activity of persons.

Abstract Society and the Society of the Future

Marx presented his earliest insight concerning sociality in terms of species being. As he progressed in economic studies, he reinterpreted this sociality in terms of the division of labor found in the industrial era—a division quite different from the primitive, natural division. The processes of abstract society found in the period of industry both distort the work activity, social interaction and growth of the individuals themselves and also conduce to a new era of free activity in which the workers will recognize their cooperative activity and its structures as their own. First, however, I present Marx's earliest approach to the understanding of sociality.

From Feuerbach Marx learned to demystify Hegel's Spirit (*Geist*) so as to see in it human potential understood as species being, his earliest conceptualization of sociality. The free, conscious activity of persons determines the species character of this being.[154] Persons partake in species being by way of their individual actions, because this activity, precisely as conscious, rises above particulars (even of itself) to the universal. The person partakes in species being, further, because his free action, along with that of all other persons, stands over against the whole of nature. Finally, the person's action contributes to the shaping of a world which stands external to him and to all other persons, for example, he may produce a product which meets the need of someone else. Human productive activity is intrinsically social.

The growing reciprocity which develops among those who work in shaping a common world actualizes the philosophical idea of species being, which the universalizing characteristics of human productive activity establishes. The development of productive forces and the progress of social structures which express this species characteristic of work activity reciprocally affect one another.

The economic analysis which commences in earnest in the *German Ideology* grows out of the anthropologial view of persons in society which Marx had developed in 1844. Studies of the *Grundrisse* clearly indicate the continuity between Marx's earlier philosophical concerns and his later economic analyses in *Capital*.[155] Certainly the continuity is not an identity but a dialectical development, the focus of which is human work activity. Class and the division of labor are two constructs which emerge out of the earlier conceptualization of sociality; both constructs are moments in the further liberation of human work activity. Marx still continues to view work activity as a key, social force in the process of the development of full, human potential.[156]

Marx indicates the perduring, yet evolving, sociality intrinsic to work when he contrasts the "natural" state of production with that based on exchange, money and the complete division of labor.[157] Use values operative at the earlier stage do exist within a particular, social framework, but they are conceptually inadequate to elucidate the special, social relations of production found at the later stage.[158] In *Capital*, Marx once again distinguishes the two dimensions of sociality: (a) a particular kind of work activity which satisfies a social want and (b) the exchangeability of such particular works on the basis of some equalization.[159] In the earlier case, a spontaneous, social division of labor emerges; in the latter, this division—focused on exchange value—is the sole form of work in society.

A condition for the social relations in the period of manufacture was the previous development of the division of labor in society as a whole.[160] In the period of capitalism, a double division of labor becomes the context of work activity; a division of labor prevails both in society and in the workshop. The circulation of commodities in exchange determines the conditions for the production of the first division. The wage contract establishes the conditions for the work activity of the second division. In the former case, anarchy characterizes the processes of exchange and, in the latter instance, the employer's will governs.[161] While these forms evolved on the strength of work activity, now work becomes reciprocally altered by its social context. Within this double division of labor, persons cannot actualize their individual freedom through work activity. In this sense, one can understand what Marx himself grasped only inchoately in 1844, "The division of labour is the economic expression of the social character of labour within the estrangement."[162] Continued work activity can overcome this estrangement, so that freedom may be actualized within the framework of new social structures which encompass the identity and security of the primitive, natural level as well as the freedom and progress of the period of exchange.

Alienation accompanies the progressive development of economic and social processes during the period of capitalism. Marx's treatment of alienation depends centrally on the alienation of human work.[163] Alienated labor is abstract labor. Social relations in the age of capital have reached such a point that goods are exchanged on the basis of a universal equivalent: abstract labor. Marx emphasizes this sociality of exchange which provides the structure for the movement of equivalents. He does so because he sees this very sociality, its harmfulness to persons notwithstanding, as a significant result of human activity.[164] Indeed, even as it wreaks severe harms on persons and society, the further development of the system of exchange advances toward the actualization of human potentials. Liberation from personal bonds and advancement of cooperation are two elements in this actualization.

The activity of persons in this system of exchange and the division of labor reshapes the very capacity to work. Before considering the progression made through this system toward a further form of sociality, I examine in terms of society two damaging effects of this alienation upon human work—the materialization of work and the separation of social relationships from the individuals. These two forms of alienation as well as the simultaneous unfolding of a new, fuller sociality proceed from the dynamism of personal work activity.

A split between the material and mental dimensions of work activity becomes socially organized.[165] One result of this is that the activity of individuals creates the social world, yet they remain unconscious of the way in which they themselves bring this about. In addition, those aspects of work which persons require for sustenance become separated from those which are creative and expressive. Further, this rift becomes socially organized with the result that even those activities which still hold potential for achieving freedom evince alienation because they are sundered from the whole of work activity.

Under alienating social relations, however, not only is work disjoined from persons, but, in addition, their own activity becomes materialized against them. The critical element lies in this: not that the work is realised outside of persons, not so much that the work is simply alien to persons—that would offer at least the chance for recognition—but, most importantly, that the realization *opposes* human labor.[166]

Whereas in the *German Ideology* Marx delineates the division of labor as that consolidation and fixation of social activity (that is, of communal work activity) in an antagonism towards workers as the division of labor,[167] in the *Grundrisse* he offers a more detailed explanation of the dynamics of this process in terms of capital. The relations of production develop to the extent that they negate the work of individuals because the system posits them as a communal or combined labor in opposition to the particular workers. Capital is the alien subjectivity of this opposition.[168] The objective conditions have developed to such an extent that capital itself acts as a subject, independent in form. Capital too produces. What it produces are its own agents of production: wage laborers; all other dimensions of these persons are left out of account.[169] Both the gap and the antagonism between workers on the one hand and the objectification in capital on the other hand grow continuously.[170] This recalls that the process of objectification does not mark an end state; indeed, this reminds one that the entire process devolves from ongoing personal action. Even as an object to capital-as-subject, labor remains "the form of activity."[171]

Commodities constitute the offspring of the sociality of exchange. The substance of commodities is the "communal substance," objectified labor.[172] Commodities, as congealed labor, are not a physical but a social magnitude: one that is derived from the relations of persons in their activity.[173] A particular, social relation among persons produces commodities, yet the relation among these socialized products comes to dominate any direct relation among persons.[174] The commodities themselves simulate the productive activity of persons and, as alien, this productivity confronts the actual productivity of persons.[175] The relation among the fetishes of commodities, their imitative activity which is productive of surplus value, and, indeed, the purchase of persons by commodities,[176] specify the parasitical inversion which originates in work activity within the particular context of exchange and the division of labor. The further evolution of productive forces and social relations, however, will reveal a state of affairs in which labor will remain communal while the society will no longer be abstract.

The circumstances of capitalist society have yet another deleterious effect: particular persons are estranged from society. The furtherance of the life of the abstract individual becomes the purpose of society in its abstract form.[177] Social production considers the person only as a provider of labor and as a consumer of goods. It measures his or her value in terms of hours of labor, hours determined in terms of the socially necessary amount.[178] This means that the individual's activity is assessed on a scale dependent upon social processes external to the needs and goals of particular, integral persons. The labor of the individual counts only in the form provided by abstract society. Persons are estranged from society because the social interaction constituted by exchange ignores persons in their fullness and freedom for the sake of processes founded upon an abstraction of the person. The actualization of personal freedom within society will require a concept of the contribution of an individual's work to social processes based on a complexity greater than that of the market process alone.

There remains, nevertheless, hope for the continued activity of the persons on whom the system depends. Just as capital grows more extensive and concentrated, so do the means for collaboration and cooperation among workers grow more elaborate and involved. Out of this complex of developments can emerge social relations which foster, on the basis of free activity, both production and recognition.

The increase of powers, knowledge and skill achieved first by manufacture and then by industrial production are available not to individual workers but to the collective laborer, that is, to the cooperative undertaking of the assembled laborers. In this, Marx's original conception of sociality as species being reaches a still more advanced conceptualization. This development takes place chiefly owing to the universal character of work and to its cooperative dimension.[179] Since the capitalist appears to be the figure in whom the universality of the work process centers and on whom the cooperation in the labor procedures depends,[180] the workers, through their activities, come to know themselves as a commodity. This realistic knowledge itself brings about a change in the workers.[181] What appear as the attributes of the capitalist, the workers recognize as the results of their own activity in cooperation with one another. From the activity of workers there derives, first, the knowledge, both scientific and practical, which results from or shapes work activity, and second, efficient and productive systems of cooperation. Thus, to both the universal and social dimensions of work activity the workers together can lay claim. The sociality which Marx had originally conceived as species being, and then reconceived in the realm of economics as the division of labor, can further unfold so that workers recognize their own free activity as the basis of production.

Marx indicates, especially in the *German Ideology* but also through some references in later works, that the developments in the forces of production and in social relations will reshape society not by a return to the primitive, natural society but beyond the abstract society to a new form of sociality characterized by specific features of each of the preceding eras. In view of the creative potential of free activity, his vision of the future society is perforce undetailed.

In the future, society will be characterized as a context where each person may freely pursue any desired branch of work, where alteration of work activity is possible and where the society itself (and not anonymous processes) regulates general production.[182] In this coming state of affairs, the community provides the situation for the actualization of personal freedom.[183] The direction of individual activity on the basis of communal needs and goals assumes the participation of each in the determination of these activities.[184] This participation makes possible the recognition of each member of society and the work of each of these individuals as integral to the whole. Thus, the entirety of society decides on the organization and application of the power which the earlier, abstract pattern of sociality bequeathed to it.[185]

The prospective sublation of the social form will incorporate the productivity and freedom of abstract society with the security and recognition of primitive social structures. The activity of working persons in overturning the system of capitalist exchange which threatens to render them merely passive, produces the society of the future.[186]

Technology as Threat and Promise

Marx's understanding of technology reaches from the past, through history, to the future. A brief outline here will specify this development. With tool-making, history commences. In the further extension of the tool—the machine—production separates itself from persons. The introduction of machines occurs within the industrial system; this results in the reduction of working persons to components of an arrangement governed for the benefit of machines. In this, machines compete with and threaten persons. The development of the machine system in the age of industry is carried out through an ever extending technology, which is the progressive organization and development of knowledge about production. This knowledge, however, is not available to the workers, although it proceeds from their own work. This inaccessibility transmutes the freedom first promised by machines into a new bondage. Nevertheless, according to Marx the era of machines will lead to a period free of their domination because of three advances: equalization, the accumulation of scientific knowledge and the development of personal work capacities. Having previewed the matter of this section, I now return to Marx's treatment of the initial stages of technological development.

Tool-making initiates human history. In the tool, a person possesses a means which frees that individual because it makes efficient use of his or her effort. Person cunningly interpose the tool, an objectification of their own individual efforts, between the procedures of his activity and the processes of nature.[187] In the machine, objectified labor now not only becomes the means of activity but also appears as the force of the productive activity.[188] Production becomes separated from the efforts of persons. With the coming of the machine, not an individual person but a complex of machines unifies a particular labor process. Persons become elements within this mechanical system.[189]

The developed forms of machines and complexes of them are distinct from workers, do not belong to them, and are beyond their control. Machines stand over against the workers as alien to them.[190] Moreover, because the efficiency of machines reduces the activity of the persons to an emulation of the mechanical, workers now face the competition of machines.

The capitalist utilizes the machines so as to extract more surplus labor from the workers.[191] Thus, machines not only are alien to and competitive with the workers, they are also a means antagonistic to them. The capitalist puts the machine to use in order to reduce the natural resistance of the workers and thus dominate them.[192] In a fetishistic perversion, the powers which devolve from the cooperative activity of workers appear as the capacity of the capitalist and of his machines.[193] The inversion goes further because it appears that persons are used and employed by machines, rather than the reverse.[194]

With the arrival of machines, persons no longer control the implements of their work activity. In addition, workers have no claim to these devices on the basis either of their particular craft roles or varied claims of ownership. Separated from these devices, working persons discover even the knowledge concretized in the machines has become withdrawn from working persons. On the basis of their own activity, yet separated from their awareness, the progressive development and organization of this knowledge, that is, technology, evolves according to the norms of the work process alone and with minimal regard for the workers themselves.

Modern industry conduces to the creation of technology: "The principle which is pursued, that of resolving each process into its constituent movements, without any regard to their possible execution by the hand of man, created the new modern science of technology."[195] Science and its application to the productive processes unfold according to principles neutral with respect to working persons. The evolution of the productive forces in this progress seems to depend on capital.[196] While the creative activity which led to the discovery of tools had characterized the outset of human history, in the age of capitalist exchange, the relationship of the forces and relations of production has separated this intellectual ferment in productive processes from the agents of production. So it appears. It is still on the strength of the activity and reflection of the workers as a social, cooperative unit that improvements and discoveries are made.[197] In order to attain a new technology, one oriented toward persons, the workers must come to a consciousness of the intellectual dimension of their own activity which is presently co-opted.[198]

The regime of machines, since they increase the productivity resulting from a given effort, makes possible a life characterized by the reduction of some signficant risks endemic to earlier periods. This same regime, however, now exposes working persons to risks concomitant with the demands specific to wage labor. Furthermore, the competition introduced by machines makes more uncertain the demand for labor; this, then, increases the risks and threat to life, contrary to that which the introduction of machinery promised.

Yet the results of this regime of machines appear negative only when regarded as a fixed result. The industrial age presents other, positive achievements. The age of technology will lead to the actualization of personal freedom in society. For example, the absorption by machines of the craftsman's skills involves the equalization of persons and machines by the elimination of a hierarchy whose power rests on a control over specialized talents. While preserving the equalization among the workers, which results from the disappearance of the hierarchy of workers with specialised talents, working persons can advance beyond the falseness of this parity.[199] Each worker has now been freed from attachment to a particular craft. Although each may be evaluated only in terms of abstract value, now each has become capable of shifting to any of a variety of tasks. Hope remains because of the further development implied in the forced equalization of their lot.

Another potentiality for the actualization of freedom lies in the scientific character of the technology which restructures society. The scientific character which results from the work of the social whole is a requirement for work to be emancipated.[200] The knowledge connected with the general work of the social whole is universal, and this universality means both that each person can come to this knowledge and that the knowledge derives from the action of the entire society. The actualization of freedom requires this flexibile access to knowledge on the part of each individual and the articulation of this knowledge within the whole of society.

Impelled by the capitalist hunt for surplus value, the thrust of industrial technology drives it to search beyond any particular discovery. The technical basis of modern industry revolutionizes.[201] On the basis of newly acquired knowledge, modern industry develops new forms of work, which require the cultivation of new skills in working persons.[202] In order to evolve in its own terms and to sustain this evolution, industrial technology must lead to the development of a worker who is fit for a variety of tasks and capable of assimilating changes in production. The full development of the entire range of capacities of the person would best provide such workers.[203]

From the economy's systematic drive to create the possibility for surplus labor and to realize profit there results one particularly significant aspect of the full development of the person. More and more technology is incorporated in instruments of production. In order to develop, produce and maintain these implements, the economic system must dedicate an increased investment of work activity to technological improvements. This research, however, leads to a universal knowledge and to the creation of workers capable of invention and discovery, involved problem-solving and complex production. The workers of the future will thus evolve as mobile and agile production workers, adept at experiment and induction, and also proficient in science and invention.[204] These twin developments in the potential of workers—the capacity for varied engagement in tasks and the sharing in a common knowledge—correspond respectively to the dialectically interacting realms of necessity and freedom. For these workers, technology no longer exploits but, rather, unfolding on the basis of their own potential as articulated in the social exchange of reseach and experiment, it represents one sphere for the achievement of freedom.

Although Marx views technology predominantly as subjugating nature and as oriented to the more efficient production of material goods and although he tends to ignore the perduring limits to human, social effort, nonetheless, he does provide a view which makes it possible to regard contemporary technology, not as an object for romantic rejection but as the potentiality for a new development in the working person. Still further, an analysis of the service component of work activity as presented in Marx's own system offers a noncontradictory corrective of his view.

Service Based in Maintenance, Reflection and Care

As today's developments in the technological and social fields transmute production, the traditionally predominant model for work, service now becomes the more significant component of work activity. The interpretation of Marx's theory offered here makes possible a comprehension of service within a society where human freedom may become actualized. In contradistinction to the idealist element in Hegel's movement of the Spirit, Marx's conception of human activity pursues a transformation of the material world. As a result, matter becomes the primary focus for any activity, even that which focuses on the benefit of another. The transformation of objects becomes the prime model for productive activity. In consequence, this type of work activity achieves a dominance within Marx's system and in his economics, to the exclusion of "non-producing" activities. The tendency toward materialism threatens to block the development of an idea of service, even though the emphasis on persons and their social interaction provided key elements toward such an idea. For Marx, praxis tends toward development through the exercise of capacities; and in this exercise praxis shapes nature, society and history.[205] I grant there are features in Marx's system which resist such an interpretation; for example, as a result of his materialist leaning, he tends to understand social relations in terms of structures and systems requisite for production. Nevertheless, other elements in his thinking can surpass these features (for example, his understanding of sociality as recognitive) so that service can become a critical, conceptual component of work. Marx's theory presents several approaches—in terms of the maintenance of workers, in terms of a transformation of reflection and in terms of care in solidarity—which contribute to an understanding of the service component of work activity.

In one of his early writings, Marx delineated the characteristics of work activity in a situation in which private property does not exist. In this setting, the production undertaken by persons results in (1) the manifestation of each of their lives in their individual activity, (2) the objectification of their particular individualities, (3) the delight in the realization that their products satisfied the need of other persons, (4) the provision of the means of mediation between the other and the species, (5) the correlative confirmation by the other ("confirmed both in your thought and in your love"), (6) the creation of a shared expression among several lives and (7) the confirmation of the workers' own communal nature.[206] In short, in the unconstrained setting, work finds its fulfillment in meeting the need of the other and in providing the means for realizing a society among the persons involved. Clearly, the individual workers gain in this situation of production; in fact, they gain their own selves insofar as they contribute to the realization of society.

In works of some years later, Marx indicates that he recognizes that use value provides a "service" in meeting a particular need of a person. He wishes, however, to draw attention to the peculiar, inverted social relation in which persons effectively "serve" the commodity, the embodiment of exchange value.[207] Marx carefully notes that the distinction between productive and unproductive labor lies not in the concrete nature of the product or the process but, rather, in the social form of production.[208] Productive labor is pre-eminently that which increases capital. Although Marx rejects the distinction of a material basis between productive and unproductive labor, he himself still betrays a materialist bias in favor of production of material goods. Because the contemporary developments in industry centered upon such production, he naturally focused his studies in this area. Service, that is, unproductive, occupations were those which had dominated the outdated, unfree economy of the craft world. His project required that he analyze the motive force of the contemporary economy, surplus value, in terms of the material products which embodied it. A service, narrowly speaking, deposited no tangible result external to the producers, and was, hence, incapable of being a vendible commodity.[209]

In addition to his rejection of "unproductive" labor, Marx was further biased toward material production out of a compassion toward the poor workers. Marx's rage on behalf of these needy was impatient of services, which promised little in terms of concrete, consumable goods.

Marx amplified his understanding of the individual worker into the image of the "collective laborer" constituted by the division of labor. This meant that he conceived of the needs of contemporary society in just the same way as he had conceived of the obvious needs of the laborer, namely, in material terms.[210] His metaphor of composition prevented Marx from recognizing the mutual work of service which could properly emerge in a conception of social and cultural work.

Furthermore, Marx's philosophical axioms dictated his concern for the material. Concerned to regard the activity of history as proper to actual persons and not to a transcendent version of Hegel's Spirit, he looked not to absolute Spirit as the focus of activity but to actual, shaped materials as the precipitate of action. Thus, persons actualize their potentials in shaping a real world and recognizing themselves in it.[211] They do this and can continue to do this by producing objects external to themselves.[212]

Yet Marx, despite this proclivity to understand production as material, offers solid bases upon which to establish a conception of service integral to work activity. Marx holds that individual consumption produces the worker himself and he contrasts this with productive consumption which results in some object.[213] The consumption by the particular worker may count as an unproductive loss, yet this consumption sustains the worker.[214] The core of service activity rests in the satisfaction of the needs of the person.

Another description of service ties it closely to the meeting of human needs:

> . . . It also does not affect the economic character of this relation [that is, between the supplier and purchaser of a service] whether the physician cures me, the teacher is successful in teaching me, or the lawyer wins my lawsuit. What is paid for is the performance of the service as such, and by its very nature the result cannot be guaranteed by those rendering the service. A large proportion of services belongs to the costs of consumption of commodities, as is the case of a cook, a maid, etc.[215]

The conclusion of this passage indicates once again the service component of work activity centers on the consumption of commodities. In this passage, however, Marx extends the range of service work organically connected with the provision of food, clothing and shelter through the addition of medical, educational and legal services. The inclusion of these other activities confirms that the working person is the focus of service: the person who is embodied, capable of developing his capacities and immersed in social structures. Furthermore, since service activity leaves no deposit of an object, this implies that it is sheer activity. In this, the service component of work activity most purely manifests its dynamism. Finally, no one of these service activities can offer a guarantee that its goal will be reached. Service activity entails the activity of a person directed to the needs of another. Since the parties are circumscribed in skills and energy, their interaction can offer no assurance of attaining the goal.

"Creativity" which figured so prominently in the early writings of Marx came to be replaced by "activity" and "work."[216] Yet the creative dimension of human efforts remains implicit in Marx's understanding of work and thereby offers another way of approaching service from within his system. The creative dimension of work becomes, in the regime of capital, structured in the research of science and technology. In these, persons in society can uncover new products, new methods, new structures, indeed, they can uncover novel aspects of themselves. In their reflection, persons can reconceive even the categories of their economic system. Indeed, it would be possible to reconceive work activity so that the priority is the service of persons, rather than the production of goods, for this assignment of a priority to service would explicate what was already available potentially in that notion of production.

Human work activity creatively reshapes the world, including the reflective processes whereby it has shaped the world. Persons can alter and reshape the very instrumentalism of reflective thought in Marx's system,[217] because it too becomes an object of reflection. This is especially so since the object of activity no longer merely implies the person, as is the case when production has priority, but now explicitly is the person with his or her needs and limits. Just as research stood apart from nature to observe it as it is for the sake of cunning manipulation, so this same attitude learned through technical research, when directed toward persons themselves, reveals both the potentials and the necessitous needs and limits simply as they are.[218] The focus of observation shifts from objects to be acted upon, to persons who act together. The reflective attitude then changes from an intrumentalism, to a readiness for interaction. Contemplation evolves from a position of superior, distant attention to a stance of caring and involved regard.[219] Whereas Smith had analyzed the productive/unproductive distinction in terms of growth in society and Marx had examined it in terms of the exploitative structures of society,[220] out of the present preoccupation with material production, there will emerge a fuller comprehension of unproductive work as activity in the service of the shaping and maintenance of society.

Marx recognizes that the growth of solidarity among the proletariat is founded upon the willingness of its members to undertake self-sacrificing acts of caring for their comrades.[221] This care begins in the desire to diminish suffering and to resist the enemy. With the movement beyond the period of capital domination, the enemy which causes suffering will no longer be seen as a particular class.[222] Death, social fragmentation and a formlessness of life threaten continually; against these, persons will deploy the habits of care in service activity which they will have learned in the earlier resistance to suffering and domination. The members of society will now comprehend their need to work in service of one another while facing the limits of mortality, in achieving recognition and in giving form to the world.[223]

In this chapter I have shown that both Hegel and Marx recognize that work activity develops the capacities of the person, confronts the limits it faces, is intrinsically social, utilizes technology progressively and is fundamentally directed toward the service of persons.

Hegel holds that work is the prime activity of persons. This activity develops their capacities, their lives in society and, in particular, their social institutions. Marx, too, emphasizes the intimate link between person and work. Through their activities persons achieve their self-realization and full freedom. In Marx's extension of the analysis of work into the economic sphere, he conceptualizes the personal capacities for work as labor power. While Marx broadens the understanding of work by studying it more analytically in its deployment within economic processes, Hegel, by contrast, examines the social structures which are founded upon, result from and provide a framework for work activity. Along with his emphasis on the individual person and in contrast to a certain passivity in Hegel's account of work activity, Marx clearly stresses the assertive and active character of work.

Persons exercise this work activity upon a field of limits. For Hegel, work activity transforms the limits constituted by the person's own embodiment, by his needs and desires, by external nature and even by death itself. Working persons do not remove these limits; they reshape them. Marx examines the limits to human activity in terms of freedom. The capitalist system arrogates personal freedom and deceives persons into a belief in their capacity to overcome limits. In the ideal for future work activity, Marx indicates that persons will put this freedom into action so that they recognize necessary limits while at the same time they undertake the creative.

Hegel's corpus presents multiple ways of examining the sociality intrinsic to work. In his early period, Hegel placed the emphasis on the recognition achieved through work. The germ of Hegel's mature presentation of the institutions which organize the political interaction arising from work activity subsists in this reciprocal recognition. In his later work he describes the abstract system of needs which forms the foundation of civil society. He then describes in detail the social institutions which proceed from work activity. Marx, too, presents several approaches to the comprehension of the sociality of work. Whereas Hegel's interpretation more fully accounts for the political interaction devolving from work activity, Marx's presentation is, of course, more systematic with regard to the economic. As he further studies economic structures and processes, he extends his early, Feuerbachian analysis, which was cast in terms of species being, through his theoretical examination of abstract society. Furthermore, in Marx's theory, on the basis of the continual transformative force of work activity, the deleterious effects of this set of social interactions lead to the cooperative society of the future. Once again in contrast to Hegel, Marx supplies at least some elements of a future vision.

Hegel describes the instrumental role which the tool plays in the human interchange with nature and social structures. That the tool endures means that the rationality which has precipitated in it is available to many others. Tools remake the world. In this process of reconstitution, working persons develop the tools still further into machines. But the use of machines against nature causes nature to take revenge on persons. Marx also treats both the evolution from tools into machines and the ills wreaked on persons when the promise, which the latter had offered, becomes subverted. Further, unlike Hegel, Marx emphasizes the social structures of the industrial system. With the rise of this system, machines assume their position of dominance. Marx is also clearer than Hegel in presenting the connection of machines to the body of knowledge which work activity incorporates in them. In addition, Marx again presents a conception of the future, here one in which workers are possessed of, have access to and control over the knowledge deposited in technological instruments.

The emphasis, found early in Hegel's work, on personal recognition conjoined with productive provision offers one approach to understanding the service element of work activity within his system of thought. In his mature works, his examination of social structures amplifies the notion of productive provison and adds the concept of the maintenance of an inclusive, common world as further elements in conceiving the service of work. Marx too furnishes several implicit approaches to an understanding of the service component of work activity. The maintenance of the worker, the creative transformation of reflection and a future, mutual care are ideas derived from Marx which indicate some understanding of the service which inheres in work activity.

At the end of the nineteenth century, Alfred Marshall at Cambridge and Gustav Schmoller at Berlin both developed systems of economic thought which they conceived against the background of the achievements of the four figures treated in the preceding two chapters. In the next chapter I examine the understanding of work as it is found in the writings of these two figures, each preeminent in the contemporary, economic thought of his nation. Thereupon, I will study work activity in the theory of Heinrich Pesch, Schmoller's student and a key figure in Roman Catholic social theory.

1. *Jenäer Systementwürfe* III, in *Gesammelte Werke*, hrsg. von der Rheinisch-Westfälischen Akademie der Wissenschaften; Bd. 8, hrsg. von Rolf-Peter Horstmann unter Mitarbeit von Johann Heinrich Trede (Hamburg: Felix Meiner Verlag, 1976), pp. 243-244.

2. H. Marcuse explains the breadth of the Hegelian understanding of work as compared to that found in the classical economic writers. He refers to Hegel's thought: ". . . Hier erscheint die Arbeit als ein Grundgeschehen des menschlichen Daseins, als ein das ganze Sein des Menschen dauernd und ständig durchherrschendes Geschehen, in dem zugleich mit der 'Welt' des Menschen etwas geschieht. Hier ist Arbeit gerade nicht eine bestimmte Tätigkeit . . . vielmehr ist sie das, worin jede einzelne Tätigkeit gründet und wieder zurückschlägt: ein Tun des Menschen als die Weise seines Seins in der Welt: wodurch er 'für sich' erst wird, was er ist, zu sich selbst kommt, die 'Form' seines Da-seins, seines 'Bleibens' gewinnt und ineins die Welt zur 'seinigen' macht." See "Über die philosophischen Grundlagen des wirtschaftswissenschaftlichen Arbeitsbegriff," *Archiv für Sozialwissenschaft und Sozialpolitik* 69 (1933): 262.

3. "A) Ich mache mich unmittelbar zum Dinge, Form die Seyn ist, in der Arbeit B) dieses meines Daseyns entäussre ich mich ebenso, mache es zu einem nur Fremden, und erhalte mich darin; eben darin schaue ich mein Anerkanntseyn an, Seyn als wissendes,—dort mein unmittelbares Ich; hier mein fürmichseyn, meine Person." *Jenäer Systementwürfe* III, p. 227.

4. *Phenomenology of the Spirit*, Translated by A.V. Miller with analysis and forward by J.N. Findlay (Oxford: Oxford University Press, 1977), Par. 194 and 195. Similarly, in the *Jenäer Systementwürfe III*, Hegel writes of "des Bürgerstandes" as "abstract" in that this activity has been separated from the natural. Indeed, these workers take work from nature and out of the consciousless state. Work has become "das Selbst des Menschen." See p. 268.

5. Alexandre Kojève, *Introduction to the Reading of Hegel: Lectures on the "Phenomenology of Spirit,"* Assembled by Raymond Queneau; edited by Allan Bloom; translated by James H. Nichols, Jr. (New York: Basic Books, Inc., 1967), p. 48.

6. "But just as lordship showed that its essential nature is the reverse of what it wants to be, so too servitude in its consummation will really turn into the opposite of what it immediately is; as a consciousness forced back into itself, it will withdraw into itself and be transformed into a truly independent consciousness." *Phenomenology of Spirit*, Par. 193.

7. L. Dupré, *The Philosophical Foundations of Marxism* (New York: Harcourt, Brace and World, 1966), p. 36.

8. *Philosophy of Right*, translated and edited by T.M. Knox (Oxford: Oxford University Press, 1952), Par. 196-198.

9. Concerning this point C. Taylor writes: ". . . Man can come to see himself in the natural environment by making it over in conformity with his own project. For in doing this we achieve another standing negation, a reflection of ourselves which endures. . . . In transforming things we change ourselves. By creating a standing reflection of ourselves as universal beings we become such beings." See *Hegel and Modern Society* (Cambridge: Cambridge University Press, 1979), p. 156. Also see the change from the savage to the civilized man, from the clumsy to the skilled person in *Philosophy of Right*, Par. 197 Zusatz. These changes result from the work activity involved in practical education.

10. Par. 194.

11. Hegel associates the development of freedom with the rise of towns and the work connected with the business class in particular. See *Philosophy of Right*, Par. 204 Zusatz.

12. R. Plant notes that the abolition of serfdom in Prussia, against the background of the French Revolution and the reforms of K. v. Stein, probably affected Hegel's thought on the freedom of the person within the context of rights secured by the State. See *Hegel: An Introduction*, 2nd ed. (Oxford: Basil Blackwell, 1983), p. 151, n. 2. I add that the abolition of serfdom may well have meant for Hegel the removal of a structural constraint inimical to the basic dynamism of work as represented in the activity of these liberated agricultural workers.

13. By definition, work activity which consists not in the creation of an object but in providing a service as such, involves another person.

14. J.M. Bernstein emphasizes the importance of Hegel's placement of the phenomenon of life on the way from consciousness to self-consciousness in the *Phenomenology of the Spirit*. See "From Self-consciousness to Community: Act and Recognition in the Master-Slave Relationship" in *The State and Civil Society: Studies in Hegel's Political Philosophy*, ed. by Z. A. Pelczynski (Cambridge: Cambridge University Press, 1984), p. 17.

15. Taylor, *Hegel and Modern Society*, p. 156.

16. *Philosophy of Mind*, Part Three of the *Encyclopaedia of the Philosophical Sciences* (1830), translated by William Wallace, together with the *Zusätze in Boumann's text (1845)*, translated by A.V. Miller, Forward by J.N. Findlay (Oxford: Clarendon Press, 1971), Par. 431.

17. *Phenomenology of Spirit*, Par. 195.

18. *Philosophy of Right*, Par. 190.

19. *Ibid.*, Par. 194.

20. Hegel speaks of yet another characteristically human feature of work—the recognition of perduring limits—in his discussion of agricultural work. These laborers provide for the future. They foresee limits and needs and so direct their efforts. Yet particularly these workers remain enmeshed both in the limits of seasonal constraints and also in the need to plan against future needs. *Ibid.*, Par. 203.

21. *Ibid.*, Par. 196.

22. A. Kojève describes the liberating activity of work, "In becoming master of Nature by work, then, the Slave frees himself from his own nature, from his own instinct that tied him to Nature and made him the Master's slave. Therefore, by freeing the Slave from Nature, work frees him from himself as well, from his Slave's nature: it frees him from the Master." See *Introduction to the Reading of Hegel*, p. 23.

23. *Philosophy of Right*, Par. 187.

24. *Ibid.*, Par. 194 Remark.

25. *Phenomenology of the Spirit*, Par. 194, emphases added.

26. *Philosophy of Mind*, Par. 432 Zusatz.

27. *Ibid.*, Par. 433.

28. Hegel draws attention to the significance of person's working activity as the basis for his recognition in the realm of the State. See *Ibid.*, Par. 432 Zusatz.

29. "Labour's 'social' character remains hidden both from classical politics (as a consequence of the fact that this classical tradition ranks moral action above production) and from modern natural law theory (as a consequence of this theory's orientation to relations of will and contract)." See M. Riedel, *Between Tradition and Revolution: The Hegelian Transformation of Political Philosophy* (Cambridge: Cambridge University Press, 1984), pp. 121-122.

30. "Die Arbeit die auf das Bedürfniss eines einzelnen ging wird in ihm a) die Arbeit eines einzelnen b) auch [wenn sie] nur auf sein Bedürfniss geht, eine Allgemeine. a) für das Arbeiten als solches itzt die Foderung vorhanden; [es] will anerkannt seyn, die Form der Allgemeinheit haben; . . . aber diss allgemeine ist für die Arbeit das wahre Wesen; und die natürliche Ungeschicklichkeit muss sich in der Erlernung des allgemeinen überwinden; die Arbeit ist nicht ein Instinct, sondern eine Vernünftigkeit, die sich im Volke zu einem allgemeinen macht, und darum der Einzelnheit des Individuums entgegengesetzt ist, die sich überwinden muss, und das Arbeiten ist ebendarum nicht als Instinkt sondern in der Weise des Geistes vorhanden, dass subjective Thätigkeit des Einzelnen doch ein andres geworden ist, eine allgemeine Regel, und erst die Geschicklichkeit des Einzelnen durch diesen Process des Erlernens wird; durch das Anders werden ihrer selbst zu sich zurückkehrt." See *Jenäer Systementwürfe* I, in *Gesammelte Werke*, hrsg. von der Rheinisch-Westfälischen Akademie der Wissenschaften; Bd. 6, hrsg. von Klaus Düsing und Heinz Kimmerle (Hamburg: Felix Meiner Verlag, 1975), pp. 319-320.

31. The activity of the single worker becomes something else: universal because it acts according to a rule and also universal because he acquires a skill which embodies this rule. K. Löwith clarifies the import of these rules: "Work turns the subjective activity of the individual into 'something other' than it seems at first glance, namely, into a 'universal', because it has been learned according to universal laws." See *From Hegel to Nietzsche: The Revolution in Nineteenth-Century Thought*, translated by David E. Greene (New York: Holt, Rinehart and Winston, 1964), p. 266.

32. The movement of self-consciousness "has itself the double significance of being both its own action and the action of the other as well." Par. 182.

33. In addition, it is important to note that property itself constitutes a social construction, as distinct from possession. S. Avineri, *Hegel's Theory of the Modern State* (Cambridge: Cambridge University Press, 1972), p. 89: ". . . Though Hegel's description of the economic process is taken . . . from classical political economy, he holds a totally different view about the basic nature of property. Property, always remains premissed on social consensus, not on the mere fact of possession." Also see R. Plant, "Hegel and Political Economy—I," *New Left Review* No. 103 (1977): 87; and also S. Benhabib, "Obligation, Contract and Exchange: On the Significance of Hegel's Abstract Right," in *The State and Civil Society*, ed. by Z.A. Pelczynski, p. 171.

34. *Philosophy of Right*, Par. 189.

35. *Ibid.*, Par. 192.

36. Jürgen Habermas states that in the reformulation of his ideas subsequent to the Jena period ". . . the dialectic of labor has been deprived of its central role within the system." See "Labor and Interaction," in *Theory and Practice*, translated by John Viertel (Boston: Beacon Press, 1973), p. 167. I suggest that Hegel shifted the location of the central dynamic of labor to place it as the mediating activity in the sphere of *Sittlichkeit* as he developed his insight into the distinction of civil society apart from the state.

37. *Philosophy of Right*, Par. 191.

38. *Ibid.*, Par. 198.

39. Plant, *Hegel: An Introduction*, p. 109.

40. A.S. Walton, "Economy, Utility and Community in Hegel's Theory of Civil Society," in *The State and Civil Society*, ed. by Z.A. Pelczynski, p. 250.

41. Bernstein, "From Self-consciousness to Community," p. 35.

42. K.-H. Ilting, "The Structure of Hegel's Philosophy of Right" in *Hegel's Political Philosophy—Problems and Perspectives: A Collection of New Essays*, ed. by Z.A. Pelczynski (Cambridge: Cambridge University Press, 1971), p. 109. Ilting refers to organic articulation and liberal division at the level of the state. I use the contrast of these models for the state as a whole, suggesting a dynamic interaction among the three levels.

43. ". . . The universal substance, as vital, exists only so far as it organically particularizes itself." See *Philosophy of Mind*, Par. 527.

44. *Philosophy of Right*, Par. 170.

45. *Philosophy of Right*, Par. 198. For an early Hegelian anticipation of the system of needs based on work—"ein ungeheures System von Gemeinschafftlichkeit, und gegenseitigen Abhängigkeit" which "als ein wildes Thier einer beständigen strengen Beherrschung und Bezähmung bedarf"—see *Jenäer Systementwürfe* I, p. 324.

46. See K.-H. Ilting, "The Structure of Hegel's *Philosophy of Right*," p. 107.

47. *Philosophy of Right*, Par. 201.

48. *Ibid.*, Par. 202. In the early *System of Ethical Life*, Hegel specifies the classes as the peasantry, the bourgeoisie, and the military nobility. See *System of Ethical Life* (*1802/1803*) and *First Philosophy of Spirit, Part III of the System of Speculative Philosophy* (*1803/1804*), edited and translated by H.S. Harris and T.M. Knox (Albany: SUNY Press, 1979), p. 152.

49. *Philosophy of Right*, Par. 207, emphases added.

50. R. Plant, "Hegel and Political Economy—II," *New Left Review* No. 104 (1977): 106.

51. *Philosophy of Mind*, Par. 534, emphases added.

52. G. Heiman, "The Sources and Significance of Hegel's Corporate Doctrine," in *Hegel's Political Philosophy*, ed. by Z.A. Pelczynski, p. 117.

53. Plant, *Hegel: An Introduction*, p. 119.

54. R. Plant, "Hegel on Identity and Legitimation," in *The State and Civil Society*, ed. by Z.A. Pelczynski, p. 231.

55. The preceding discussion has emphasized the benefit that the corporation provides the working person vis-à-vis economic processes. G. Heiman clarifies part of the economic background of Hegel's corporative notion when he indicates that Justinian's *Digest* (see 3.4.1) permitted corporate bodies mostly for economic purposes. See "Sources and Significance," p. 116.

56. *Philosophy of Right*, Par. 297 Zusatz.

57. Heiman, "Sources and Significance," p. 126. Heiman responds to one criticism of Hegel's idea, by remarking that the corporation does not dominate persons because it is concerned only with their occupational lives, not with their private, religious and philosophic interests. See p. 134. Although he thereby underplays the significance of work in the life of persons, especially given that the corporations are concerned with the education of their members, nevertheless against the question of domination, his point is well taken. See *Philosophy of Right*, Par. 252.

58. *Philosophy of Right*, Par. 290 Zusatz.

59. R. Plant notes that Hegel avoids, by this construction, the dangers to which the French Revolution succumbed as a result of removing intermediate and sectional groupings. See *Hegel: An Introduction*, p. 172.

60. *Philosophy of Right*, Par. 289.

61. *Philosophy of Mind*, Par. 432 Zusatz.

62. R. Plant, *Hegel: An Introduction*, p. 112.

63. Paul Chamley, "La Doctrine Economique de Hegel et la Conception hégélienne du Travail," *Hegel Studien* Beiheft 4 (1965): 152.

64. *Jenäersystementwürfe* I, p. 300.

65. *Ibid.*, p. 320.

66. *System of Ethical Life*, p. 113.

67. H.S. Harris presents this point with particular clarity: ". . . When he was making the spade itself Adam was realizing his rational capacities, he existed as a universal subject. . . . The making of tools sets the boundary between nature and culture. Labor here becomes rational self-expression (craftsmanship) instead of being a mechanical expense of energy (toiling)." "Introduction," in Hegel, *System of Ethical Life*, p. 33.

68. *Ibid.*, p. 117-118; *Jenäersystementwürfe* I, p. 321; *Jenäersystementwürfe* III, p. 225: ". . . er braucht blosse Bewegung, und diese findet er in der äussern Natur; oder die reine Bewegung ist eben das Verhältniss der abstracten Formen die äussre Thätigkeit.—Maschine."

69. This idea may be found early and late in Hegel's writings: *System of Ethical Life*, p. 117; *Philosophy of Right*, Par. 198; and *Philosophy of Mind*, Par. 526.

70. *Jenäersystementwürfe* I, p. 321.

71. *Ibid.*, p. 323.

72. *Ibid.*, p. 324.

73. G. Lukács, *The Young Hegel: Studies in the Relations Between Dialectics and Economics*, translated by Rodney Livingstone (London: Merlin Press, 1975), p. 413.

74. Hegel, having read his Smith, expresses his admiration for the system of interdependence which works for the provision of goods for all of society: "When men are thus dependent on one another and reciprocally related to one another in their work and the satisfaction of their needs, subjective self-seeking turns into a contribution to the satisfaction of everyone else. . . . The compulsion which brings this about is rooted in the complex interdependence of each on all, and it now presents itself to each as the universal permanent capital which gives each the opportunity *by the exercise of his education and skill*, to draw a share from it and so be assured of his livelihood, while what he thus earns *by means of his work* maintains and increases the general capital." See *Philosophy of Right*, Par. 199. I set aside for the present an examination of the difference between the mechanisms indicated at opening and the closing of this passage—it remains unclear whether the mediation of needs will be accomplished by way of market processes or by a permanent, general capital which constitutes a fund governed by other processes—in order to re-emphasize how the system of needs is established upon individual work efforts. See my emphases in the passage cited. The same stress on work as the basis of the exchange (as well as a similar ambiguity concerning the mechanism of need satisfaction) may be found in *Philosophy of Mind*, Par. 524.

75. "The business class has for its task the adaptation of raw materials, and for its means of livelihood it is thrown back on its work, on reflection and intelligence, and essentially on the mediation of one man's needs and work with those of others." See *Philosophy of Right*, Par. 204. Notice that this work of individuals Hegel conceives as a task in the service of the whole.

76. "The universal class [the class of civil servants] has for its task the universal interests of the community." *Ibid.*, Par. 205. An adumbration of the universal class may be found in the *Phenomenology of Spirit*, Par. 503, where the noble consciousness sacrifices his individual satisfactions for the sake of efforts on behalf of the universal, as represented by the ruling power, in a heroism of service.

77. G. Lukács, *Young Hegel*, p. 408.

78. P. Chamley, *Economie, Politique et Philosophie chez Steuart et Hegel*, Annales de la Faculté de Droit et des Sciences Politiques et Economiques de Strasbourg, XII (Paris: Librairie Dalloz, 1963), p. 53.

79. Heiman, "Sources and Significance," p. 130.

80. Jean-Guy Meunier rejects the widely accepted phenomenological interpretation of Marx's presentation concerning work. "Le concept de travail est corrélatif au concept de production; il s'agit d'un procès. Il doit être intégré dans le champ opératoire de ce concept. Le travail définit un procès de production. Il appartient à l'une des formes de ce procès et non pas à l'ordre de la manifestation, de la subjectivité humaine." (p. 70.) "Le travail n'est pas l'expression de l'Homme mais l'expression de l'activité générique de l'homme, l'activité de l'homme en tant qu'être générique." (p. 71.) Yet one need not understand the phenomenological interpretation of Marx as individually and ahistorically subjective. The processes of production are driven by persons in their collaboration. These persons and this cooperation are transformed over time through their own activities. From these result the changing forms of the division of labor. As will be presented below, the activity of work articulates the relationship between persons and society not only in Marx's early work but also in his mature writing. See Jean-Guy Meunier, *Genèse du Matérialisme dans les Écritures de Jeunesse de Karl Marx* (Ottawa, Canada: Éditions de l'Université d'Ottawa, 1981).

81. *Capital*, Vol. I, translated by Samuel Moore and Edward Aveling, edited by Frederick Engels (New York, International Publishers, 1967), p. 178.

82. *Grundrisse: Foundations of the Critique of Political Economy*, translated with a foreword by Martin Nicolaus (New York: Vintage Books, 1973), p. 364: "This appropriation, by means of which living labour makes instrument and material in the production process into the body of its soul and thereby resurrects them from the dead, does indeed stand in antithesis to the fact that labour itself is objectless, is reality only in the immediate vitality of the worker. . . ."

83. *Capital*, Vol. I, 1967 edition, p. 177.

84. Bertell Ollman, *Alienation: Marx's Conception of Man in Capitalist Society*, 2nd ed. (Cambridge: Cambridge University Press, 1976), pp. 98-99.

85. *Capital: A Critique of Political Economy*, Vol. I, introduced by Ernest Mandel, translated by Ben Fowkes, "Appendix" translated by Rodney Livingstone (Harmondsworth: Penguin, 1976), Appendix, p. 998.

86. ". . . Human activity produces both a real freedom and a free reality. . . ." L. Dupré, *The Philosophical Foundations of Marxism* (New York: Harcourt Brace and World, 1966), p. 216.

87. *Grundrisse*, translated and edited by David McLellan (New York: Harper and Row, 1971), p. 124.

88. H. Klages agrees with H. Freyer concerning the work of production as the action of freedom. See Helmut Klages, *Technischer Humanismus: Philosophie und Soziologie der Arbeit bei Karl Marx*, Soziologische Gegenwartsfragen, Neue Folge (Stuttgart: Ferdinand Enke Verlag, 1964), p. 109.

89. "The very great difference is . . . whether the means of production employ the workers, or whether the workers as subjects employ the means of production . . . ." See *Theories of Surplus Value*, 3 Pts. (Moscow: Progress Publishers, 1963-1971), Pt. 2, p. 580.

90. "[Estranged Labour]," *1844 Manuscripts*, in Karl Marx and Frederick Engels, *Collected Works*, Vol. 3 (New York: International Publishers, 1975), pp. 276-277.

91. Although the free and liberating activity of praxis will be examined still more closely below—exactly as it provides a critical force in the sphere of economic activities—it merits attention to observe here how the Marxian understanding of productive activity as praxis falls prey to the temptation to see production as the primary, or even the sole, mode of praxis. When this occurs, Marx improperly narrows the scope of his project to reconcile the human person with his activity in society and he illegitimately generalizes the economic activity predominant at a particular period. See Louis Dupré, *Marx's Social Critique of Culture* (New Haven: Yale University Press, 1983), pp. 13 and 87. This attention will provide a vantage from which to grasp the significance of free work activity, apart from Marx's unduly narrowing, conceptual moves, which were contradictory to his own project.

92. *Capital*, Vol. I, 1967 edition, pp. 46-47, n. 1.

93. *A Contribution to the Critique of Political Economy*, translated by S.W. Ryazanskaya, edited by Maurice Dobb (New York: International Publishers, 1970), p. 36.

94. "So far therefore as labour is a creation of use-value, as useful labour, it is a necessary condition, independent of all forms of society, for the existence of the human race; it is an eternal nature-imposed necessity, without which there can be no material exchanges between man and Nature, and therefore no life." See *Capital*, Vol. I, 1967 ed., pp. 42-43.

95. *1844 Manuscripts*, in *Collected Works*, Vol. 3, pp. 241-242.

96. *Wage Labour and Capital*, in Karl Marx and Frederick Engels, *Collected Works*, Vol. 9, (New York: International Publishers, 1977), p. 225.

97. "The medium through which estrangement takes place is itself practical." See *1844 Manuscripts*, in *Collected Works*, Vol. 3, p. 279.

98. *Grundrisse*, 1973 edition, p. 462.

99. *Capital*, Volume I, 1976 edition, "Appendix," p. 1013.

100. *1844 Manuscripts*, in *Collected Works*, Vol. 3, pp. 237-238.

101. *Grundrisse*, McClellan edition, p. 81.

102. *Theories of Surplus Value*, Vol. 2, p. 234.

103. *Capital*, Volume I, pp. 338-339.

104. *German Ideology*, in Karl Marx and Frederick Engels, *Collected Works*, Vol. 5 (New York: International Publishers, 1976), p. 87.

105. G.A. Cohen, *Karl Marx's Theory of History: A Defence*, (Princeton: Princeton University Press, 1978), p. 129.

106. E. Mandel, *The Formation of the Economic Thought of Karl Marx: 1843 to "Capital,"* translated by Brian Pearce (New York: Monthly Review Press, 1971), pp. 46 and 70.

107. M. Blaug, *Economic Theory in Retrospect*, 3rd ed. (Cambridge: Cambridge University Press, 1978), p. 242.

108. *Capital*, Vol. I, 1967 edition, pp. 46-47, n. 1.

109. Marx indicates that this same distinction would have helped Ricardo grasp the social, as opposed to the physical, character of capital. See *Theories of Surplus Value*, Pt. 2, p. 400.

110. *Capital*, Vol. I, 1967 edition, p. 167. The following citation emphasizes particularly the tie between these capacities and the person: "Man himself, viewed as the impersonation of labour-power, is a natural object, a thing, although a living, conscious thing, and labour is the manifestation of this power residing in him." See p. 202 of the same edition and volume of *Capital*.

111. F. Engels, in his introduction to *Wage Labour and Capital* notes, ". . . This labour-power has grown up with his [that is, the worker's] person and is inseparable from it." See *Wage Labour and Capital* and *Values, Price and Profit*, Introduction by Frederick Engels (New York: International Publishers, 1933 and 1976), p. 10.

112. *Capital*, Vol. I, 1967 edition, pp. 183-184.

113. This point is emphasized by Marx and Engels in *Wage Labour and Capital*, "But labour is the worker's own life activity, the manifestation of his own life." See *Collected Works*, Vol. 9, p. 202. I have drawn some attention to this intimate connection between the capacities for work activity and the person himself. Perhaps the following makes my motivation clearer. M. Blaug writes, "Marx was very proud of the distinction betwen labor and labor-power, which he thought cleared up Smith's confusion between embodied labor and commanded labor. But what he really discovered was the Walrasian distinction between the service flow of labor and the stock of labor resources, and it is perfectly true that this distinction is peculiar to a nonslave economy." See *Economic Theory in Retrospect*, p. 283. The Walrasian distinction omits both the focus of activity in the person and the connection of the person to the world by way of his/her own capacities. Marx's distinction incorporates the human person into his understanding of the economic system.

114. *Capital*, Vol. I, 1967 edition, p. 177.

115. *Grundrisse*, 1973 edition, p. 267 and *Theories of Surplus Value*, Pt. 3, p. 178.

116. *Capital*, Vol. I, 1967 edition, p. 519. These dimensions are dependent on the person who embodies labor power. So even when Marx must simplify his analysis by excluding these from further consideration, he does this at risk of obscuring a central insight.

117. *Ibid.*, p. 38.

118. *Ibid.*, p. 58.

119. The intention I have indicated above, to focus on the common elements in human work activity, does not rely on this abstraction found in the capitalist system. My isolating of the common factors does not abstract from instances of exchange value produced within the capitalist industrial system but, rather, from instances of human productive activity (broadly understood as the provision of goods and services) in all economic systems. In this other abstraction from the concrete work activities of persons, through which I seek the central elements of work, I develop a project which Marx himself initiated in his emphatic, conceptual joining of work activity to the human person.

120. *Capital*, Vol. I, 1967 edition, p. 537.

121. *Ibid.*, p. 538, emphasis added.

122. In the *Contribution to the Critique of Political Economy*, labor *time* as materialized in commodities is described as both the substance and standard of exchange value. See p. 30.

123. *Capital*, Vol. I, 1967 edition, p. 195.

124. *Theories of Surplus Value*, Pt. 2, pp. 396-397.

125. When Marx describes the determination of the value of labor power as a commodity, he means, of course, its exchange value. The market system compels a quantification of the value of this personal capacity. See *Capital*, Vol. I, 1967 edition, pp. 170-171. This requisite valuation can be carried out only through the elision of the difference between the (re-)production of persons and that of objects and services.

126. G.A. Cohen, "The Labour Theory of Value and the Concept of Exploitation," *Philosophy and Public Affairs*, 8, No. 4 (Summer, 1979), p. 354.

127. R.J. Bernstein emphasizes the significance for praxis that "the object produced is understood as a congealed form of the activity." On the strength of this understanding Marx extends his earlier philosophical insights into his economic interpretation. See *Praxis and Action: Contemporary Philosophies of Human Activity* (Philadelphia: University of Pennsylvania Press, 1971), p. 60, n. 62.

128. *German Ideology*, in *Collected Works*, Vol. 5, p. 52.

129. *Grundrisse*, 1973 edition, p. 325.

130. *Grundrisse*, McClellan edition, pp. 124 and 142.

131. Alfred Sohn-Rethel, *Intellectual and Manual Labour: A Critique of Epistemology* (London: Macmillan, 1978), p. 56.

132. *Theories of Surplus Value*, Pt. 1, p. 282.

133. *1844 Manuscripts*, in *Collected Works*, Vol. 3, p. 291.

134. *Capital*, Vol. I, 1967 edition, p. 635, emphases added. Capitalist accumulation not only requires this reserve army; it also produces it. See *Ibid.*, p. 630.

135. R. Tucker indicates that the drive to infinite self-enrichment on the part of Hegel's spirit reappears in Marx's thought as the self-aggrandizing of capital. "Hegel's *Logic*, which he reread while working on *Capital*, remained for him always the 'money of the spirit'." See *Philosophy and Myth in Karl Marx* (Cambridge: Cambridge University Press, 1961), pp. 213-214.

136. Klages, *Technischer Humanismus*, p. 173.

137. *Capital*, Vol. I, 1967 edition, p. 487.

138. *Theories of Surplus Value*, Pt. 1, p. 283.

139. *Capital*, Vol. I, 1967 edition, p. 190.

140. Dupré, *Marx's Social Critique of Culture*, p. 177.

141. *Grundrisse*, McClellan edition, pp. 142-143.

142. *Ibid.*, p. 145.

143. *Ibid.*, pp. 75-76.

144. *Grundrisse*, 1973 edition, p. 361.

145. *German Ideology*, in *Collected Works*, Vol. 5, p. 42.

146. *Capital*, Vol. III, translated by David Fernbach (Harmondsworth: Penguin, 1981), p. 959. The two sentences immediately preceding those cited above elucidate the connection with needs and with free activity: "This realm of natural necessity expands with his development, because his needs do too; but the productive forces to satisfy these expand at the same time. Freedom, in this sphere, can consist only in this, that socialized man, the associated producers, govern the human metabolism with nature in a rational way, bringing it under their collective control, instead of being dominated by it as a blind power; accomplishing it with the least expenditure of energy and in conditions most worthy and appropriate for their human nature."

147. Klages, *Technischer Humanismus*, p. 114.

148. Marcuse, "Über die philosophischen Grundlagen," p. 283.

149. A. Heller indicates how needs, and centrally the need for work activity, conduce to the development of freedom, "Needs are simultaneously passions and capacities (the passion and the capacity to appropriate the object) and thus capacities are themselves needs. The capacity for objective activity is thus one of the greatest needs of man." See *The Theory of Need in Marx* (London: Allison and Busby, 1976), pp. 41-42.

150. "The free labourer . . . sells himself and, indeed, sells himself piecemeal. He sells at auction eight, ten, twelve, fifteen hours of his life, day after day . . . to the capitalist." See *Wage Labour and Capital*, in *Collected Works*, Vol. 9, p. 203.

151. In *Theories of Surplus Value*, Pt. 2, p. 408, he writes that the length of the worker's life is, "the normal duration of his labour-power." Also see *Capital*, Vol. I, 1967 edition, p. 641.

152. "A machine which does not serve the purposes of labour, is useless. In addition, it falls a prey to the destructive influence of natural forces. Iron rusts and wood rots. . . . Living labour must seize upon these things and rouse them from their death-sleep, change them from being mere possible use-values into real and effective ones." See *Capital*, Vol. I, 1967 edition, p. 183.

153. *German Ideology*, in *Collected Works*, Vol. 5, p. 43. Also see *Capital*, Vol. I, 1967 edition, pp. 171-172.

154. "Conscious life activity distinguishes man immediately from animal life activity. It is just because of this that he is a species being. Or it is only because he is a species-being that he is a conscious being, i.e. his own life is an object for him. . . . In creating a world of objects by his practical activity, in his work upon organic nature, man proves himself a conscious species-being, i.e. a being that treats the species as its own essential being, or that treats itself as a species being." See *1844 Manuscripts*, in *Collected Works*, Vol. 3, p. 276.

155. See, for example, Mandel, *Formation of the Economic Thought*, here p. 180.

156. Mandel, *Formation of the Economic Thought*, p. 162.

157. *German Ideology*, in *Collected Works*, Vol. 5, p. 63.

158. *Contribution to the Critique of Political Economy*, p. 28.

159. *Capital*, Vol. I, 1967 edition, p. 73.

160. *Ibid.*, p. 353.

161. *Ibid.*, pp. 354-356.

162. *1844 Manuscripts*, in *Collected Works*, Vol. 3, p. 317.

163. It is interesting to note Marx's use of Steuart, from whose *Inquiry into the Principles of Political Oeconomy* he quotes, "Labour which through its alienation creates a universal equivalent, I call industry." The quotation is taken from Vol. 1 of the Dublin, 1770 edition, pp. 361-362 as cited by Marx in *Contribution to the Critique of Political Economy*, p. 58.

164. See *German Ideology*, in *Collected Works*, Vol. 5, p. 41 where Marx emphasizes the connection between activity and social relations, including that of market exchange, as a corrective to Feuerbach's approach.

165. Marx indicates that this split marks the true division of labor. See *German Ideology*, in *Collected Works*, Vol. 5, pp. 44-45.

166. *Grundrisse*, 1973 edition, p. 454.

167. *German Ideology*, in *Collected Works*, Vol. 5, p. 47.

168. "Capital therefore is the existence of social labour—the combination of labour as subject as well as object—but this existence as itself existing independently opposite its real moments—hence itself a particular existence apart from them." See *Grundrisse*, 1973 edition,

p. 471. The Hegelian construction of this dynamic appears here and even more clearly on p. 470.

169. *Capital*, Vol. I, 1967 edition, p. 571.

170. *Theories of Surplus Value*, Pt. 2, p. 416.

171. *Grundrisse*, 1973 edition, p. 302.

172. *Ibid.*, pp. 271-272.

173. *Theories of Surplus Value*, Pt. 3, p. 129.

174. *Capital*, Vol. I, 1967 edition, p. 72.

175. *Theories of Surplus Value*, Pt. 3, p. 245.

176. *Capital*, Vol. I, 1976 edition, "Appendix," p. 1003.

177. *1844 Manuscripts*, in *Collected Works*, Vol. 3, p. 276.

178. Marx criticizes Ricardo for concentrating only on the quantitative dimension of exchange value and for omitting its qualitative aspect. See *Theories of Surplus Value*, Pt. 2, p. 504.

179. *Capital*, Vol. III, p. 175.

180. *Capital*, Vol. I, 1976 edition, "Appendix," p. 1053.

181. Georg Lukács, *History and Class Consciousness: Studies in Marxist Dialectics* (Cambridge: MIT Press, 1971), p. 169.

182. *German Ideology*, in *Collected Works*, Vol 5, p. 47.

183. "Only within the community has each individual the means of cultivating his gifts in all directions; hence personal freedom becomes possible only within the community." See *Ibid.*, p. 78.

184. *Grundrisse*, McClellan edition, pp. 74-75.

185. More specific images of the future society Marx sees limned in the Paris Commune as well as in the Rochdale collectives. Reference to the Rochdale system as a system beyond alienated sociality may be found in *Capital*, Vol. III, pp. 178-179. Also, Marx mentions the association of workers and cooperative factories in *Theories of Surplus Value*, Pt. 3, pp. 104, 264, 273, 505, and 574.

186. R.W. Miller presents a useful insight concerning Marx's rejection of a tendency to passivity in Hegelian thinking. See *Analyzing Marx: Morality, Power and History* (Princeton: Princeton University Press, 1984), p. 90.

187. *Capital*, Vol. I, 1967 edition, p 179.

188. *Grundrisse*, 1973 edition, p. 694.

189. *Grundrisse*, McClellan edition, p. 133. Also see *Theories of Surplus Value*, Pt. 1, p. 411.

190. *Theories of Surplus Value*, Pt. 1, p. 391.

191. *Grundrisse*, McClellan edition, p. 138.

192. *Capital*, Vol. I, 1967 edition, p. 403.

193. *Theories of Surplus Value*, Pt. 1, p. 389.

194. *Ibid.*, p. 390 and *Capital*, Vol. I, 1967 edition, p. 310.

195. *Capital*, Vol. I, 1967 edition, p. 486.

196. *Grundrisse*, McClellan edition, p. 135; *Theories of Surplus Value*, Pt. 1, p. 390; and *Capital*, Vol. I, 1967 edition, p. 423.

197. *Capital*, Vol. III, pp. 198-199.

198. Sohn-Rethel writes, "With a technology dependent on the knowledge of the workers, the capitalist mode of production would be an impossibility." See *Intellectual and Manual Labour*, p. 123.

199. *Capital*, Vol. I, 1967 edition, p. 420.

200. Klages, *Technischer Humanismus*, p. 90.

201. *Capital*, Vol. I, 1967 edition, p. 486.

202. *Capital*, Vol. I, 1976 edition, "Appendix," p. 1034.

203. *Capital*, Vol. I, 1967 edition, p. 488.

204. Klages, *Technischer Humanismus*, p. 128.

205. Allen W. Wood, *Karl Marx* (London: Routledge and Kegan Paul, 1981), p. 29. See also Mihailo Marković *From Affluence to Praxis: Philosophy and Social Criticism* (Ann Arbor: University of Michigan Press, 1974), pp. 64-65.

206. "Comments on James Mill," in *Collected Works*, Vol. 3, pp. 227-228.

207. *Contribution to the Critique of Political Economy*, p. 37 and *Theories of Surplus Value*, Pt. 1, p. 403.

208. *Theories of Surplus Value*, Pt. 1, p. 157.

209. *Ibid.*, pp. 404-405.

210. *Capital*, Vol. I, 1967 edition, pp. 508-509.

211. Wood, *Karl Marx*, p. 39.

212. "Labour is only productive so long as it is producing its antithesis." See *Grundrisse*, McClellan edition, p. 79, n. 1. Such production allows, then, the dialectic of history to continue. The "provision for others" motif found in his early writing, Marx again emphasizes in the *Grundrisse* in conjunction with the establishment of an external reality which reacts upon the workers. See *Ibid.*, p. 100.

213. *Capital*, Vol. I, 1967 edition, p. 183 and 573. Notice that the (re)production of the person constitutes him once again with his needs and limits.

214. *Wage Labour and Capital*, in *Collected Works*, Vol. 9, p. 214. It is of some interest that Marx at one point understands production to include labor which ". . . produces, trains, develops, maintains or reproduces labour power itself." Since the acceptance of this as part of the definition of productive labor ". . . would open the flood-gates for false pretensions to the title of productive labour," Marx leaves this dimension of production out of account. These functions concerning labor power constitute the service of human capacities, that is, of persons. One supposes that Marx's philosophical interests—and perhaps his distance from the Ricardian socialists—led him to reject this element from his definition of productive work. See *Theories of Surplus Value*, Pt. 1, p. 172.

215. *Ibid.*, Pt. 1, p. 405. For Marx's sharp comment on Nassau Senior's attempt to consider the medical care of a child as production, see p. 288 of this same volume.

216. Ollman, *Alienation*, p. 102.

217. Jürgen Habermas, *Knowledge and Human Interests*, translated by Jeremy J. Shapiro (Boston: Beacon Press, 1971), p. 44.

218. George G. Brenkert, *Marx's Ethics of Freedom* (London: Routledge and Kegan Paul, 1983), p. 109.

219. In this light, see Eugene Kamenka's comment that Marx viewed the proletariat in terms of its suffering and deprivation—characteristic of his materialistic approach—rather than in terms of its character and capacity for freedom. Hence, a concern for useful production results. Another approach, one more oriented to service, would recognize the potential for free change on the part of the proletariat and stimulate the actualization of this liberty. See *The Ethical Foundations of Marxism*, 2nd edition (London Routledge and Kegan Paul, 1972), p. xiv.

220. I. Gough, "Marx's Theory of Productive and Unproductive Labour," *New Left Review* 76 (Nov.-Dec., 1972): 66. Gough's article provides a solid analysis of many of the other issues concerning this distinction, for which there is no further space in my presentation.

221. R.W. Miller carefully examines the workings of this dynamic in *Analyzing Marx*, pp. 71-72.

222. The optimism implied in this capacity to expand one's care and service beyond one's own group meets with criticism, for example, by Kamenka in *The Ethical Foundations of Marxism*, p. 107. Although R.W. Miller raises the question whether the concern will be shown to strangers, his own response is insufficient. See *Analyzing Marx*, pp. 73-76. I note, first, that the constraints of space have perhaps had the effect of making the optimism appear naive. That extensive efforts, setbacks and other vicissitudes will mark the process of the expansion of

concern, I have no doubt. Secondly, this checkered development itself highlights the struggle for recognition and social form. The reconciliation necessary in this struggle itself constitutes a particular work of service in the face of very human limits. ". . . It is not unconstrained intersubjectivity itself that we call dialectic, but the history of its repression and re-establishment." See Habermas, *Knowledge and Human Interests*, p. 59.

223. Carmen Sirianni writes of "the terror of death, the dread of insignificance, and the need for cosmic recognition" as the deepest sources of domination. See "Production and Power in a Classless Society: A Critical Analysis of the Utopian Dimensions of Marxist Theory," *Socialist Review* Issue 59, 11, No. 5 (Sept.-Oct., 1981): 73-74 n.

Chapter 3
The Turn of the Century: Marshall, Schmoller and Pesch

Alfred Marshall (1842-1924), the leading figure in British economics at the turn of the century, continued the main line of development initiated by Smith and Ricardo, generalized about the evolution of economic processes as a whole and strongly influenced the development of the marginalist approach to economics. In addition, Marshall was influenced by Hegel's ideas concerning organic social change and the ever more complex development of needs and capacities. Furthermore, this leading economist was markedly concerned about workers. His first published essay was entitled "The Future of the Working Classes" (1873). He contributed to the Royal Commission on Labour and was especially responsible for those sections which treat trade unions, minimum wage, and the irregularity of employment.[1] Indeed, Marshall faults Ricardo and the other classical economists for their deficient view of the worker:

> This did little harm so long as they treated of money and foreign trade, but great harm when they treated of the relations between different industrial classes. It led them to regard labor as a commodity without throwing themselves into the point of view of the workman; without allowing for his human passions, his instincts and habits, his sympathies and antipathies, his class jealousies and class adhesiveness, his want of knowledge and of the opportunities for free and vigorous action.[2]

I shall now examine how Marshall advances to a fuller understanding of the workers and their activity and how he nevertheless fails to maintain his attention toward the ordinary worker.

The Worker's Personal Progress

For Marshall, economics is the study of wealth and, more importantly, the study of the human person—in particular, the study of how that person is molded through his ordinary work.[3] His interest in the supply of labor as a productive factor focuses on how the worker's energy, activity and character are at the base of technical economic problems.[4] Although he frequently understands work as a "discommodity" and often emphasizes the increasing marginal disutility of effort for the laborer, Marshall's view on work cannot be reduced to this single view. ". . . Much of the work by which people earn their living is pleasurable in itself, and there is truth to the contention of the socialists that more of it might be made so."[5] Both the pleasure and discommodity of the work experience depend not solely on the task in itself but, rather, depend significantly on the physical strength, character, and will of the worker.[6] Despite the general validity of diminishing marginal utility, this economist finds in intellectual work a significant area in which marginal utility increases; even with regard to physical work, some segments of the work process may not only result in pleasure, but may actually lead to its increase.[7] The attitude toward a job depends on the particular worker, his prior development, and his changing relation to the task.

Marshall directs his attention to the effect of work activity upon the worker and asserts that this result need not leave the worker in a coarse and rough condition.[8] The needs of the workers give rise to their activity and this activity leads, in turn, to higher needs and activities.[9] The capacities and powers of individuals evolve through their own work activity. In taking this view, Marshall expressly challenges many of his British predecessors in economics. They saw the person as a constant quantity; he recognizes that both the patterns of work activity and the person behind them are liable to transformation.[10] Further, in this transformation the workers are altered not merely by being worn down; rather, through their work they achieve their potentials. When the institutions of society have been arranged so that work is fully revealed, then it will stand manifest as "the healthy energetic exercise of faculties," as "the aim of life itself."[11]

Marshall understands that work activity will develop persons in terms of work skills, intellectual capacity and virtuous character. Through their activity and through their collaboration, workers gradually learn the skills of an industrial people, namely:

> To be able to bear in mind many things at a time, to have everything ready when wanted, to act promptly and show resource when anything

> goes wrong, to accommodate oneself quickly to changes in detail of the work done . . . to have always a reserve force which will come out in emergency. . . .[12]

Work adapts persons to the demands of industrial production and this adaptation is benign, for it develops some of their higher capacities. Marshall regards education as an aid to making workers more skilled and more responsible for their tasks.[13]

Apart from those skills necessary to their occupations, the development of workers also includes their intellectual development. Work activity, when stunting forms are excluded, will stimulate new paths of thought, increase knowledge, elevate tastes and conduce to more comprehensive interests.[14] Furthermore, work brings about the development of "character." This encompasses virtues such as regard for others and self-respect. It is upon such character that the progress of society itself is founded.[15]

The development of the person through work initially depends on the foundation laid within the family: "Thoughtful parents bring up their children to the most advantageous occupations to which they have access. . . ."[16] Two processes are interwoven here: education and the selection of an occupation. Persons do not weigh the costs and benefits only of particular tasks; rather, they determine the advantages of a particular occupation relative to the efforts which must be expended to attain it. Further, parents make this same estimate on behalf of their children.[17] For Marshall it is a moral principle that children should not be brought into the world unless they can be given at least as good an education as that received by the parents. Education determines the sort of occupation; the occupation is central to the development of personal capacities; the education provided by the parents demonstrates their regard for the person who is their child.

Partly on the basis of the wage which is offered, persons select their individual occupations. In accordance with Smith's analysis, Marshall asserts that the uncertainty of success in an occupation, the inconstancy of employment and the social rank of the occupation determine the wage. Each of these indicates that wages mark the significance of work in the development of the person. Marshall extends Smith's analysis when he notes a difficulty in the comparative evaluation of professions: prior commitment on the part of the individual to a particular occupation is required, for it is only on the basis of this condition that the development of talents and tastes—the chief elements of success—may be properly assessed.[18] So intimate an activity is employment that the possible maturation of potentials requires an antecedent, personal investment. Quite in contrast to this situation, Marshall observes that the very waiting for work as it is borne by trained but unemployed persons causes more strain than the work itself would.[19] Implicitly then, the proper use of human effort lies in work. Those who fail to develop as a result of inactivity deserve compensation. Conversely, a higher wage, corresponding to a higher social rank, represents an inducement to "persons in a lower grade" to bring up their offspring to this higher level of education, where more skills can be exercised.[20]

Through this analysis, Marshall has conceived of parents as entrepreneurs of human capital "on the look-out for the most advantageous occupations for their children."[21] He expressly speaks of the most valuable capital as that which is invested in human beings and that the most significant element of this results from parental, in particular, maternal care.[22] He considers wages to be analogous to the interest returned on capital.[23]

In the long-run, the return on this investment cannot fall to the parents. The ultimate beneficiary is society.[24] The education for and choice of occupation undertaken by the parents thus stands as an ethical concern not only as an expression of regard for the child as person, but also as a responsibility to maintain and better society. The failure to invest well in this human capital will have a long-run and cumulative negative impact on society.

Marshall utilizes a notion of freedom and self-development derived from Hegel[25] in order to present the significance of work activity for the person. Parents undertake free choices in determining occupations for their children. In the course of the education offered by the parents and during work itself, working persons come to learn: independence and the habit of choosing one's course; self-reliance; deliberation and promptness of choice; and the shaping of one's course toward distant aims. These traits of character proper to modern industrial life describe the ideal, free individual. Yet such individuals could only describe a minority in Marshall's society.[26] In the ideal world, Marshall would have this freedom practiced fully by each individual. In the actual world, he sees the entrepreneur as the exemplar of this freedom.

In the eyes of Marshall, employees, employers and professionals are all workers.[27] Among all workers, entrepreneurs exercise their freedom most fully. Marshall indicates the special position accorded to those persons active in business when he says that the truly liberal, general education prepares the mind to engage its finest potentials in the work of business.[28] Below I will examine further the significance which Marshall attaches to the freely working entrepreneur, whom he holds to be chiefly responsible for economic progress. I now turn, however, to an examination of the way in which Marshall sees work as directed against the limits which weigh upon human efforts.

Efficiency Directed to Raise the Standard of Life

When a person satisfies one of his wants by direct work, this exemplifies the simplest case of equilibrium.[29] The working person remains the center of the interrelation of production and consumption, even on the broadest level of the entire economy.[30] Work produces the means for sustenance, those for the satisfaction of wants and the requisites for the development of physical, mental and moral activities.[31] Marshall here advances the British economic tradition beyond its strongly hedonistic interpretation of producing activity. Sustenance remains the ground upon which human wants emerge and seek satisfaction—wants such as those which Marshall calls "conventional necessaries" and habitual comforts—but he goes still further and includes, in the goals of production, activities and even their improvement. Human needs are not static and constraining, but dynamic and stimulating. The standard of life does not rise solely as a result of the availability of more goods for sustenance and want-satisfaction:

> . . . The true key-note of economic progress is the development of new activities, rather than of new wants. . . .
>
> The term "standard of life" is here taken to mean the standard of activities adjusted to wants. Thus a rise in the standard of life implies an increase in intelligence and energy and self-respect; leading to more care and judgment in expenditure. . . .[32]

The needs of working persons are not so structured that they remain capable only of a change in the quantity or variety of what they consume. Rather, their work activity progressively augments the very activities which each person undertakes.

The rational use of human efforts calls for efficiency. For Marshall, industrial efficiency requires both the training of the population and its vigor. Obviously the supply of food and the discoveries of medical science do much to secure this vigor. In addition, Marshall believes that hopefulness, freedom and change are other conditions requisite for vigor.[33] This is to say that the work environment of persons must both permit and stimulate these three conditions so that their vigor may develop fully and so that they may work efficiently with each other.

Workers face both the physical limitations imposed by nature and those other restrictions, significantly dependent on work conditions, which inhibit the exercise of freedom. Marshall recognizes the embodied limits which constrain work activity. He realizes that work-fatigue can contribute to aging and that a low wage-rate may result from a marginal utility for money which is extremely high as a consequence of the fear of hunger.[34] More broadly, because of these special, embodied limits, he differentiates labor from other commodities in terms of his concept of the long period of production:

> . . . Since labour is slowly produced and slowly worn out, we must take the term 'long period' more strictly, and regard it as implying a greater duration, when we are considering the relations of normal demand and supply for labour, than when we are considering ordinary commodities.[35]

The first two of his "peculiarities" of work likewise demonstrate Marshall's theoretical grasp of the significance of embodied limits. These two particular characteristics examine the obligations entailed in the work contract: labor is sold, but the worker remains his/her own property; once it has been sold, the worker must present him/herself where those services are delivered.[36] The first circumstance argues for the special treatment accorded the worker as an agent of production. The second is the basis for the particular attention which must be paid to work conditions and for the importance of one's work associates, as Marshall himself notes.

Other elements of Marshall's examination of labor do not, however, give an adequate treatment of the embodied limits of the working person. When he writes that an unemployed person is likely to suffer not because he is a wage-earner but because he lacks a reserve, he fails to remember that this worker is a person who is capable of suffering, not an establishment which may be sustained by a reserve fund.[37] When Marshall, in his treatment of the costs of production, excludes considerations both of the expenses of education and of the costs of the exhaustion which work exacts of health and strength, he betrays a tendency to equate labor with other productive inputs.[38] Indeed, although he indicates that workers do not often notice that the pay necessary for them to work "the last, most distressing hour" gives them a claim to the producers' surplus, he fails to mark how each hour is connected with those which have preceded in the embodied history of the working person.[39] Finally, Marshall denotes perishability as the third peculiarity of labor power, which he stipulates as the irrecoverability of the time spent unemployed.[40] Marshall is at pains to show that other material agents of production are perishable in the same sense. In so doing he ignores both the significant difference found in the close relation of the labor sold to the embodied provider himself and ignores the unique developmental role he himself has assigned to personal work. Neither of these characteristics has a parallel in the case of material agents of production. Marshall advances beyond the position of his predecessors in his greater understanding of the aids to vigor and the embodied limits faced by the worker, yet he often fails to attend fully to the significance of the worker's embodied limits. The social sphere in which the working person experiences these limits and undertakes his efforts will be the next subject to be examined in Marshall's understanding of work.

Individual Efforts Toward Social Progress

I now present Marshall's treatment of the significance of institutions, principles, and workers' groups in his understanding of society.

In an important way, work activity constitutes society, for the demand for work depends on the national dividend, which, in turn, derives through the cycle of social processes from work.[41] Although the sundry agents of employment are rivals amongst themselves, they are also collaborators in that they mutually constitute the field of employment.[42] The total supply of labor and the collective demand for it do not constitute society. Marshall's theoretical depiction of the perfect market mechanisms must not be mistaken for his image of society. Indeed, he himself saw equilibrium analysis as only preparatory to a fuller, organic treatment of society.[43]

Social customs and institutions are integral to his understanding of economic phenomena in society conceived as an organism. Skills needed in work, particularly that activity performed with machines, often represent a general learning which society provides the worker.[44] An industry concentrated in a particular locality will often benefit from the special exchange of knowledge and skills among the workers in that area.[45] The family represents both a set of rearing customs and an institution; the family is the key locus for training in the skills and habits required for work in society

The significance of institutions for Marshall's view of work in society is expressed in Book IV of the *Principles* where he lists organization along with land, labor and capital as an agent of production. His examination of the growth, transformation, and functioning of both capital and labor groups attests to his earnest adherence to an organic view of society. And finally, Marshall holds that the organization of a well-ordered state proceeds from motives many of which are independent of those connected with the pursuit of wealth; yet the institution of the state is itself critical in attaining economic prosperity.[46] Marshall specifies still further the organic character of this society along with its structures, when he indicates that these effects of custom are neither short term nor merely perduring, but, rather, cumulative either for good or ill.[47]

Ethical principles represent another element in the construction of Marshall's concept of society. In fact, Marshall is convinced that progress in such principles is required for the advancing condition of a nation's industry.[48] The full development of persons includes the unfolding of their moral capacities. Individual progress effects social progress. The ethical principles which higher society articulates and follows, will, in turn, contribute to the still further advance of other persons. In particular, both working and the social situation of work activity mold the character of a person.[49]

In the ideal society persons will be interrelated by a system of duty alone:

> . . . Where everyone shall work for the public weal with all his might, expecting no further reward than that he in common with his neighbours shall have whatever is necessary to enable him to work well, and to lead a refined and intellectual life. . . .[50]

On the way toward this ethical and ideal society of work, Marshall believed in a gradual amelioration of social conditions, dependent on the virtuous sense of social duty on the part, especially, of the employing classes. On the part of the laboring classes, this improvement depended on the virtue of unselfish sacrifice undertaken for the sake of the improved education of their children.[51]

Ethical principles, customs and institutions contribute to the stucturing of society in Marshall's view. In the constitution of society, the organizations of working persons occupy a position of clear importance. Following Smith and Mill, Marshall describes the disadvantages for working persons in their attempts to negotiate the terms of labor. Such a disadvantageous position in bargaining lowers both the wages and the efficiency of the worker.[52] This latter reduction causes a cumulative loss for society as a whole. Marshall favors the creation of groups of workers—trade unions—for the redress of this worker disadvantage and the remedy of this social loss. To mistake the concern for wages as the main argument for unions would be to mistake Marshall's thrust. Unions benefit workers by training them to be persons and citizens.[53] The proper level of wages, attained free from any disadvantage in bargaining, provides a means for the achievement of full personhood and citizenship.[54] Unions have the capacity to secure the development of the skills, understanding, business power and inventive resources of the workers.[55] Unions together constitute an institution wherein it is possible for persons to be encouraged to develop their potentials so that each does not see him- or herself as a "producing machine."[56]

The union situation offers more than the training of the individual, for persons themselves are thereby guided to participation in a social group. In the well-run trade union, each member learns direct responsibility, the ordering of public matters, the discharge of public functions, judgment, self-control and knowledge of his colleagues, just as do members of small republics or village democracies.[57] Workers learn an even more weighty political lesson when the members acquire an understanding of how to cooperate with other unions.[58]

Marshall also refers to another grouping of workers which surpasses the union organization. Marshall describes several types of cooperative associations in which laborers collectively own the capital. These cooperatives improve on unions because they enable working persons to partake in the direction of production.[59] The responsibility for a still broader range of the production process stimulates a greater development of personal skills and knowledge.[60] Marshall presents a normative subsidiarity when he emphasizes decentralized organization, self-help and the absence of all interference by the state and others in the exercise of individual freedom within the cooperative movement.[61]

The individual workers collaborating within a union or cooperative come to a wholeness of their own character while contributing to the responsible life of customs and institutions which structure their larger society.[62] Marshall holds that both the interaction of workers among themselves while at work and the stimulus to the development of skills, knowledge and practical judgment which the business of each day presents are central to the process of bringing ethical and industrial phenomena into relationship.[63] Since a certain wholeness can be achieved through worker organizations, these groups are critical in evaluating the ethics of economic processes.

Although Marshall extols workers' organizations as a factor in the wider economic world, his view of technology as another striking feature of the modern scene does not cohere with the priority he places on the personal development that he sees fostered in institutions of workers' organizations.

Technology as an Entrepreneurial Tool

According to Marshall, the social separation between the masters and the workers increased especially in the later part of the eighteenth century as a result of the introduction of a broad range of important, new, mechanical inventions.[64] At the same time, as has been frequently observed, a growing subdivision of labor accompanied the improvement in machines. Marshall greatly respects the organizational skill of the person engaged in business who is capable of ordering and directing the multiple, particular tasks which constitute the division of labor. Since those persons dedicated to business concern themselves with the total production process and since they now stand at a distance from their employees, they are inclined to see the employees at most as equivalent to the machines because they too are elements in the course of production. Marshall, despite his repeated emphasis on the significance of the development of the working person's potential, shares this view which equates worker and machine.

Oddly enough, Marshall clearly sees the danger of such an equation: ". . . Men and machines stand on much the same footing: but while machinery is a mere implement of production, man's welfare is also its ultimate aim."[65] Nevertheless, he does compare the working person to a machine.[66] Like a machine, the person must be employed constantly within a narrow range of work in order to maintain economy in production.[67] Likewise, when workers must undertake several tasks, it is necessary that each of these assignments require of them the maximal use of their personal skills and abilities. Yet this requirement no longer devolves from the worker's need for development; but rather, as in the case of the machine, each job must extract from the worker the maximal output. The acquisition of highly specialized machinery is an advantage afforded to large establishments; this same advantage applies equally well to highly specialized skill.[68] Here once again Marshall reckons the advantage in terms of productive efficiency, not personal development. He has implicitly extended the concept of work to machine capital, while narrowing the concept of work. Both capital and labor "work" in the sense that both machines and working persons partake in a productive process directed and evaluated by the businessperson. At the same time as this extension has been accomplished, the concept of work has been narrowed with a resultant loss of its personal elements.

Not only are working persons equated with the machine, but they must also compete with it. The principle of substitution determines the judgment of businesspersons as they seek the most efficient production process:

> The efficiency of human agents of production on the one hand, and that of material agents on the other, are weighed against one another and compared with their money costs; and each tends to be applied as far as it is more efficient than the other in proportion to its money cost. A chief function of business enterprise is to facilitate the free action of this great principle of substitution.[69]

In this evaluation concerning which productive agent to substitute, the persons engaged in business compare the money cost of inputs in terms of time. First, they weigh the uncertainty of human life against the readily and accurately determined length of machine activity. Second, they consider the usually longer earning life of the person in contrast to the briefer service period of most machines.[70] A third approach by way of time is also used in this evaluation: Marshall conceives capital as labor-with-waiting.[71] Yet capital cannot be conceived simply as labor. It results from labor, true; but it is something else. To call it a type of labor illegitimately implies that it has organic potentials to develop and that it may responsibly collaborate with working persons. The waiting indicated here is not a virtuous patience similar to self-control. It is rather the delay sustained by entrepreneurs because of their choices of productive inputs, a delay in the achievement of a productive process. Marshall's conception of education as an investment in human capital and his distinction between the short and long run periods of production, especially as this applies to working persons, easily led both to the equating of men and machines and to the conception of their competition within the process of the efficient substitution of inputs.[72]

I reiterate that Marshall's concept of the relation between technology and the human person remains ambiguous. Despite his tendency to stress the equation of and competition between labor and machinery, he still distinguishes between labor and machines. He recognizes that capital, including machinery, "consists in a great part of knowledge and organisation."[73] Since only working persons can discover and provide both of these, the source of labor power is clearly distinct (at the minimum) from that force provided by machine capital. Humans are significantly different from machines, additionally, owing to the way in which technology shapes them. The direction of technological change brings about the rising relative importance of general, at the expense of specialized, personal abilities.[74] Marshall also holds that the operation of intricate, costly machinery elicits a development of the intelligence and responsiblity in the worker.[75] The products of advanced technology thus contribute to the development of personal character. This second Marshallian argument for the distinctiveness of human productive inputs rests both on the development which cannot be paralleled in machines, and on the unique contribution of this character development to production. The rising standard of life which is the goal of production depends more importantly on the progress of the skills, intellect, and ethical character of working persons, than on the investment in machine capital.[76]

Lastly, Marshall hints at another critical difference between persons and machines. He notes that some outcomes of technology—such as factories, houses, mine shafts, and railway embankments—last longer than do the persons who construct and use them.[77] Yet these are exceptions to the general rule about machine perishability. Implicit here is this significant difference: although persons and many machines may substitute for one another in processes of production, some very large and long-perduring tools are not easily replaced by working persons. These lasting tools often only indirectly contribute to production, while shaping a world within which persons may meet and advance their needs. The world of things, tools, and machines may be said to shape persons—but this analogous expression depends upon the deliberate, active response by persons in work. Only working persons shape a world—of perduring implements, customs, and institutions—to be at the service both of their production and their lives. Although these long-term achievements of production and technology easily become the focus of attention at cost of the regard for mortal persons, their significance arises only from continued personal activity. The differences between machines and persons reveal the distinctiveness of the latter. I now present Marshall's contributions to an understanding of service.

Entrepreneurship and Education

Marshall realizes that many of the most worthy benefits produced in society do not themselves contribute directly to the production of material wealth.[78] What the nature of these "nonproductive" benefits might be, he indicates when he writes that "the desire for the exercise and development of activities, spreading through every rank of society" induces the pursuit of science, literature, and art as well as a demand for those who teach in these areas.[79] These various forms of service work provide "nonproductive" benefits to society, since the progressive development of persons and society requires the full actualization of potentials. The higher levels of these potentials demand further work activities. That is, they no longer require the satisfaction of sustenance and wants through material goods but, rather, call for exercise in activities which offer services to the members of society.

The Marshallian theoretical world restricts work as service activity to two particular forms: entrepreneurship[80] and education. For Marshall, the entrepreneur represents a "nonproductive" worker whose activity yields the highest benefit to society. The entrepreneur, "like the skilled artisan, renders services which society needs. . . ."[81] The services which the entrepreneur contributes do not proceed simply from his work as a manager, organizer, and investor. Marshall regards the businessman also as providing services to his workers in offering them high wages and proper training and in attending to their happiness and culture.[82] In his leadership both in the processes of production and in social care, the entrepreneur provides critical services to society.[83]

The other significant service activity in Marshallian society is education, which falls as an obligation on both parents and society.[84] Parents are obligated to give children a better education than their own, and society is to rescue each child from that ignorance in which he acts as a mere "producing machine."[85] Education represents one factor in the "production of human beings."[86] That is, it is one of those aforementioned social benefits which do not result in a material commodity. According to Marshall, education benefits society particularly because it remedies the problem of the feeblest and most ignorant class of workers. One must carefully note that the education which Marshall understands here is not mere training in some technical skill. He is concerned for workers, yes, but not solely in their capacity as wage laborers:

> . . . A good education confers great indirect benefits even on the ordinary workman. It stimulates his mental activity; it fosters in him a habit of wise inquisitiveness; it makes him more intelligent, more ready, more trustworthy in his ordinary work; it raises the tone of his life in working hours and out of working hours.[87]

The benefits of education exceed the advantages it brings through the worker's activity to the production process. Insofar as it is thus "nonproductive," education represents a chief means for the service of persons.

I have presented Marshall's understanding of work in terms of the effort to raise the standard of life through both personal and social progress. While he clearly believes that the organizations of workers plainly assist them in their advancement as persons and citizens, his position on the role of technology in this human progress remains ambiguous. Further, while the businessman serves society by energetically directing its evolution, education also represents a critical, ethical and social service.

The complexity of human satisfaction, the pattern in which society and its economic processes alter from period to period, the integration of ethics and economics, and a pre-eminent concern for the worker—these central concerns of Marshall also characterize the work of another school of economic thought contemporary to Marshall's, namely, the work of the Later German Historical School. Marshall knew of the work of these scholars and praised it.[88] The influence of the Germans on him enabled Marshall to advance beyond his classical forebears, especially in terms of the historical examination of economic institutions.[89] At this point, I direct my attention to the concept of work as used by Gustav Schmoller, professor at the University of Berlin and the acknowledged leader of the Later German Historical School. His work represents the achievement of German economics at turn of the century and provides the basis for twentieth-century Roman Catholic social teaching—especially as articulated by Heinrich Pesch and Oswald v. Nell-Breuning.

Gustav Schmoller (1838-1917), professor and rector at the University of Berlin, leader of the Later German Historical School, co-founder and chairman of the *Verein für Sozialpolitik* (Association for Social Policy), member of the Prussian *Staatsrat* (state council), representative in the *Herrenhaus* (upper house of the legislature), directed the course of German economics during the period of the Prussian empire. Although Thomas Carlyle, Auguste Comte, and John Stuart Mill markedly influenced the Later Historical School,[90] the Earlier Historical School and, in particular, its leading figures—Wilhelm Roscher, Bruno Hildebrand, and Karl Knies—were its source of inspiration. The thought of Hegel exercised influence on these three economists, especially in terms of a grasp of social reality as an organic totality[91] and in terms of the progressive development of history.[92] The earlier historical economists emphasized the connection between ethics and economics as well.[93]

Schmoller's school claimed its study to be an historical-ethical economics: "historical" in opposition to the theoretical claims of classical economics; "ethical" in contrast to the procedures of scientific socialism.[94] Unlike the British enterprise of articulating unchanging and universal economic laws, Schmoller and his colleagues sought to understand the causes and patterns in the development of particular social organizations. In their study of historical progress, they advanced beyond the position of the Earlier Historical School because they did not see capitalism as a final stage for society.[95]

From the time of his dissertation on economic views during the Reformation, Schmoller emphasized the interrelation of economic activity and ethical action.[96] Since, unlike the classical economists, he understood human motivation to be constituted by many factors, including the ethical, and while he still recognized and analyzed the technical factors of production, Schmoller could posit an intimate relation at the level of individual workers between ethics and economics. Influenced by those German economists who were affected by the German humanist ideal, he viewed economics as concerned not merely with the production of goods, but with the exercise and development of human powers for the ends of persons.[97]

Schmoller does not conceive that the relevance of ethics to economics remains simply on the personal level. Economic studies demonstrated the deleterious impact of the systems of society on persons; this indicated that reform must take place in these systems. The historical procedure of Schmoller and his economist-colleagues indicated, further, that reform could take place: just as technology had introduced profound changes in production, so too new social institutions and services could be devised.[98] For Schmoller and his colleagues the *Verein für Sozialpolitik* was a critical means for shaping public opinion, the political environment and social legislation in order to bring about reform. They conceived of their detailed studies of particular social, political, economic, legal and administrative institutions as the matter upon the basis of which their suggestions for reform would be drawn. For Schmoller, the pre-eminent agent of reform was the state.

A concern for the workers informed the project of the Later Historical School and, especially, that of Schmoller. Alerted by the increasingly popular socialist views to the plight of the workers in the period of belated German industrialization, Schmoller began to address this complex of issues early in his studies. His first publication treated the "worker question."[99] The union question formed a chief point of discussion at the first meeting of the *Verein für Sozialpolitik*.[100] Throughout his career, in historical studies, contemporary descriptions and policy recommendations, Schmoller paid heed to the lot of the worker and sought to remedy the situation. To the Later Historical School must be credited a significant contribution towards the amelioration of the workers' situation and of their integration into society and state.[101]

Persons Actively Engaged Toward Their Totality

In contrast to the classical economists, Schmoller considers the breadth of the range of motivations which impel the person. Indeed, he focuses not, as did the classicals, on sovereign, natural laws but, rather, on the free activity of persons. Schmoller's attention to the complexity of the person counterbalances his treatment of large institutions and historical processes. Central to persons is their capacity for ethical judgment. Schmoller also holds conversely that ethics is one basis for all social and economic organization.[102]

Schmoller founds his ethics on a double principle of totality which he derives from Johann Gottlieb Fichte. The stress in this ethical system lies not upon pleasure in itself, as the simple motive toward the good, but upon human action. An ethical action is one in which the lower natural drives are unified with the higher impulse to self-determination.[103] Furthermore, such actions must reflect an attention to the social whole which structures the range of objective goods sought through the various social institutions.[104] To rectify the Fichtean error which delimited the private sphere alone as the realm of ethics and to distinguish himself from both materialistic and idealistic ethics, Schmoller establishes his empirical ethics, in continuity with the tradition of Hume and Smith, upon the moral feelings of the individual and upon the evaluation of these from the perspective of society.[105]

Within this framework, Schmoller conceives the person as oriented toward happiness, understood broadly as a manifold of pleasures derived from varied social activity wherein one's capacities are exercised.[106] Within the theoretical, ethical construct of totality, Schmoller can evaluate the motivation through profit as one of the desires of the person. Schmoller explicitly advances beyond the Earlier Historical School in his contention that the profit motive is not a primitive impulse but has developed with the rise of individualism. Modern individuals pursue profit for the sake of their individual happiness. The development of this drive represents an advance insofar as it stimulates the discipline of work.[107] The profit motive can also be justified insofar as it encourages the further development of both individuals and their personal freedom and insofar as it stimulates the striving and self-assertion of communities. To secure these aims and to prevent the dangers of greed, injustice and mistreatment of the weak, society must establish ethical means and social frameworks.[108]

Within the ethical construct of totality, the chief force is human work activity. Work, "that human activity which pursues ethical-rational aims with lasting effort"[109] is the prime activity through which individuals become themselves and society maintains itself. Persons can find much pleasure, establish a sense of their own selves, and even discover a lasting, inner balance through the very activity of work.[110] Through work activity, persons learn the virtues which characterize the working class.[111] As they work, persons come to realize their own several reactions and feelings, come to rank them and to realize that the action conduces to lasting happiness in terms of their higher feelings.[112] They learn to delay their gratification for the sake of future returns.[113]

Workers learn to forget themselves as each focuses on the object of his or her particular endeavor. They also learn that the higher values are those of society. As their work brings them to recognize the importance of the social, it brings them in the end to properly ethical goals; they come to work for the welfare of society.[114] So significant is work for society, that Schmoller believes that one can psychologically derive the economic system better from the virtue of industry than from the profit motive.[115] In Schmoller's system, work reveals its ethical significance since it is the school of virtues and the basis of social organization.[116]

Schmoller does not emphasize the burden character of work because he views it first of all as a process of personal and social formation, that is, as a central form of education. In this light, the emphasis no longer lies upon the negative dimensions of work activity solely as imposing costs but, rather, on these same dimensions as calling forth a response from persons, which leads in turn to their more complete formation. The circumstances of work, however, can be so distorted, that such a formation of the total person does not take place. Yet, unless members of each class learn the broad range of virtues, and do acquire more than technical knowledge and skills, the unity of society will be dissolved.[117] Despite the many distressing factors of the modern labor situation, Schmoller continues to regard work as integral to the educative formation of person and society.

The Needs of Workers

Persons undertake work activity to provide for their needs. In so doing they develop their desires. Furthermore, society itself, formed through the work of persons, influences the development of these needs. Schmoller, having studied the rapid evolution of contemporary German society, pays particular heed to the needs of those workers affected by these profound social changes.

The basic needs, in Schmoller's view, constitute a far larger field than the range which is commonly included under sustenance. He includes not solely physiological but also customary requirements for both physical and intellectual vitality, in addition to the requisites for social interaction.[118] These needs arise from the complex and varied feelings of the person. Pleasure and displeasure—and the recollection of these—provide the stimulus for human action intended to satisfy the entire range of needs.[119] The pursuit of necessities commonly delimits the realm of economics. Yet more varied feelings and higher goals than the basic provision of necessities stimulate human work activity. Nor are the demands from the feelings finally determinative; work activity becomes ethical insofar as it undertakes a shaping of these needs into a totality.[120] In fact, the activity undertaken by persons instills new drives in them. The development of their internal life interacts in a complex fashion with the development of society. The modern broadening of needs raised the capacity for work and thereby enriched persons in terms of internal and external goods.[121]

Needs do not center solely on the individual. The complexity of motives and goals is found both in activity directed toward self-constitution as well as in the work expended for the sake of others along with oneself, for example, for the sake of one's family.[122] Needs are social from yet another perspective: the drives and needs of persons today result from the formative activity of earlier generations.[123] Still further, needs have become socialized in that the members of higher cultures do not meet needs individually, directly and immediately; but rather, they cooperate through the indirect means of exchange and complex institutions to provide for needs in advance, so that together they might better avoid neediness.[124]

New needs which remain unmet constitute the shadow side of social evolution. When the economy shifted the recompense for work to the money-wage, many workers did not learn soon enough how to foresee their needs and how to plan for their satisfaction. Their day-to-day perspective did not meet the new requirement for a foresight extending over far longer periods.[125] The needs now appeared to be the same as they had been; the changing socio-economic structure nevertheless reconstituted those needs, placing new demands on the workers. But more than the education to new habits was required: because the workers had no reserve for security, they were more exposed to need.[126] The new economic system deprived them of their former security, the certainty of maintaining their place under the old, near-feudal arrangements.

The economic revolution reconstituted the need for work itself. Whereas one engaged in the house or farm system of work in varied labors depending on the season of the year and, with the passing of time, depending on one's own age, now the economy demanded that a single worker produce a single type of labor. When the demand for that labor decreased, even seasonally, one could become unemployed. When a young man, not surprisingly, had made no provision for his later years when he was no longer a desired employee, he lacked work and support altogether.[127] The uncertainty and dependence of the worker in the new work relation indicate new needs which, Schmoller believes, must be met under the direction of the state. Economic crises afflict these workers who do not have the wherewithal to adapt themselves to the rapidly changing circumstances for which they are in no way responsible.[128] Through legislation and enforcement to assure good wages and some property ownership, the state must take the lead to preserve these workers, some of its own members, until they gain the wherewithal to endure economic upheavals and to participate more fully in the life of society.

The Progressive, Ethical Community

For Schmoller economics examines primarily neither individual actions nor developments in technology. Rather, the focus of economics rests upon the interaction of individual and society as influenced by technology, custom, law and ethics.[129] Society is an ethical community which undergoes a progressive development. Institutions and the division of labor structure this society, while the formation of various worker organizations marks its further progress.

Schmoller draws from Fichte once again in his conception of society as an ethical community of persons. This community is ethical because the individual can only realize objective values insofar as he enters into the community. In addition, the community is ethical because the individual directs his efforts not to his own immediate end, but to those of others, either directly or indirectly. Finally, this community is ethical because concord, not conflict, characterizes the interaction of its members.

Insofar as Schmoller adopts an organic view of society in opposition to the mechanical model favored in classical economics, he commits himself to understanding the structure of society not as an agglomeration of individuals but, rather, as an anatomy of institutions. These organs manifest their descent from the Hegelian structures of objective spirit. The institutions of society assume a particular signficance for Schmoller: they confront the members of society as the most important outcomes of the ethical life.[130] These institutions are ethical because they organize the otherwise conflicting efforts of persons and direct them toward particular ends sought by many. The institutions efficiently structure otherwise individual work activities for the attainment of personal and group goals.

The institutions of capital (the corporations) and the institutions of labor (the workers organizations) are especially important in Schmoller's understanding of ethical community. These organizations structure and direct individual work efforts. The work activity of the individuals contributes to the fulfillment of individual and group duties vis-à-vis the whole of society. Persons participate in the process of organizing society's efforts in particular by carrying out the obligations of the various offices of self-administration. In this way they become capable of responsibility once they learn what governance entails.[131] In appearance, they are caught in a web of obligations, yet, by collaborating with their fellow citizens in work activity which is coordinated in order to continue to satisfy needs, working persons actually secure their freedom in society.

Both ethically and structurally, the organizations of the ethical community constantly progress. Custom—a constitutive element of ethics according to Schmoller—is reformed and purified throughout history. By means of custom human persons construct within nature a second world—a world of culture which includes the economy.[132] In addition, in the course of historical development they improve the law—another element of ethics. Indeed, a progressive evolution of the knowledge and feelings enables ethical judgment itself to improve.[133] This progress of the ethical occurs in conjunction with technical and institutional developments. As the capacity for ethical imagination deepens, the development of the ethical community progresses toward an ever more comprehensive society.[134] Ethical progress contributes to the development of total, social progress as well because, in their developed ethical sense, individuals dedicate their efforts to the whole.[135]

Persons must make this progress real on the basis of technical and social conditions. Beyond the family and the community, firms and cooperatives are further institutions which have been established alongside the earlier, more limited groups for the attaining of economic ends.[136] Ethical standards perfect themselves by way of and in terms of the culture and the institutions in which they are embedded.

The division of labor constitutes the social interaction of the elements in the ethical community. The division of labor coordinates the work activity of individuals, on the basis of which institutions and society itself progresses. In examining Schmoller's understanding of the coordination of work activity, it is important to note that he underplays the importance of the technical (or factory) division of labor for the sake of emphasizing the significance of the social division.

His treatment of the division of labor indicates that Schmoller understands that the technical division obviously brings about a more efficient productive process by the reduction of energy expended and the development of skills. Yet even as he considers these economic advantages, Schmoller desires the attainment of solidarity. Through the freedom of work choice and the more just distribution of income, he wants to secure the promised strengthening in the feelings of solidarity—within society and not solely within the working classes—which ought to result from the division of labor.[137]

The social division of labor presupposes a natural grouping with a common language, a common set of ideas, common needs, a shared system of exchange and a unity of custom, law and morality.[138] Within this framework, the division of labor takes place as a process. This is to be stressed: the organic character of Schmoller's concept of society is dynamic and progressive.

The process whereby an individual person organizes all his or her work and activities for the efficient meeting of his or her needs provides the model for the social allotment of work activity.[139] As determined by its work assignment and as specified by the virtues proper to each type of activity, each of the various work groups in society collaborates with the others, with a special contribution from those persons engaged in business.[140] Schmoller realizes that certain difficulties result from this treatment. The specialization in training which results from the particular work assignments—not to speak of the still narrower training in skills concomitant to the technical subdivision of a work process—makes his own humanist ideal of educative formation of the person appear merely utopian.[141]

He also recognizes that certain extreme malformations of the division of labor do occur, but that these can be limited through proper counterweights and controls.[142] Schmoller regularly argues that the state and its bureaucracy will establish these countermeasures. The persons who occupy roles in these social organizations are themselves part of the division of labor.[143]

Schmoller assigns a particular importance to the regulative work of the state government; an inadvisable underestimate of the significance for the whole of society of the average person's work activity results. The process which the division of labor makes up restricts the progressive range of the workers' contributions; they remain passive functionaries in a system directed by others. The process of the social division of labor takes place through the contributions within each sphere of work, but the system truncates possible interplay among spheres—an example of such an interaction would be a competition among workers from all spheres for the purpose of demonstrating suitability for a position in government. Further, Schmoller's emphasis on state direction of the economy de-emphasizes the progressive contribution of ordinary work.

Does Schmoller offer any institutional corrective to the bureaucratic, centralist distortion of the division of labor? Specifically, what sort of role does he envision for groups of workers within society? He declares that a healthy democracy must reduce the income inequalities and class egoism which could result from imbalances in the division of labor.[144] He indicates that both the democratization of the state and of the economy accompany the improvement of the totality of social institutions.[145] How might such a democratization, with a more prominent role for the workers, be institutionalized?

Schmoller first responds to this challenge by highlighting the modern firm as a locus of social cooperation.[146] The company becomes a means for social peace, an organ of the ethical community, dependent on the cooperative relations between owners and workers. Schmoller also describes larger social units, *Berufsvereine* (occupational associations), comprising both employers and employees. He intends these institutions to bring about social peace through a mutual education of the parties.[147] Through these institutions, Schmoller hopes that the relationship between workers and employers will be re-established on a sounder footing where conflict will be obviated, duties observed, and common interests recognized.[148]

As a result of the stormy labor conflicts in Germany at the time, Schmoller plays down the role of labor unions as organs which can cooperate in social progress. However, he does examine workers' organizations with some approval. Society's progress has developed the workers to such an extent that the workers too can actively contribute to society by way of the articulation of their groups into the totality of society.[149]

Schmoller also mentions other forms of workers' organization. As early as 1864, based on the successes of Schulze-Delitzsch, he sets forth the idea of cooperatives.[150] Later, in 1872, he claims the freedom of workers includes their ability to set up organizations modeled on the guild structure.[151] As a variant of worker organization, in "*Über Gewinnbeteiligung*" from 1890, he argues the advantages for profit-sharing, including the benefit of reduced employer-worker conflict.[152] In the same piece he goes still further: writing of the requirement of peace in the workforce and in the plants, Schmoller sees an advantage for the firm in a limited degree of shared management for a certain range of workers, namely, for the higher-level employees.[153] However, when he returns to the theme of profit-sharing at the end of his life, he no longer mentions even this limited notion of shared management.[154]

Although Schmoller steps hesitantly toward a valorization of independent workers groups, for the most part his centralist and bureaucratic approach to the social questions posed by the rise of the "working classes" inhibited his appreciation of what the workers in their own activity could contribute to the reformation of society in order to address these societal ills. Schmoller criticizes Fichte's failure cohesively to integrate ethics and economics because of his assigning too great a role to external regulation.[155] Yet Schmoller himself ignores the contribution which could proceed from the work activity of each person in favor of centralized direction of the economy and legislative measures against social ills. Thus he submitted to a romantic view of society where each person worked in a well defined position. Save in the conception of the work of the overseeing bureaucrats, the German, humanist view of work as the independent creative activity of each in shaping the common world, was lost.

The Differential Benefits of Technology

Schmoller presents a more refined understanding of the nature of technology, demonstrates a better grasp of the effect of machines on working individuals and offers a subtler, more comprehensive presentation of the positive and negative impact of technology on society than has been found in the theorists I have treated earlier.

Each type of work activity involves both intellectual and mechanical elements.[156] Insofar as a particular activity involves more than technical means, it is liable to ethical evaluation.[157] Because of the close union of mechanical and intellectual means, the entirety of the work process must be examined from an ethical perspective, without endangering the evaluation of a procedure in terms of its effectiveness. Not only its external means, but also the methods in which these are wielded, represent technology.[158] Tools, weapons, apparatuses, machines—these all exist as projections of human organs into nature, but not solely as material projections. They also represent the inner, intellectual processes which persons have discovered and deployed for particular ends.[159] The creation of a tool transcends the passing solution of a work problem to become a part of the world which persons change by the invention of tools.[160] Technology is constituted by a tradition; within this tradition, workers hand down the implements, their use, the estimate of their net benefit and the ideas which stand behind them. Persons together in society recall the tools' use and significance.

Schmoller presents modern technology as equivocal in its effects on individual workers. He clearly recognizes that the *Teilarbeit* (piece or partial work), which too often characterizes work activity as a result of modern technology, leads to the ruin of persons. In this light, the modern, mechanical processes which shape particular work activities substitute poorly for the older, outmoded means.[161] Indeed, as a whole, technological progress has not been concerned with advancing the subjective happiness of persons. As a consequence of this omission, persons suffer individually not only because of their particular labors but further because of a situation in which quiet and traditional ways are lost while competition rises.[162] Although he recognizes these problems, Schmoller maintains a hope that further technical development will remove these difficulties.[163]

He also affirms the positive benefit for the individual of this new technology. It stimulates the worker's sense of order and precision, his reflective ability and his desire for further technical knowledge. Further, this type of work stimulates the development of habits such as economy, forethought, cleverness and diligence.[164] Human rationality, decisiveness and adaptability are those capacities which, while stimulated by the work with machines, distinguish working persons from them.[165] The rational capacity of the worker also figures in another significant way. In some sense, the worker at the machine participates in the knowledge which the inventor has deposited in the machine.[166] The worker then partakes in a knowledge entrusted to society. Thus this work also stimulates the workers in their own capacities to alter, develop and discover still more advanced knowledge and applications.

Schmoller likewise studies the larger, social impact of technology. The striking developments of modern technology derive largely from the substitution of machines for tools. Although there are clear-cut variations in the percentage of effort which has been spared by use of machines, one can easily recognize how society as a whole has been markedly altered by the vast quantitities of power which are now available, by the simplification of work processes and by the shortening of the periods of work. This means, Schmoller says, that planning and action in preparation for crises and disturbances may now be more easily undertaken for society as a whole.[167]

The introduction of advanced technology has altered the relationship of persons with one another in society. On the one hand, persons are now more individual—thus there is the possibility that they may act and relate as free citizens. On the other hand, one of the most significant results of this technology has been the achievement of better arrangements for interaction in state and society.[168] Advances in technology make possible a more highly developed society, but they do not guarantee it. Social progress also requires a parallel development of the intellectual and ethical strengths of society. Schmoller hopes such a development may be the case, particularly that the freedom of the worker and his integration into society may be secured.[169]

The social impact of technology has had a varied benefit. Some technical processes have been introduced which produce more but only at a greater cost. Improvements have been developed in key consumer items, but others have been neglected—often for a net loss. Further, the advance in the wealth of persons has lagged behind the increase in the rate of production.[170] New technologies, contrary to their promise, have often depressed wages or made unemployment more likely. Indeed, persons often must work harder in their same fields because of the introduction of advanced machines. Additionally, when one looks at the high cost of food and housing, one realizes that technical advance has occurred more swiftly in some fields than others and that those areas in which advance has occurred are not necessarily the most significant in terms of human need.[171]

Technological progress also benefits groups in society disparately. Because the employment of advanced machines causes some groups to rise at the expense of others, the social and institutional changes wrought by modern technology exacerbate class differences so that questions concerning ownership and the power of control come to the fore.[172] Although society to date has not established progress within a comprehensive new order in accord with ethical guidelines, Schmoller continues to hope for the rectification of present social wrongs.[173]

Centralized, State Provision of Service

Throughout his life, Schmoller was consistently concerned for the lower classes—in contemporary Germany, these were comprised of the workers—so that they would come to take part in all human goods.[174] His attention was directed toward the workers, in part through his familiarity with the earlier, romantic economists who recognized the significance of these weaker members for the whole of society. Additionally, Schmoller's concept of the ethical community depends not only on Fichte, but also, in terms of this attention, again to the lowest classes, on Lorenz v. Stein. In v. Stein's thought, the contemporary situation of the workers contradicted the very idea of the state.[175] Deprived, passive and isolated, the workers deserve attention from society under the direction of the state which ameliorates their situation and integrates them into both society and itself. This very concern constitutes the primary notion of service in Schmoller.

In this light, Schmoller's insistence both on reform and on the contribution of economic study to it constitutes one element within his overall emphasis on the service of the workers. By its study of historical and progressive changes, economics serves. The economist must carefully avoid stating economic laws as laws of nature, the immutability of which might guarantee the permanence of unjust conditions.[176] The historical economist understands the past so that in the present he may recommend to the state how progress might be furthered into the future.[177] The service of the economist contributes to service on the part of the state toward its citizens.

As technological, social and ethical progress has made possible more extensive and effective groupings of persons, so does the state become necessary as an agent which can correct the egoistic tendency of groups.[178] The activity of reform directed by the state on behalf of the workers is integral to Schmoller's concept of service. The workers are not simple objects of assistance. The workers are potentially full members of society whose structures the state helps to readjust so that they may achieve this potential.

One particular type of work activity, which other economists would term "unproductive," assumes especial significance in this state service directed toward the workers. The members of several, different classes who partake in a mutual work for the common good undertake this particular activity. A variety of professionals, most of whose work is "unproductive," constitute an extended officialdom which concerns itself with the public affairs.[179] The particular services of these individuals coalesce into a neutral, controlling ministration, one of the chief aims of which is to ameliorate social ills among the lower classes by realigning social structures.

The reshaping of society requires the newly institutionalized care of the weak and the poor—such as children, the elderly, the chronically ill, widows and orphans—in the form of aid or insurance. This care, Schmoller asserts, is a duty which society owes its members.[180] The local community often administers this care the best, because in the community natural interests and sympathy often facilitate the provision of such services.[181] Where local means do not suffice, centrally organized service structures must be introduced. These facilitate the cooperation among state officials, employers and workers and also prevent the emergence of grossly unequal treatment of those in need.[182] Of particular interest to Schmoller is *Arbeiterversicherung*, insurance for the protection of the worker weakened through illness and accident or for the assistance of his survivors in the event of his demise.[183] This new centralized insurance institution serves the worker by maintaining him and his family and thereby benefits the whole of society.[184]

Beyond provision for the needy and weak and beyond the guarantees of insurance, Schmoller envisions yet one more "unproductive" work activity as service—education understood in the sense of *Bildung* (formative education), that is, the total formation of the person. Especially in view of his progressive view of personal capacities, one can assert that the deepest root of Schmoller's sociopolitical thought lies in his desire to elevate the lower, working classes by means of education.[185] While he naturally refers to the educative role of the parents of future workers, he also points out both the educative dimension of the the work contract for young workers and the role which unions may play in the education of their members.[186] Schmoller believes that the system of schools and the obligation of attendance have together been the chief reason for the rise of the workers, because these have brought about substantial developments in their intelligence and self-awareness. Indeed, such education has made possible both political parties and the organizing of the economy and, thus, has made social peace more likely.[187] The particular work activity of education, which may be offered in many situations but pre-eminently in the state-organized school system, develops the worker and reshapes society.

Schmoller's turn-of-the-century, central European considerations concerning work activity encompass (1) human needs which are extensive and socially delimited; (2) the worker as located in a double totality of personhood and society; (3) a society whose divisions are centrally regulated for the sake of a common ethical progress; (4) a comprehensive description of technology in its human invention, in close association with the person, and in a conjoint relation with the whole of society; and (5) a service inclusive of reform, of an equitable care for the workers and of education. Schmoller and his teaching influenced a broad and powerful array of colleagues and students. To the world of these academics and politicians and to a single figure in particular I now turn.

Heinrich Pesch, S.J. (1854-1926), an economist who was also well schooled in theology and whose work was very influential for the tradition of Catholic social teaching, was deeply acquainted with contemporary social questions and especially with the situation of the workers. While a student at Bonn, Pesch had studied and worked closely with the later leader of Catholic social reform in Germany, Georg v. Hertling. During his four years of study in England, he gained a firsthand familiarity with the plight of the workers in Lancashire, Liverpool and Manchester. When Pesch returned to Germany, especially during his time in Mainz, the work of Wilhelm Emmanuel v. Ketteler, the former bishop of that see and initiator of the social movement in German Catholicism, significantly influenced Pesch's outlook. Later, in the course of private instruction in economics under the tutelage of Rudolf Meyer, a member of the circle of Karl v. Vogelsang,[188] Pesch was able, through his observations in northern Bohemia, to augment his knowledge concerning the situation of labor.

In the latter half of the nineteenth century, the ferment of ideas which arose in response to the plight of the workers and which profoundly stimulated Catholic social teaching also invigorated the thinking of Pesch. Ketteler, to whom Leo XIII acknowledged an indebtedness for central ideas of the encyclical letter *Rerum Novarum* (*The Condition of Labor*, 1891), had argued for an ethical reconstruction of society for the benefit of workers on the basis of corporative, self-governing work organizations.[189] In the last quarter of the century, the influence of the Austrian v. Vogelsang was extended through the studies and conferences of the *Union de Fribourg* and the *Freie Vereinigung katholischer Sozialpolitiker.*[190] Both of these groups recommended a new corporative social order established upon vocational-cooperative groups;[191] this program markedly influenced the system offered by Pesch. The development of this teaching did not remain a theoretical, internal, ecclesiastical discussion. The *Reichstag* (Parliament) efforts of Franz Hitze, a proponent of v. Vogelsang's corporatism, and those of v. Hertling were bringing Catholic socio-political ideas—heavily informed by the newly rising Neo-Thomism in Catholic philosophy—to effect in national social legislation.[192]

Having already begun his writing on social and economic questions in *Stimmen aus Maria Laach*, Pesch decided to return to studies with the express intention of learning the ethical economics of the Later Historical School.[193] In the course of these two years (1901-1903) in Berlin, he maintained contact with his friend and former fellow student Leopold Wilhelmi. Wilhelmi's role as president of the Office of Imperial Statistics, his contribution to laws benefiting workers, and his founding of the *Reichsarbeitsblatt*, all enabled him to foster Pesch's understanding of the situation of German workers.

While at the University, Pesch, in addition to his study of German social legislation (with K. Bornhak) and worker's insurance (with L. v. Bortkiewicz), also had contact with H. Dade (a follower of Roscher), Adolf Wagner and Gustav Schmoller. He participated in Schmoller's course, "The Contemporary Situation of the Workers."[194]

While distancing himself from the extreme emphasis on state-monarchical socialism to which some of its members (e.g., Wagner) adhered, Pesch shared the concern of the Later Historical School of finding a way between the extremes of Manchester liberalism and socialist revolution. At the same time, Pesch criticized Schmoller for a failure to give proper weight to theoretical analysis in economics and, especially, in explaining the relationship between custom (*Sitte*) and ethics (*Sittlichkeit*).[195] Pesch shared with the Later Historical School a concern for the organic nature of social life, a recognition of a role for the state in economic affairs, a grasp of the historical analysis needed in economic study, a realization of the complexity of human motivations in economic activity, a rejection of the exclusive dependence on liberal economics, a desire to still class strife, an eagerness to raise the lower classes, a desire for social reform and a concern for the role of ethics in economics.

The Free Subject as the Chief Cause of Production

The goal of social reform was the elevation of the circumstances and status of the working person. In the analysis of the person, Pesch adopted Wagner's understanding[196] of the complex of motives which lead a person to work: material self-interest, the desire for recognition and honor, joy in activity itself and the wish to fulfil one's duty all contribute to the desire for work.[197] The sense of independence, even the merely economic, instigates work on the part of some persons.[198] Pesch clearly indicates that self-interest includes the inner development of the person; thus economics must examine how the person achieves his fullness not only in material but also in ethical matters.[199] Work possesses a double, ethical purpose which consists in the person's completing of him- or herself, including the attainment of the material goods requisite to this end, and the indirect satisfaction of the needs of others through the person's work contribution.[200] Effective social reform designs institutions which enable persons to achieve these personal and social goals of work in their complex interaction.

Work is a nearly universal human activity which is intimately related to the person who engages in the effort. Because work exercises the highest capacities of the person, all who are capable are obliged to work.[201] The personal effort of work always involves the intellectual-spiritual (*geistig*) side of the person and devolves, to be sure, partly from necessity, but also and importantly, from human freedom.[202] By way of the person's embodiment, the intellectual-spiritual powers, including freedom of action, enter into the basic activity of economic life. Human purpose informs the entire economy through work activity.[203]

Because of the individual's intellectual-spiritual nature, which permits him or her to stand apart from the present conditions of control or subordination, the human person is a subject which shapes (*das gestaltende Subjekt*) and is the chief cause of social progress.[204] The work of the person enables each to shape the world; in this way the individual can be said to govern it. Where the person is active in production, there the individual is the sole subject of production. Everything else is object or means.[205] Work activity expresses the subjective nature of the person. The person "is never a mere object or tool but always and everywhere subject and end of the economy and of economic activity.[206] Even when working for a firm, workers remain producing subjects and ends in themselves. The worker is no mere means of production; the person's work cannot be conceived in terms of the thing-character of a commodity; neither can the individual's work activity be conceptualized as an obligation owed, strictly speaking, to society.[207]

Although Pesch's concept of the person begins with individuals in their needs and desires, he also recognizes the social drives and needs which partially constitute the person as well as the duties consequently owed to society.[208] As free subjects, working persons are responsible, particularly for integrating their individual capacities and activities into society.[209] This consciousness of their obligation to society depends on the freedom of the individual and on the natural impulse to partake in society. In fact, these social duties themselves betoken the responsibility of the person to shape society rather than let its shape be determined by heteronomous mechanisms. Work activity is the central means by which individuals serve the common good.[210]

In seeking to avoid both the individualism of classical economics and the dominance of the group found in Marxist or state socialism, Pesch focused on the importance of the individual-within-society and highlighted the significance of work activity for the relation between person and society. On the one hand, Pesch specifies the freedom to be secured for persons in their work as the freedom of type, place and manner of work.[211] On the other hand, he recognizes that work is embedded in a sociality of history and locality and that reason, not desire, constitutes the freedom of workers with the result that the activity of each must be coordinated with that of the others.[212]

Pesch argues still further for a certain social limitation to freedom: first, persons are prone to error and, second, inequality of power may lead to oppression of the weaker.[213] Freedom is not an axiom, but a problem to be resolved. Such difficulties as the two just mentioned require the higher, regulating principle of justice as a corrective to the false developments of individual freedom.[214] While individual freedom may be corrected through social regulation, social groupings themselves can constitute spheres where the freedom of persons may expand beyond individual action alone.

Although individual freedom is subject to social regulation, that regulation is itself restricted by the rights of the person. By nature, the person has a right to shape and govern the world.[215] Based on their individual personhood, which places each above the world of objects, persons possess natural rights which secure their individual development and independence.[216] Among these rights, Pesch speaks of the right to work, by which persons put their individual capacities into action.[217] Here I have considered the priority of working persons in the exercise of their freedom in a social setting. I will examine several, special work settings at greater length below.

Persons in society must regulate the ordinary work setting within the framework of the wage contract so that the free activity and rights of the person are secure. The social regulation of the work contract redirects the free action of the employer within the social bond established through the contract so that the entire personhood of the worker might be secure in its activity and development in itself and for others.[218] Both materially and formally the employed worker as a partner to a contract is as free as the employing person and is equal to him or her.[219] Not capital and labor, but persons bind themselves to cooperation through a contract in which capital is a means and persons the subject and goal of production.[220] The wage itself does not constitute a recompense for the product of work activity; rather it must be a recognition for the personal activity of the worker.[221] Because of the personhood of the employed worker and in view of the inequality in power between those who control capital and those who offer labor, the social regulation of law must prevent dehumanizing, reifying and patriarchal misuses of freedom from distorting the work community based on contract.[222]

Pesch explores the unique role of the person both in terms of activity as an expression of that individual's free personality, and in terms of the more specifically economic category of production. Two factors interact in production: the material world and human work activity.[223] Personal activity, in its intellectual-spiritual and embodied components, is the *causa efficiens principalis* (principal efficient cause) of production.[224] More specifically, human work activity is not the sole source of production, but the human person is the sole subject or producer. This subject calls forth and directs and cooperates with the other effective powers of production.[225] As a consequence of this, the production process and the goods created must not damage any aspect of the well-being of consumers, workers or society itself.[226] More generally, as a result of his position, Pesch is able, first, to counter some socialist claims that labor is the unique source of production; second, to oppose the extreme capitalist position which asserts that capital alone makes production possible; and, finally, to maintain, along with a sound recognition of the function of material reality, his declaration of the pre-eminence of the working person .

The Natural Limits Which Constrain Work Activity

Pesch recognizes that the human person is needy both on the physical and the intellectual-spiritual side and that both these elements must be developed. Although the work activity of the person can contribute to the meeting of these needs, this can be accomplished only in society. Yet this social facilitation of the satisfaction of needs also means that social processes may expose persons to the grim consequences of their needs without recourse to any protection.[227] Against any such threat, Pesch declares that the individual's personhood establishes that person's right to existence.[228] Further specified, this is a right to work and to the fruits of one's work.[229] Society must design processes to meet the obligations resulting from these rights.

Pesch's handling of the just wage reflects his concern to guarantee the right to existence. There are several ways of ascertaining the justice of the wage in terms of the limits which the working person faces. Echoing Smith, Pesch recognizes that the wage must compensate for the time and effort expended in the task as well as in the education for the task. The wage must also provide compensation for the dangers which threaten to limit any further activity on the part of the worker.[230] Another way in which limits contribute to the wage determination is that they place demands upon the worker even apart from the work process. The just wage must allow the individual to meet physical requirements as well as those necessities for the development of higher, personal capacities.[231] Indeed, the wage should be such that the worker can participate in the overall progress of society. In a still further and extended sense, the wage is determined by personal limits: workers are free persons who, constrained by social obligations and finite in terms of time, energy and skill, must choose their actions. The recompense for work must reflect, in part, the specific responsibilities which the worker assumes in the course of a particular task[232] and the responsibility which that individual must assume as a particular member of society, for example, for familial members.[233] Whether in compensation for efforts which expose the person to danger, in provision for full, personal development or in recompense for responsibility, work activity in itself must be remunerated, for it is the deployment of free and limited personal action.

The personal capacity for work is a natural and limited good which enables workers to meet their needs to a degree. Social interaction, coordination and supplementation often facilitate the attainment of the requisites for their full existence.[234] Since workers are persons-in-society, they are responsible not only for economizing their efforts toward need-satisfaction through their cooperation in society, but also for the prudent deployment of their individual efforts in order to meet the needs of society, because human needs are also social needs. Furthermore, increasing population density, more advanced cultural interaction and the ever extending range which accompanies the imperative to govern the earth impel persons increasingly to coordinate their work activity.

This economic action intends to meet both present needs and those which progressively develop; it aims at achieving the common good, which consists in the material, spiritual and ethical benefits to society and its members. The end—even of complex, social interaction—remains the provision of benefits to persons in society and to society insofar as it benefits persons. When Pesch indicates that the five norms which guide the provision for needs are the economic, national-economic, hygienic, aesthetic and ethical principles,[235] he thereby emphasizes that human needs are: material, social, embodied and, in addition, required by the higher human capacities. Thus, the limits which face the working person proceed from the totality of that individual's personhood. The social work of production must arrange for the satisfaction of the needs that devolve from the personhood of each member of society. Economic value depends on the social evaluation of a good in its capacity to meet the needs of persons.[236] More explicitly, neither technical efficiency nor the physical items fabricated nor profitability but, rather, the contribution to the satisfaction of human and social needs is the proper measure of production.[237] Pesch conceives society, then, not as a field of competition, but rather, as a work-community of cooperative effort for common provision.[238]

Beyond the understanding of limits implicit in his treatment of remuneration and of social production, Pesch distinguishes himself from other economists by altering the concepts of nature and, consequently, of its limits on economic activity. In so doing, he attempts to correct their views of the relation between ethics and economics. The earlier, classical economists had, in their materialist presuppositions, seen economics only as a branch of the natural sciences and had understood self-love as the chief motivation of the economic activity.[239] In contrast, Schmoller and others of his school had emphasized the signficance of the social and historical sciences for an adequate approach to economics. Pesch held that in this the German historical economists had relativized ethics in the course of progress so that the study of ethics was reduced to empirical observations.[240] Both streams of the tradition of economics saw nature as the a- or irrational force against which persons struggle in their efforts to dominate it.

For Pesch, however, nature is patterned according to a rational, objective law. The values indicated by this law are accessible to personal reason.[241] In this understanding of nature, "ought" is an integral part of the empirical "is."[242] Economics, as the study of nature understood in the sense of the proper action of human persons upon material things and processes, must be established upon this natural law. The science of economics examines, in part, the free activity of persons, which activity must accord with natural law, lest the attainment of the end of economic activity, the common good, be impeded.[243] To be sure, Pesch does not say that the natural law indicates particular economic policies; rather, he affirms the need for more precise determination and application by ordinary social and legal regulation.[244] The natural law does provide positive and negative guides for economic action. This natural law is another source of limits which both constrain and stimulate human activity. For Pesch, unlike the laws of nature to which the classical economists adverted and unlike the mechanicist interpretation of the progress of history, this limit is expressly ethical; it bounds the way to the personal and social exercise of potentials.

Society's System of Work

For Pesch, the personal activity of work both actualizes social cooperation[245] and differentiates groups through which society is supported and in which working persons can realize their own potentials. The system[246] of work in society comprises both the division of labor (*Arbeitsteilung*)—based upon practical rationality[247] —and the unification through work (*Arbeitsvereinigung*). Through both, persons jointly complete one another.[248] Pesch establishes social interaction upon a common and personal activity. Society is not a conglomerate of individuals, nor is it a socialism where the means of production are ideally socialized. The ideal, rather, is the socialization of producers, that is, working persons who collaborate.[249] The system of work in solidarity proceeds neither from the individual nor from the social group as primary but, rather, from the person-in-society.[250]

Pesch is able to treat the person-in-society as the basis of a developing system because he focuses on human work activity. The working person finds completion in society, exerts him/herself in common with his/her like, attends to the well-being of others and to the interest of the whole of society.[251] Work activity mediates between person and society because it both proceeds intimately from persons themselves and because it achieves its ends in society. In that Pesch explains this social system as a work system, he highlights two points: 1) the working person is the subject of production, superior to all other productive factors (as has been discussed in an earlier subsection) and 2) the common good and its progress depend on creative production through work activity.[252] Persons are elements of society, ends in themselves and the principal agents of social activity and progress.

In the achievement of this system, freedom is restrained so that personal ends might better be reached by allying persons one with the other and uniting their diverse strengths. Further, the joint emphasis on individual and society conceptually permits and actually encourages the subsidiary groupings of persons toward limited ends which advance the common good. Prominent in such groupings are those organized around particular, cooperative work activities.[253]

Several guidelines result from this view of the person in society. In each and every work relationship, the worker must be treated as a person. To the degree that the work relationship is one of authority, employer and worker are bound by reciprocal duties and rights. Of critical significance for the social order itself is that all workers in an enterprise—management, white or blue collar—are collaborators. Similarly, all participants in a venture are members of that firm in the sense that they are not to be dismissed at whim. The individual worker, furthermore, must be a recipient of the benefits produced by the joint enterprise.[254]

The individual workers are bound together through several groupings which are subordinate to that community which society as a whole constitutes. A group of workers makes up the individual factory or firm; those active in the same industry share a social bond beyond company lines; unions provide another organization for representing the social ties resulting from work activity.[255] Each worker, in his or her particular line of work, is united within a sub-group of society. However, the workers as citizens recognize that they direct their work activity to the common good and thereby unite themselves in a social *Arbeitsge-*

meinschaft (community of work), in which even competition is evaluated in terms of its utility in advancing the common good.[256]

Moral philosophy, Pesch holds, offers central principles toward a well-grounded science of society. One of these is the principle of solidarity. The system of work will become fully social when citizens fully satisfy the obligations resulting from solidarity.[257] One must carefully distinguish two aspects of Peschian solidarity: the factual and the ethical. A three-fold solidarity of workers as persons in general, as citizens and as co-workers (*Berufsgenossen*) brings about an actual socialization of working persons. In fact, modern developments in industry, finance, technology and communication have advanced the solidaristic interaction of persons in society. The world of work has already created a society where interactions more markedly evidence solidarity. The prescriptive dimension of solidarity cannot, however, be conflated with the factual and thereby be overlooked in its distinctive claim. Pesch, appraising the actual social ties, holds them to be only the material presupposed in a complete organic state—which requires the form of an inner bond.[258] This internally united nature of society becomes realized progressively through the more complex and modern forms of social and economic interaction, not as a result of chance but, rather, as the product of human reason and decision. The higher human capacities shape the matter provided by the modern patterns of interaction into the form of a proper society, namely, one informed by solidarity. To proceed toward this solidary society and to rectify the distortions which emerge are two tasks which ethically oblige members of society.

For Pesch solidarity is an ethical duty which encompasses a subordination and cooperation for the purpose of the whole society which, in turn, indirectly benefits each.[259] Solidarity, in addition, prescribes a fullness and continuity of the interaction. The subordination of this system differs from subjugation, for solidarity advocates a system which more readily accomplishes the potential of each person.[260] Just as the achievement of solidarity does not depend on technical developments alone, so too it is not brought about by natural, even social, drives or instincts. Reason and choice shape a society in solidarity.[261]

By the prescription of solidarity, Pesch wishes to avoid both the atomistic fragmentation of liberal social theory and the conflicts toward which that theory conduces when individuals band together for momentary advantage in the struggle with others.[262] Solidarity also avoids the distortion of centralized socialism, because it appeals to the activity and freedom of the working person to produce for the social good and to act responsibly for participation in this social project.[263] Thus, solidarity spurs persons to shape a system which achieves both order and freedom, through the organic integration of individuals and their groups in the whole of society. While modern achievements in technology, marketing and the division of labor provided the epochal stimulus for new social organization, developments in organic groups of workers will take on particular significance in the further advance toward the accomplishment of solidarity.

The social system of work, guided by the principle of solidarity, will be achieved through the institution of work groups. These groups are intermediate bodies between the working person and the state. The modern economy—and, consequently, society in general—has been significantly shaped by the rise of organizations of working persons which were originally creations of private, contractual obligation and have now come to a public standing. Incorporated companies and labor unions are important examples of this development. The organic conception of society utilizes the phenomenon of these institutions in order to build a multi-level and coordinated vision of society. Just as novel patterns of cooperation in work activity originally constituted these new organizations, so Pesch's version of the organic society retains, indeed highlights, the centrality of work activity.

German social philosophy long held to variations on the insight that the state could be organically integrated on the basis of intermediate bodies which were constituted by sundry types of work activity. Fichte, for example, established the state on the basis of functional groups determined by social positions (*Stände*) and the type of trade practiced within each division.[264] In an earlier chapter, I presented Hegel's version of the organic state which assigned a significant role to corporations and work classes.

Contemporaneous with developments of such theory throughout the century, the problems of the working classes were waxing in stark force. Along with other leaders in Germany, several Catholic leaders, such as Franz v. Baader and Ketteler, were of the opinion that subordinate and semi-independent work organizations would avoid revolution and articulate the workers into society. Toward the end of the century, when, under the leadership of Hitze and Hertling, reformism marked the Catholic contribution to social policy, proponents of the ideas of Vogelsang still asserted the need for social transformation on the pattern of *Stände* groupings.[265] While the reformist group attempted legislatively to ameliorate the difficult position of the worker under an increasingly concentrated capitalism, the group which favored intermediate work groups—corporations—often relied on a medieval guild model as a pattern for social transformation in continuity with German traditions.

Within the general tradition of German socio-political thought and the specific heritage of Catholic thought concerning the organization of society, Pesch favored a social system of work composed of occupational groups. The legally defined occupational groups enable the workers in the sundry types of related work to have their interests represented so that the common good may be reached cooperatively.[266] Further, both the human necessity of mutual assistance and the natural drive to form groups lead to the establishment of these occupational organizations.

The natural basis for the groups makes it possible that they may maintain a necessary independence in their mediation between the state and the individual. The occupational groups then have their own natural claim to juridical existence, so long as they contribute to the common good.[267] The groups are organs of civil society which function for the sake of the common good. Although the work activity of each person has a social character and although each must fulfil certain duties in society, personal work is not a function in society as such.[268] Workers assist the whole by way of their work activity in the occupational group; at the same time the group shields them from the demands of the whole society and, at still another remove, from possible impositions by the state.

The vocational groups provide secure bases upon which the state can be established in order to complete the integration of society.[269] Yet, I re-emphasize, the state cannot regard these groups as its organs, for they are properly structures of civil society.[270] The state can introduce mandatory vocational groups, but only under two restrictions: a) Such groups may comprise only a small portion of those in society as a whole. b) The state must proceed according to subsidiarity; that is, it must first attempt to stimulate and encourage the initiative of the members of a possible vocational group to establish their own organization.[271]

The occupational group not only protects the working person from the state, it is also a means whereby workers can mutually prevent arbitary and harmful actions on the part of their fellows.[272] The occupational group provides a structure which protects workers from domination and stimulates them to the exercise of freedom.

The guild model, which some characteristics of the occupational group evoke, may inhibit an appreciation of this new organizational form insofar as it appears solely as an expression of romantic nostalgia. On the contrary, insofar as it selects features of the medieval model—cooperation between managers and actual producers, worker voice in the production process, opportunity for the workers to comprehend the entire pattern of production—these intermediate groups represent loci where freedom can grow in range, responsibility and cooperation. This structure secures both the order requisite for societal production and the recognition accorded the personal exercise of freedom.

A vocational group must be distinguished from a producers' or consumers' cooperative, for these Peschian groups structure the social system of work that organizes the entirety of society and production for the public good. Each group is an organization of executive, managerial and production workers; that is, natural interests and economic function determine its vertical organization.[273] These vocational groups enable work activity to be more productive and regulate it better for the good of the whole.

Pesch maintains that the social system of work established through vocational groups will bring major benefits to society. Class conflict will be overcome because employers and workers will recognize their common interests within each group and their common purpose in society as a whole.[274] The structured cooperation of the social system of work remedies the unfair competition between large and small firms and useless, anarchic competition.[275]

The power invested in concentrations of capital can be regulated since all persons—owners, entrepreneurs, managers, service and production workers—must follow the regulations of the vocational group and societal position (*Stand*) within which they work.[276] Within the individual firm, membership of all in the corresponding vocational order will give weight to the claim that all workers, whether management or production, are co-workers. The labor contract establishes a quasi-partnership in which the ordinary worker acknowledges the directive instructions of management, while not submitting to any claims of commanding authority on the part of owners or managers.[277]

The vocational groups also offer particular benefits to the individual workers. Out of common interests and related occupations within a group, a common attitude, sensibility and set of customs develop to unite, not just the means and efforts toward an economic end, but also the working persons themselves.[278] The association of this group of workers provides a way for securing employment, income and just working conditions for its members,[279] while the social order founded on the vocational groups insures the civil freedom of its members.[280] When the space and structure for each worker's initiative, economic independence and free activity is secure, this civil freedom is guaranteed.[281] Within this world of free activity and civil freedom, vocational groups enable the working person to grow in responsibility as he grows in knowledge about the economic means and goals of his group.

The social system of work can assure freedom to its members because it establishes a set of organic relationships as the social dimension of their work activity. This system offers a field in which they may collaborate with other persons in the provision of mutual and social goods. Thus, working persons, instead of losing themselves in the empty satisfactions of capricious choice, can freely invest their efforts in projects which encourage their own growth.

As this concern for the freedom of the individual attests, Pesch by no means intends for the vocational group to subsume ordinary working people under a patriarchal tutelage of owners and managers. He insists on other structures which, in conjunction with the free activity and growing responsibility of the worker within the vocational group, underscore the centrality of the working person in his view of the economy. The liberty to band together in unions, Pesch maintains, is a natural right.[282] In his declaration that work activity proceeds from the person on the basis of the person's choice, that this work naturally inclines toward collaboration, but that labor is not undertaken as a socially determined obligation, Pesch outlines an understanding of work in which he maintains the workers' freedom to strike.

The centrality of the worker in the advancing socio-economic system also leads Pesch to indicate still further structures which would surpass the vocational group; indeed, these developments move the social system of work far beyond any medieval vestiges of static or patriarchal structures. The quasi-partnership established through the work contract between the managerial and production co-workers indicates a movement toward profit-sharing plans.[283] The workers would face no threat in such an arrangement, for they do not share in the risk of the firm. This is so because the contract does not establish them as full partners. Pesch holds out the possibility of a still further advance in the work relationship: co-ownership. Pesch believes that a system of insurance could protect the workers—who as a rule hold only minimal reserves against loss—in their participation in the risk of the firm. Moreover, co-ownership would bring a greater allegiance on the part of the workers to the challenges which the firm faces and a greater understanding of how their own needs fit into the other demands on the firm.[284]

One final arrangement represents Pesch's understanding of the fullest role which the worker may assume in the firm: workers' participation. Article 165 of the Weimar Constitution of 11. August, 1919 declared that all employees are entitled to participate in the regulation of wage and work conditions and of the development of the firm's production.[285] With approval, Pesch interprets this as insuring the direct participation (that is, not only indirectly through public institutions) of all workers in the shaping and direction of the whole economy, beginning with their own economic units.[286] "Co-administration, participation in the returns and ownership must be made possible for the workers," so that they, no longer restricted to the role of wage laborer, may be stimulated to increase production.[287] Consultation with ordinary workers will conduce to the most economic utilization of work activity.[288] The constitutionally protected restructuring of the work relationship, in conjunction with the system of *Betriebsräte* (works councils), will prevent the proletarianization of the people.[289]

This early, comprehensive system of workers' participation involved (as had been characteristic for the German, organic approach) several levels for worker interaction with employers: in the firm, at the district level, and in national institutions. Pesch does not explore all of these in depth. The *Betriebsräte* were the consultative bodies of workers and other employees in the firm. Here, Pesch calls for parity of representation.[290] On the board of the firm (*Aufsichtsrat*), the workers are to have a limited representation.[291] In addition, because each firm belonged to a group of companies constituting one branch of business, the unions maintained a valid role because they could represent workers throughout the industry.[292] Rather than advocating ambitiously comprehensive schemes, Pesch supports a moderate approach to *Betriebsparlamentarismus* (deliberative and legislative processes in the firm) carried out by means of committees in the factories and firms.[293] As the ill fortune of the grander Weimar plans for worker participation demonstrated, the times were not yet ripe for a more comprehensive structure which could protect and promote the central position of the working person in society.

A Deficient Analysis of Technology

Since the human person is the sole producer in work activity, Pesch regards tools and machines only as instrumental causes.[294] So much does Pesch focus on the person and the freedom of his activity that he specifies one characteristic of work activity as the freedom to select the means used in one's occupation.[295] The means provided through technology not only are instruments for use in a specific task, but also, when taken as a whole, they are instruments for the most comprehensive provision for human needs.[296]

These tools and machines both facilitate the attainment of specific as well as general human ends, and they also result from the application of human, rational capacities to work activities so that human control over the world may be extended. Although Pesch realizes that the process of discovery which results in more advanced technical processes and devices has undergone a radical change away from the activity of the individual worker to a separate and planned occupation on the part of selected workers,[297] he does not examine the more extensive impact of technology on society, economic processes and the workers. Pesch continues to emphasize the centrality of the person in work activity, but does not present a comprehensive understanding of the connection between the implements of work and persons in society.

Subsidiarity and the Service of the Weakest

Clearly for Pesch, work activity not only benefits each person who undertakes this activity, but also serves one's fellows in society through the provision of goods and services and through the exercise of the social virtues, justice and love.[298] Pesch emphasizes, beyond the concerns for justice found in many of the thinkers I have already considered, that love is required for the maintenance and development of society, because it supplies that which lacks in actions undertaken from the perspective of justice alone.[299]

Pesch carefully and concretely specifies this understanding of the practice of love connected with work activity. Work activity most frequently occurs within the employer-employee relationship. This relationship must, of course, secure recompense to the worker, because the work of each is to contribute to the development of his or her own person. This relationship configures the social interaction of work on the pattern of hierarchical subordination; yet the employer-employee structure must recognize that work activity implies, in addition, a mutual, responsible service which is focused on workers because of the dangers they face and the weaknesses they experience. That is, the employer must set up not only a system for recompense and control, but also one which serves the workers by showing them actual care.[300] This is not the case yet, but it can become so. The practice of love has an even wider scope. Pesch holds that persons direct their productive activity toward personal and social welfare as well as toward the culture of the society and, further, toward the progress of this society's welfare and culture.[301] Production is itself an activity of culture which works the advance of this culture. In particular, the actualization of values in society advances this culture. Indeed, the extension of society's concern to still more members of that society marks this development. Work activity provides the motive force whereby this practice of values and the extension of concern take place.

The weakest members of society have the greatest claim on the service component of work. Thus as a minimum, the members of society must not configure the system of work so that it serves those who already possess much.[302] Nor must the conquest and defeat of some characterize the processes of economic life, because then their spiritual, intellectual, and material capabilities are unavailable to society.[303] The service of work, as structured in the entire social system, must protect the weak.[304]

Pesch carefully notes that state action does not provide the only instance of the service which the system of work must offer. To be sure, where other means are not sufficient, the state must step in to serve those in need. The principle of subsidiarity indicates, however, that the state must maintain a framework in which individuals and their ordinary groupings can act responsibly for the good of the whole.[305] This subsidiary activity includes action in the face of threats and weakness. Decentralized and participatory efforts protect persons in their weaknesses within the economic system and stimulate the protection of their interests in common with those of others.[306] These same efforts make it possible for a society to manifest and exercize care and concern as it strives progressively to incorporate the social virtues.

The subsidiary action of persons who work together serves both themselves and society as a whole. Such groupings prevent the dangers which result from extremely centralized and bureaucratic power lodged in the authority of the state.[307] In addition, the decentralized cooperation of co-workers helps integrate more members of society into its life, thus preventing the threat to society which results when groups are excluded.[308] The *Berufsstände* serve society in that they depend on, stimulate and cultivate the capacities for initiative and commitment on the part of many persons.

Political participation by workers in terms of their work activity is the completion of the service component of work activity. Distributive justice requires that workers—craftspeople, factory workers and farmers—be integrated into the common life through the guarantees provided by a vehicle for the representation of their interests and secured by bodies of law which protect them in their particular work activities.[309]

Similar to the view of Hegel, Pesch maintains that workers' associations serve the common good through the education to political action which they furnish working persons. These workers are trained to be conscious of the social purpose of their own activity and are exercised in a sense of the law.[310] The work organizations which Pesch has in mind are not class-based associations of proletarians. They are, rather, organizations of citizens which together encompass all the members of society.[311] As an ensemble, these groups of workers can include all members of society because they are founded upon the common feature of work activity. These work groups can contribute to the progressive and cultural development of society because the central activity of work produces the requisites for welfare which make possible any development on a higher level and because the collaborative activity of deliberation and decision activates the higher potentials of each participant for the benefit of the entire society. Political participation on the basis of work activity exercises progressive and unitive values in society.

I have examined how Pesch emphasizes the singular importance of working persons who together constitute the sole subject of society. They give shape both to objects and to society itself. Pesch's view of human limits which must be met so that persons may develop and may bear responsibility was also considered. Persons-in-society meet these limits as they collaborate in an array of work-based groupings which make participation possible. Although Pesch's understanding of technology is severely limited, his theory of service is rich. It emphasizes that it is in work that society engages in the practice of value and the extension of concern. The service component of work leads to political participation. The social system of work evolves so that associations of workers integrate all members of society into a productive collaboration and responsible citizenship. Just as I have indicated Pesch's roots in the German Historical School, so now I turn to the intellectual descendant of Pesch, Oswald v. Nell-Breuning, whose understanding of work activity has enabled him to articulate still more clearly the role of workers' participation in the contemporary economy.

1. J.M. Keynes, "Alfred Marshall 1842-1924" in *Alfred Marshall: Critical Assessments*, ed. by J.C. Wood, Vol. I, (London: Croom Helm, 1982), p. 44.

2. A. Marshall, "The Present Position of Economics" in *Memorials of Alfred Marshall*, ed. by A.C. Pigou, (London: Macmillan and Co., 1925), p. 155. Also see Alfred Marshall, *Principles of Economics*, 2 vols., 9th (Variorum) ed. with annotations by C.W. Guillebaud (London: Macmillan, 1961), Vol. 1 (Text), p. 762.

3. Marshall, *Principles*, p. 1.

4. Talcott Parsons, "Wants and Needs in Marshall" in *Alfred Marshall: Critical Assessments*, ed. by J.C. Wood, Vol I., p. 185.

5. *Principles*, p. 23.

6. *Ibid.*, pp. 194 and 527.

7. *Ibid.*, pp. 141-142.

8. A. Marshall, "The Future of the Working Classes," in *Memorials*, ed. by A.C. Pigou, p. 103.

9. Parsons, "Wants and Needs in Marshall," in *Alfred Marshall: Critical Assessments*, ed. by J.C. Wood, Vol. I, p. 180. Also see Walter W. Haines, "The Psychoeconomics of Human Needs: Maslow's Hierarchy and Marshall's Organic Growth," *Journal of Behavioral Economics*, 11 (1982): 103.

10. Keynes, "Alfred Marshall: 1842-1924," in *Alfred Marshall: Critical Assessments*, ed. by J.C. Wood, Vol. I, p. 75.

11. Marshall, "The Future of the Working Classes" in *Memorials*, ed. by A.C. Pigou, p. 115.

12. *Principles*, pp. 206-207.

13. *Ibid.*, p. 718.

14. Marshall, "The Future of the Working Classes," in *Memorials*, ed. by A.C. Pigou, p. 111.

15. Marshall, "Fragment" (1922), in *Memorials*, ed. by A.C. Pigou, p. 367.

16. *Principles*, p. 143.

17. *Ibid.*, p. 541.

18. *Ibid.*, p. 578.

19. *Ibid.*, p. 556.

20. Marshall, Letter to Bishop Westcott (1900), in *Memorials*, ed. by A.C. Pigou, p. 391.

21. *Principles*, p. 549.

22. *Ibid.*, p. 564.

23. *Ibid.*, p. 660.

24. Marshall in G.J. Stigler, "Alfred Marshall's Lectures on Progress and Poverty" in *Alfred Marshall: Critical Assessments*, ed. by J.C. Wood, Vol. IV, p. 160.

25. John Dennis Chasse, "Marshall, the Human Agent, and Economic Growth: Wants and Activities Revisited," *History of Political Economy*, 16 (1984): 384.

26. Jacob Viner, "Marshall's Economics in Relation to the Man and His Times" in *Alfred Marshall: Critical Assessments*, ed. by J.C. Wood, Vol. I., p. 254.

27. *Principles*, pp. 250 and 546. See also p. 293, "Looking at businessmen from one point of view, we may regard them as a highly skilled industrial grade. . . ."

28. *Ibid.*, p. 208.

29. *Ibid.*, p. 331.

30. *Ibid.*, p. 139.

31. *Ibid.*, p. 173.

32. *Ibid.*, p. 689.

33. *Ibid.*, pp. 197 and 691.

34. *Ibid.*, pp. 262 and 335.

35. *Ibid.*, pp. 573-574. Consider also the fifth peculiarity of work, which describes the long preparation and the slow return resulting from this training, p. 570.

36. *Ibid.*, p. 560 and 566-567.

37. *Ibid.*, p. 664. See also Marshall's treatment of the fourth peculiarity of work on p. 567.

38. *Ibid.*, p. 552.

39. *Ibid.*, p. 527.

40. *Ibid.*, p. 567.

41. *Ibid.*, p. 697.

42. *Ibid.*, p. 665.

43. *Ibid.*, p. 323.

44. *Ibid.*, p. 205.

45. *Ibid.*, p. 271.

46. *Ibid.*, p. 247.

47. *Ibid.*, pp. 559-560.

48. See, for example, Alfred Marshall, *The Early Economic Writings of Alfred Marshall—1867-1890*, ed. by J.K. Whitaker, (London: Macmillan for the Royal Society, 1975), Vol. 2, p. 374.

49. *Ibid.*, Vol. II, p. 374 and *Principles*, p. 565.

50. *The Early Economic Writings of Alfred Marshall—1867-1890*, Vol. 2, p. 341.

51. J.K. Whitaker in *The Early Economic Writings of Alfred Marshall—1867-1890*, Vol. 2, p. 343. Several times Marshall attempted the formulation of particular principles, as distinct from general virtues: The general adoption of an action ought not be harmful (p. 127). An action if generally adopted should not cause greater injury to the agent than the benefit of adopting it himself (p. 366). That policy of a group which injures the common good is morally wrong (A. and M.P. Marshall, *The Economics of Industry*, 2nd ed. (London: Macmillan and Co., 1881), p. 211). No gain should be desired "which would involve a very much greater loss of happiness to others." *Ibid.*, p. 426.

52. *Principles*, p. 569.

53. A. and M.P. Marshall, *The Economics of Industry*, 2nd ed., p. 425.

54. "Freedom to combine with others in his own occupation would have widened his horizon, and given him larger matters to think about. . . . Thus the early struggle for the principle that workmen should be free to do in combination the counterpart of anything which employers were free to do in combination was in effect an effort to obtain conditions of life consistent with true self-respect and broad social interests." *Principles*, p. 703. Also see how leisure, secured indirectly by obtaining a proper level of wages, contributes to the twin goal of personhood and citizenship (p. 720).

55. A. Marshall, "Fragments on Trades Unions," in *The Early Economic Writings of Alfred Marshall—1867-1890*, Volume II, p. 351; and A. and M.P. Marshall, *The Economics of Industry*, 2nd ed., p. 425.

56. A. Marshall, "The Future of the Working Classes," in *Memorials*, ed. by A.C. Pigou, p. 105.

57. A. Marshall, "Some Features of American Industry," in *The Early Economic Writings of Alfred Marshall—1867-1890*, Volume II, p. 364.

58. *Ibid.*, p. 366.

59. A. Marshall, "A Fair Rate of Wages" (1887), in *Memorials*, ed. by A.C. Pigou, p. 225.

60. A. Marshall, "Cooperation" (1889), in *Memorials*, ed. by A.C. Pigou, pp. 228-229.

61. *Ibid.*, p. 249; and A. and M.P. Marshall, *The Economics of Industry*, 2nd ed., p. 218.

62. Marshall speaks of this harmony as the first principal factor of ethical growth in society and he understands this harmony as Hegel's objective freedom. See "Some Features of American Industry" in *The Early Economic Writings of Alfred Marshall—1867-1890*, Volume II, p. 375.

63. *Ibid.*, p. 377.

64. A. and M.P. Marshall, *The Economics of Industry*, 2nd ed., p. 188.

65. *Principles*, p. 265.

66. *Ibid.*

67. Notice that this Marshallian phrase substitutes for the concept "industrial efficiency," which I have examined earlier.

68. *Principles*, p. 283. Marshall carries the equating of men and machines so far that machines become personified. This occurs, for example, when he speaks of the wastefulness in keeping the untiring iron and steel of machines resting in idleness. See *Ibid.*, p. 695.

69. *Ibid.*, p. 662.

70. *Ibid.*, pp. 570-571.

71. *Ibid.*, p. 541.

72. D.A. Walker, "Marshall on the Long-Run Supply of Labor" in *Alfred Marshall: Critical Assessments*, ed. by J.C. Wood, Vol. III, p. 533.

73. *Principles*, p. 138.

74. Walker, "Marshall on the Long-Run Supply of Labor," in *Alfred Marshall: Critical Assessments*, ed. by J.C. Wood, Vol. IV, p. 532.

75. *Principles*, p. 258.

76. Chasse, "Marshall, the Human Agent and Economic Growth: Wants and Activities Revisited," p. 388.

77. *Principles*, p. 571.

78. *Ibid.*, p. 67.

79. *Principles*, pp. 88-89.

80. "By 1890, and increasingly thereafter, it was the restless, pioneering, farsighted entrepreneur . . . on whom chief responsibility for the economic and moral progress of the world came to be placed." J.K. Whitaker, "Some Neglected Aspects of Alfred Marshall's Economic and Social Thought," in *Alfred Marshall: Critical Assessments*, ed. by J.C. Wood, Vol. I, p. 476.

81. *Principles*, p. 664.

82. *Ibid.*, pp. 565-566.

83. Marshall, in "Some Features of American Industry," sets forth a pattern of ethical advance, one which he finds characteristic of the America he visited as a young man, and for which he found a theoretical account in Hegel's subjective freedom. This pattern, I believe, finds echo in the work of the mature Marshall in his characterization of the entrepreneur: "The other factor I would describe as the education of a firm will through the overcoming of difficulties. This will does not glide carelessly into conformity with the conditions by which it is surrounded, but submits every particular action to the judgment of reason. . . . Such a society [i.e., one which is shaped by such individuals] will be an empire of energy, of strong but subdued enthusiasm, of grand ideals." See *The Early Economic Writings of Alfred Marshall—1867-1890*, pp. 375-376.

84. Marshall, "The Future of the Working Classes" in *Memorials*, ed. by A.C. Pigou, p. 114.

85. *Ibid.*, p. 117.

86. Marshall, "A Fair Rate of Wages," in *Memorials*, ed. by A.C. Pigou, p. 225.

87. *Principles*, p. 211.

88. Marshall, "The Present Position of Economics," in *Memorials*, ed. by A.C. Pigou, p. 165; and *Principles*, pp. 783-784. Notice that Marshall was familiar with the work both of the Earlier and the Later Historical Schools. It is of some interest that on the last page indicated here, Marshall makes reference to A. Wagner's list of economic motivations. One of these motivations is the craving for occupation and the pleasure of activity.

89. T. Parsons, "Economics and Sociology: Marshall in Relation to the Thought of His Time," in *Alfred Marshall: Critical Assessments*, ed. by J.C. Wood, Vol. I, p. 219; G.F. Shove, "The Place of Marshall's *Principles* in the Development of Economic Theory," *The Economic Journal*, Vol. 52, December, 1942, p. 309; A.J. Youngson, "Marshall on Economic Growth" in *Alfred Marshall: Critical Assessments*, ed. by J.C. Wood, Vol. IV, pp. 100-101.

90. Fritz Völkerling, *Der deutsche Kathedersozialismus*, (Berlin: Verlag die Wirtschaft, 1959), pp. 39-41, 44 and 49.

91. Albert Müssiggang, *Die soziale Frage in der historischen Schule der deutschen Nationalökonomie* (Tübingen: J.C.B. Mohr (Paul Siebeck), 1968), p. 225. Also see Arno Friedrichs, *Klassische Philosophie und Wirtschaftswissenschaft: Untersuchungen zur Geschichte der deutschen Geistesleben im neunzehnten Jahrhundert* (Gotha: Friedrich Andreas Perthes, 1913), p. 565.

92. Hans Freyer, *Die Bewertung der Wirtschaft im philosophischen Denken des 19. Jahrhundert* (Leipzig: W. Engelmann, 1921; reprint ed., Hildesheim: Georg Olms Verlagsbuchhandlung, 1966), p. 122.

93. In this too see Friedrichs, *Klassische Philosophie und Wirtschaftswissenschaft*, p. 568.

94. Wilhelm Kromphardt, "Die Überwindung der Klassenkämpfe nach Gustav von Schmoller, in *Schmollers Jahrbuch*, hrsg. von A. Spiethoff, 62. Jahrgang, II. Halbband (1938): 717-718.

95. Freyer, *Die Bewertung der Wirtschaft*, pp. 129-130.

96. Bernhard Pfister, "Schmoller," in *Staatslexikon*, 1932 ed., Bd. 4, Sp. 1284-1286.

97. Freyer, *Die Bewertung der Wirtschaft*, p. 30.

98. G. Schmoller, *Über das Maschinenzeitalter in seinem Zusammenhang mit dem Volkswohlstand und der sozialen Verfassung der Volkswirtschaft* (Berlin: Julius Springer Verlag, 1903), pp. 23-24.

99. "Die Arbeiterfrage" in *Preussische Jahrbücher* 14. Jahrgang (1864): 393-424 and 523-547; 15. Jahrgang (1865): 32-63.

100. Völkerling, *Der deutsche Kathedersozialismus*, p. 54.

101. H. Delbrück, G. Schmoller und A. Wagner, Über die Stumm'sche *Herrenhaus-Rede gegen die Kathedersozialisten—Schreiben an den Geh. Justizrath Professor D. Dr. Hinschius* (Berlin: Verlag von Georg Stilke, 1897), pp. 4-5; and Müssiggang, *Die soziale Frage in der historischen Schule*, p. 1.

102. *Grundriss der allgemeinen Volkswirtschaftslehre*, Erster Teil (Leipzig: Duncker und Humblot, 1908), p. 41.

103. *Zur Litteraturgeschichte der Staats- und Sozialwissenschaften* (Leipzig: Duncker und Humblot, 1888; reprint ed., New York: Burt Franklin, 1968), pp. 86-87.

104. Schmoller, *Zur Litteraturgeschichte*, p. 93.

105. *Grundriss*, I, pp. 71, 50-51 and 42-43.

106. *Ibid.*, I, pp. 20-22.

107. *Ibid.*, I, pp. 34-35.

108. *Ibid.*, I, p. 36-37.

109. ". . . Jede menschliche Thätigkeit. . . welche mit dauernder Anstrengung sittlich-vernunftige Zwecke verfolgt," My translation. *Ibid.*, I, p. 38.

110. *Ibid.*, I, p. 29.

111. G. Schmoller, *Die soziale Frage: Klassenbildung, Arbeiterfrage, Klassenkampf* (München und Leipzig: Verlag von Duncker und Humblot, 1918), p. 201.

112. *Grundriss*, I, p. 21.

113. *Ibid.*, I, p. 39.

114. *Ibid.*, I, p. 42.

115. *Ibid.*, I, p. 38.

116. G. Schmoller, *Über einige Grundfragen der Socialpolitik und der Volkswirtschaftslehre* (Leipzig: Duncker und Humblot, 1898), pp. 45-46.

117. *Die soziale Frage* (München und Leipzig: Verlag von Duncker und Humblot, 1918), p. 619.

118. *Ibid.*, p. 292.

119. "Der ganze Umkreis menschlicher Gefühle, der niedrigen wie der höheren, erzeugt so Bedürfnisse. Der Mensch hat sinnliche, ästhetische, intellektuelle, moralische Bedürfnisse. . . . Im weiteren Sinne ist Bedürfnisbefriedigung der Zweck alles menschlichen Handelns, nicht bloss des wirtschaftlichen, denn zu allem Handeln geben Lust- und Unlustgefühle und die Erinnerung an sie den Anstoss." See *Grundriss*, I, p. 23.

120. *Über einige Grundfragen*, p. 47.

121. *Grundriss*, I, p. 26.

122. *Ibid.*, p. 29.

123. *Ibid.*, p. 27.

124. *Ibid.*, p. 552 and also p. 338. In opposition to the Darwinian view of a struggle for existence, Schmoller emphasizes those characteristics of human life—feelings of sympathy, customs, morals, law—which reduce conflict within groups. See p. 65.

125. "Das Leben vom Tag zum Tage blieb das alte, während die neue Wirtschaftsverfassung Vorsorge für Wochen, Monate, Jahre forderte." See *Die soziale Frage*, p. 333.

126. *Ibid.*, p. 302.

127. *Ibid.*, pp. 432-433 and 373.

128. *Ibid.*, pp. 430-431.

129. *Über einige Grundfragen*, p. 46.

130. "Die gesellschaftlichen Institutionen und Organe treten uns als das wichtigste Ergebnis des sittlichen Lebens entgegen. Es sind die Krystallisationen desselben." *Grundriss*, I, p. 61.

131. Kromphardt, "Die Überwindung der Klassenkämpfe," pp. 730-731. Kromphardt refers to Schmoller's "Demokratie und soziale Zukunft" in *Soziale Praxis*, 22. Bd., 1912/1913, p. 148.

132. "Durch die Sitte baut der Mensch in die Natur, eine zweite Welt, die Welt der Kultur hinein. Und zu dieser Welt der Kultur gehört auch die Volkswirtschaft." See *Über einige Grundfragen*, p. 47.

133. *Grundriss*, I, p. 43.

134. ". . . Jedes Prinzip sittlicher Wertschätzung von Handlungen baut sich auf bestimmten, materiell-technischen, gesellschaftlichen, und psychologisch-geschichtlichen Voraussetzungen auf. Die ethische Vorstellungswelt erstreckt sich von der sinnlichen Lust des individuellen

Lebens durch zahllose Glieder hindurch bis zur Menschheit, zum Weltganzen, zur Ewigkeit. Das Gute hat kein ruhendes, sondern ein sich stetig vervollkommendes Dasein. Der nie ruhende Sieg des Höheren über das Niedrige, des Ganzen über das Einzelne macht das Wesen des Guten aus." *Ibid.*, p. 44.

135. Müssiggang, *Die soziale Frage in der historischen Schule*, p. 230.

136. *Grundriss*, I, pp. 552-554.

137. *Die soziale Frage*, p. 76. Also see p. 636.

138. *Ibid.*, p. 78.

139. *Ibid.*, p. 71.

140. "Wir mögen unser Beamtentum und seine grossen Tugenden, unsere liberalen Berufen mit ihrem Idealismus, unser Bauerntum mit seinem kernhaften Muskeln und schlichten Gemütseigenschaften, unsere aufstrebende Arbeiterwelt mit ihrem Bildungstrieb, ihrer technischen Tüchtigkeit, ihrer aufopfernden Vereinstätigkeit noch so hoch schätzen, sie bedürfen als Ergänzung der ganz anders gearteten, aber nicht minder wertvollen psychischen und gesellschaftlichen Kräfte der Geschäftswelt ebenso, wie diese ohne jene anderen gesellschaftlichen Kräfte und Tendenzen nicht glücklich wirken könnte." *Grundriss*, I, p. 554.

141. *Die soziale Frage*, p. 82.

142. *Ibid.*, p. 83.

143. Several problems present themselves as a result. First, the corrective view from the top implies that the work problems which result from the technical division of labor—here consigned to the farther reaches of the social division—will be overlooked. Second, the view from the top will have no vantage from which to see its own role in the social division as liable to reform. Since the social division of labor asserts the advantage for society of specialization on the basis of talents and abilities, this applies as well to government. An aristocracy will arise which will be responsible for the adjustment of the process of the social division of labor. See *Die soziale Frage*, pp. 183-184. The danger posed by such an aristocracy reveals itself when the inegalitarian tendency of Schmoller's social system is clearly seen: ". . . Die längst widerlegte Aufklärungshypothese, dass alle Menschen von Natur gleich, nur durch gesellschaftliche Einrichtungen ungleich geworden seien"—see *Ibid.*, pp. 157-158. This inegalitarianism forces even more questions because Schmoller writes, ". . . Der gesellschaftliche Fortschritt und die Arbeitsteilung ist nicht möglich ohne dass immer wieder zeitweise ihr einzelne Individuen und Klassen geopfert werden." *Ibid.*, p. 83. Working persons are assigned to different works because they are differentially suited to these works. These assignments are placements in lifelong cooperative roles in society. The differences in works then too easily become differences in the treatment of persons.

144. *Ibid.*, p. 130.

145. *Ibid.*, p. 318.

146. *Grundriss*, I, p. 550.

147. *Die Soziale Frage*, pp. 504-505.

148. Müssiggang, *Die soziale Frage in der historischen Schule*, p. 133.

149. "Die selbstbewusste Organisation des Arbeiterstandes an sich ist der Ausdruck der weltgeschichtlichen Tatsache, dass die Menschheit eine Kulturhöhe erreicht hat wie nie früher, eine Kulturhöhe, die auch die unteren Klassen nicht mehr zum passiven Fussgestell der oberen, sondern zu einem selbstbewussten aktiven Gliede des Gesamtorganismus machen will und kann." *Die soziale Frage*, p. 641.

150. Müssiggang, *Die soziale Frage in der historischen Schule*, p. 131.

151. "Rede zur Eröffnung der Besprechung über die soziale Frage in Eisenach den 6. Oktober, 1872" in *Zur Social- und Gewerbepolitik der* Gegenwart: Reden und Aufsätze (Leipzig: Verlag von Duncker und Humblot, 1890), p. 12.

152. *Zur Social- und Gewerbepolitik* p. 455.

153. "Das geschieht aber wenn wir in jedem einzelnen grösseren Geschäft der Angestellten durch Arbeiterausschüsse einen gewissen Anteil an der Verwaltung und innerer Ordnung der Geschäftsverhältnisse und durch eine entsprechende Gewinnbeteiligung neben dem festen Lohn ein Interesse am Gedeihen des Geschäfts geben." *Ibid.*, p. 461.

154. *Die soziale Frage*, pp. 239-240.

155. *Zur Litteraturgeschichte*, p. 77.

156. *Grundriss*, I, p. 39.

157. *Ibid.*, p. 60.

158. *Ibid.*, p. 190.

159. *Ibid.*, p. 192. Also see *Über das Maschinenzeitalter*, p. 12: "Denn die letzte Ursache des Fortschrittes liegt in der Rationalisierung der Arbeitsprozesse auf grund naturwissenschaftlicher Erkenntnis. . . ."

160. "Wenn der Mensch, wie der Affe, einen Stein zum Öffnen einer Frucht, einen Stock zum Schlagen brauchte, so hatte er noch kein Werkzeug; erst dann konnte man davon sprechen, wenn er diesen Stein, diesen Stock stetig bei sich führte, wenn die Erinnerung an den Nutzen dieses Hülfsmittel die Unbequemlichkeit der Aufbewahrung, des Mitschleppens überwand." See *Grundriss*, I, p. 192.

161. *Grundriss*, I, p. 226.

162. *Über das Maschinenzeitalter*, p. 19.

163. *Grundriss*, I, p. 227.

164. *Ibid.*, p. 227 and *Über das Maschinenzeitalter*, p. 21-22. Also see *Ibid.*, pp. 26-27 where

Schmoller asserts that the most advantageous work agreements have been made by workers in those trades where technological advances have most stimulated their intelligence.

165. "Für einen sehr erheblichen Teil aller Arbeitsprozesse muss der Mensch jeden Moment seine Hand andern Bedingungen anpassen, muss er sich jeden Moment etwas veranderte Ziele setzen. Das kann die Maschine nicht, sie hat keine Seele, kein eigenes Leben, sie überträgt nur die Anstösse, die der Mensch in ihr entzündet, sie konzentriert gewisse Naturkräfte; aber sie kann den Menschen in der Wirtschaft doch nur partiell ersetzen." *Über das Maschinenzeitalter*, p. 15.

166. ". . . Die grosse mechanische Anstrengung werde dem Menschen überhaupt abgenommen, der bedienende Maschinenarbeiter werde gleichsam der hohen geistigen Fähigkeit und Kraft teilhaftig, über die der Erfinder des vervollkommneten technischen Progresses, der Maschine verfügte." See *Ibid.*, p. 13.

167. *Grundriss*, I, pp. 221 and 228; *Über das Maschinenzeitalter*, pp. 11 and 15-16.

168. *Über das Maschinenzeitalter*, pp. 20-21. The modern company represents one of these new establishments in the social sphere. See *Ibid.*, p. 29. Also note *Grundriss*, I, p. 225: "Die wichtigste sociale Folge der Grosstechnik ist die Entstehung eines breiten Lohnarbeiterstandes."

169. *Grundriss*, I, pp. 229-230 and *Über das Maschinenzeitalter*, p. 25.

170. *Über das Maschinenzeitalter*, pp. 17-18; and Grundriss, I, pp. 223 and 226.

171. *Grundriss*, I, pp. 223-226.

172. *Über das Maschinenzeitalter*, p. 21; and Grundriss, I, pp. 224-225.

173. *Grundriss*, I, p. 228.

174. His goal, as stated in his 1864 article on the *Arbeiterfrage*, was to bring "unaufhaltsam immer mehr Glieder zur Teilnahme an allen Gütern der Humanität." See Müssiggang, *Die soziale Frage in der historischen Schule*, p. 132.

175. For the romantic economists, see Freyer, *Die Bewertung der Wirtschaft*, pp. 46-47. For v. Stein, see Kromphardt, "Die Überwindung der Klassenkämpfe," p. 720.

176. *Über einige Grundfragen*, p. 55.

177. Freyer, *Die Bewertung der Wirtschaft*, p. 131.

178. "Je grösser die Gemeinwesen wurden, desto weniger genügte die Sorge von Vereinen, Genossenschaften, Gemeinden, desto mehr mussten diese Veranstaltungen im Gesamtinteresse gemacht, gerecht gehandhabt, von den egoistischen Sonderinteressen einzelner Geschäfte, Orte, und Klassen befreit werden." See *Grundriss*, I, p. 340.

179. "Erst die letzten Jahrhunderte haben nun . . . einen Kreis von Juristen, Beamten, Offizieren, Geistlichen, Lehrern geschaffen, die, häufig aus allen Kreisen der Gesellschaft sich ergänzend, doch gleichmässig auf den Universitäten gebildet, teils durch Besitz, teils durch Besoldung wirtschaftlich sichergestellt, ihr ganzes Leben den öffentlichen Geschäften widmen.

". . . Sie bilden mit den Rechtsanwälten, Ärtzen, Künstlern, Journalisten eine Art neutraler Sphäre gegenüber den eigentlich kämpfenden, ausschliesslich um Besitz ringenden Klassen." See *Die soziale Frage*, p. 629.

180. *Ibid.*, p. 331. Distinct from support of the poor are the actual payments resulting from a system of insurance. These are due as a person's right. See *Ibid.*, p. 413.

181. *Ibid.*, pp. 341-342.

182. The mandatory insurance scheme and the centralized job placement agency provide two examples. See *Ibid.*, pp. 426 and 504.

183. *Ibid.*, pp. 367-374. The Prussian law which instituted this insurance was the model for what has since become an expected condition of modern work life.

184. "Der wichtigste Grund für die immer dringlicher werdende Durchführung aller dieser Zweige der Arbeiterversicherung ist die Wirkung auf die Lebenshaltung. Mögen die einzelnen Arbeiter und ihre Familien vor allem durch die Not des Augenblicks schwer leiden und daher der Versicherung so dringlich bedürfen, für die ganze Gesellschaft ist noch wichtiger, dass die gesamte Arbeiterschaft ohne die Versicherung in ihrer ganzen Lebenshaltung immer wieder bedroht ist: alle die erwähnten Übelstände reissen die dadurch Betroffenene notwendig immer wieder und immer stärker auf die abschüssige Bahn einer Gewöhnung an schlechteres Leben. . . ." See *Ibid.*, pp. 377-378.

185. Otto v. Zwiedineck-Sudenhörst, "Gustav von Schmollers Gewerbepolitik," in *Schmollers Jahrbuch*, hrsg. von A. Spiethoff, 62. Jahrgang, II. Halbband (1938): 527.

186. *Die soziale Frage*, p. 198 and *Über das Maschinenzeitalter*, pp. 28-29.

187. *Die soziale Frage*, pp. 317-318.

188. Vogelsang was a publisher and editor. He was an early leader in the efforts in the German and Austrian region to reform social conditions on the basis on the basis of Christian social ethics.

189. Ralph H. Bowen, *German Theories of the Corporative State: With Special Reference to the Period 1870-1919* (New York: Whittlesey House, McGraw Hill Book Company, Inc., 1947), p. 89.

190. "In 1882 he helped to organize the *Vereinigung katholischer Sozialpolitiker*, a joint Austro-German committee to study social reform topics and make regular reports to the annual German Catholic congresses. . . . The results of the first two conferences at Haid (Bohemia) and Salzburg in 1883 offered conservative, totalist interpretations of the social question, rejecting the justice of the wage contract and demanding instead a corporatist reorganization of large industry and the elaboration of a 'social contractual relationship' (*Gesellschaftsvertrag*) between industrial workers and capitalist employers." John W. Boyer, *Political Radicalism in Late Imperial Vienna: Origins of the Christian Social Movement 1848-1897* (Chicago: University of Chicago Press, 1981), p. 173.

191. Karl Hümmer, *Der ständische Gedanke in der katholisch-sozialen Literatur des 19. Jahrhunderts* (Würzburg: Buchdrückerei Bavaria, 1927), pp. 94-100. "Vogelsang's ideal society was one in which horizontal class divisions would be replaced by vertical, hierarchical divisions of various broad economic sectors, the three most important being large industry, the small craft trades and shops, and agriculture." Boyer, *Political Radicalism*, p. 177.

192. Bowen, *German Theories of the Corporative State*, pp. 90 and 96; and Clemens Bauer, "Wandlungen der sozialpolitische Ideenwelt in deutschen Katholizismus des 19. Jahrhunderts," in *Sektion für Sozial und Wirtschaftswissenschaft der Görres-Gesellschaft*, hrsg. von Jakob Strieder and Johannes Messner, Heft 2 (Paderborn: Verlag von Ferdinand Schöningh, 1931), pp. 35 and 38-39.

193. Heinrich Pesch, ["Selbstdarstellung"] in *Volkswirtschaftslehre der Gegenwart*, Bd. I, hrsg. von Felix Meiner (Leipzig, Felix Meiner Verlag, 1924), p. 196.

194. Franz H. Mueller, *Heinrich Pesch: Sein Leben und seine Lehre* (Köln: Verlag J.P. Bachem, 1980), p. 42.

195. ["Selbstdarstellung"], p. 197.

196. This is an understanding which Marshall also accepts.

197. Heinrich Pesch, *Lehrbuch der Nationalökonomie*, 4. Bd., *Allgemeine Volkswirtschaftslehre III, 1 und 2*, Erste und Zweite Auflage (Freiburg im Breisgau: Herder, 1922), pp. 417-418.

198. *Ibid.*, IV, p. 399.

199. *Lehrbuch der Nationalökonomie*, 3. Bd., *Allgemeine Volkswirtschaftslehre, II*, (Freiburg im Breisgau: Herdsche Verlagshandlung, 1913), p. 99; and *Lehrbuch der Nationalökonomie*, 1. Bd., Grundlegung (Freiburg im Breisgau: Herdersche Buchhandlung, 1905), p. 370.

200. Heinrich Pesch, *Liberalismus, Socialismus und christliche Gesellschaftsordnung*, 2. Auflage (Freiburg im Breisgau: Herdersche Verlagshandlung, 1901), 1. Theil, p. 438.

201. *Lehrbuch der Nationalökonomie*, 2. Bd.: *Allgemeine Volkswirtschaftslehre I*, 4. und 5. unveränderte Auflage (Freiburg im Breisgau: Herder, 1925), p. 218.

202. *Ibid.*, IV, p. 339; and *Lehrbuch der Nationalökonomie*, 5. Bd., *Allgemeine Volkswirtschaftslehre III, 3-5*, Erste und Zweite Auflage (Freiburg im Breisgau: Herder, 1923), p. 630.

203. Gustav Gundlach, "Solidarist Economics" in *Social Order*, 1, No. 4 (April, 1951): 183.

204. *Lehrbuch*, V, p. 611.

205. *Ibid.*, IV, p. 526.

206. The person "ist niemals blosses Objekt oder Werkzeug sondern immer und überall Subjekt und Ziel der Wirtschaft und wirtschaftlichen Tätigkeit. . . ." See *Ibid.*, I, p. 17.

207. *Ibid.*, V, p. 563; *Ibid.*, IV, p. 330; *Liberalismus*, I, p. 448 and 602.

208. *Liberalismus*, I, p. 72.

209. *Lehrbuch*, II, 1925, p. 280. See also: ". . . Freiheit ohne Verantwortlichkeit, wie diese aus der sozialen Zusammengehörigkeit erwächst, ist eben Willkür." —*Ibid.*, III, 1913, p. 78 and (2.-4. Auflage) 1926, p. 9.

210. *Ibid.*, IV, pp. 382-383.

211. *Ibid.*, III, 1913, pp. 80-86 and 1926, pp. 11-17.

212. *Ibid.*, III, 1913, pp. 97-99 and 1926, pp. 27-29. The appeal of reason for coordination of the personal exercise of freedom Pesch explicitly bases on Kant.

213. *Ibid.*, III, 1913, p. 98 and 1926, p. 28.

214. *Ibid.*, II, 1925, p. 267.

215. *Liberalismus*, I, p. 48.

216. *Lehrbuch*, II, 1925, p. 265.

217. *Ibid.*, III, 1913, p. 57; not found in 1926 edition.

218. "Das Arbeitsverhältnis in eine richtige oder erträgliche Beziehung zum ganzen Leben des Arbeiters zu setzen, zu seinem körperlichen, geistigen, religiösen und sittlichen Leben, wie zu seinem Familienleben, dazu bedurfte es einer Beschränkung der Macht des Betriebes, wie sie eben durch die unveräusserlichen Rechte der menschlichen Persönlichkeit des Arbeiters, das Wohl seiner Familie und auch durch das Interesse der Gesamtheit gefordert wurde." See *Ibid.*, III, 1913, pp. 275-276 and 1926, p. 211.

219. *Liberalismus*, I, p. 595.

220. *Lehrbuch*, II, (Erste Auflage) 1909, p. 208 and 1925, p. 259.

221. *Ibid.*, V, p. 629.

222. *Ibid.*, III, 1913, pp. 289-294 and 1926, p. 228-230.

223. "Ökonomische Produktion ist . . . diejenige menschliche Tätigkeit oder Wirksamkeit, durch welche im Dienste unserer Bedarfsversorgung mit äusseren Gütern neue Substanzen bzw. substantielle Zusammensetzungen gewonnen werden, oder wodurch einem vorhandenen Dinge, Stoffe, Wesen, sei es eine akzidentelle Form, sie es einen neuen Zustand verliehen wird." See *Ibid.*, IV, p. 304.

224. *Ibid.*, IV, p. 322.

225. *Ibid.*, IV, pp. 510 and 518.

226. *Ibid.*, IV, p. 315.

227. *Lehrbuch*, IV, pp. 510 and 518.

228. *Liberalismus*, I, p. 169.

229. *Ibid.*, I, p. 174, n. 1.

230. *Lehrbuch*, V, p. 644.

231. *Ibid.*, V, pp. 610-611. Here Pesch follows the thinking of Lorenz v. Stein.

232. *Ibid.*, V, p. 644.

233. *Ibid.*, V, p. 637.

234. *Liberalismus*, I, p. 47; and Jacques Yenni, "Pesch's Goal of the Economy" *Social Order* 4 (April, 1951): 172.

235. *Lehrbuch*, II, 1925, p. 278.

236. *Ibid.*, I, pp. 41-45.

237. *Ibid.*, IV, p. 314.

238. *Ibid.*, II, 1925, p. 220.

239. *Ibid.*, I, 424.

240. *Liberalismus*, II, pp. 345-348.

241. Here, as elsewhere Pesch acknowledges his dependence on neo-Thomists such as Luigi Taparelli. *Lehrbuch*, I, p. 20.

242. *Liberalismus*, I, pp. 47-48.

243. *Lehrbuch*, I, p. 423.

244. *Liberalismus*, I, p. 125.

245. Work is ". . . wirtschaftlich das stärkste Bindemittel unter den Menschen." —*Ibid.*, IV, p. 421.

246. Pesch describes a "*soziales Arbeitssystem*," see *Lehrbuch*, II, 1909, pp. 190-228. Later, he describes it as a "*solidaristisches Arbeitssystem*," see *Ibid.*, II, 1925, pp. 213-284. In this way he removes a source of confusion between his system and socialism; he thereby highlights a central principle of his social conception.

247. *Ibid.*, IV, p. 422.

248. *Liberalismus*, I, p. 430; *Lehrbuch*, I, 430 and IV, 431.

249. "Vergesellschaftung der Produzenten"—*Lehrbuch*, II, p. 207.

250. *Ibid.*, II, 1925, p. 238. Here Pesch acknowledges an indebtedness to Viktor Cathrein, a major neo-Thomist moral philosopher.

251. "Der arbeitende Mensch inmitten der Gesellschaft, wo er die Ergänzung findet, deren er fähig und bedürftig ist, der arbeitende Mensch inmitten der Gesellschaft, der mit seinesgleichen arbeitet, *unitis viribus* in Arbeitsteilung und Arbeitsgemeinschaft, der Rücksicht nimmt auf das Wohl anderer, der sein eigenes Interesse dem höheren Interesse der Gemeinschaft zu unterordnen versteht." See *Ibid.*, II, 1925, p. 216.

252. *Ibid.*, III, 1913, p. 63.

253. *Ibid.*, II, 1925, p. 269.

254. "Sie sind Glieder der Unternehmen, auch in dem Sinne, dass die Unterordnung nicht nur für die Bereicherung ihres rechtlichen Inhabers, sondern zum materiellen Wohle aller in ihr Arbeitenden tätig sein muss." See *Ibid.*, III, 1913, p. 182; 1926, pp. 113-114.

255. *Ibid.*, III, 1926, p. 114.

256. *Ibid.*, IV, p. 437; III, 1913, p. 90 and 1926, pp. 20-21.

257. *Ibid.*, II, 1925, p. 218-219.

258. "Die sittlich-organische Gesellschaftsauffassung . . . erblickt in dem äusserem Zusammensein lediglich die Voraussetzung oder die 'Materie' speziell der staatlichen Gesellschaft, nicht deren 'formales' Wesen. Ihrem Wesen nach ist die Gesellschaft eine innerlich verbundene Gemeinschaft." See *Ibid.*, I, p. 372.

259. "Unterordnung unter ein gesellschaftliches Ganze mit Rücksicht auf den Zweck des Ganzen, auch positives Zusammenwirken für diesen Zweck, der mittelbar die einzelnen beglückt—das ist die Solidarität als Grundsatz, als sittliche Pflicht." See *Ibid.*, I, pp. 30-31.

260. *Ibid.*, I, pp. 371-372.

261. *Liberalismus*, I, pp. 435-436.

262. *Lehrbuch*, III, 1913, p. 822; 1926, p. 743.

263. Clemens Ruhnau, *Der Katholizismus in der sozialen Bewährung: Die Einheit theologischen und sozialethischen Denkens im Werk Heinrich Peschs* (Paderborn: Ferdinand Schöningh, 1980), p. 181.

264. Bowen, *German Theories of the Corporative State*, pp. 28-31.

265. Clemens Bauer, "Wandlungen der sozialpolitischen Ideenwelt," pp. 41-42.

266. "Er fördert die berufsgenossenschaftliche Organisation auf öffentlich-rechtlicher Grundlage, weil er auch für unsere Zeit in dieser Organisation ein wirksames Mittel erblickt für die Vertretung der Standesinteressen in Harmonie mit dem Wohle der Gesamtheit. See *Lehrbuch*, I, p. 383." Also see *Liberalismus*, I, p. 137.

267. *Liberalismus*, I, p. 547.

268. *Ibid.*, I, p. 509.

269. *Ibid.*, I, p. 93-94.

270. *Ibid.*, I, p. 510.

271. *Ibid.*, I, p. 577-578.

272. Richard E. Mulcahy, "Economic Freedom in Pesch," *Social Order* 1, No. 4 (April, 1951): 168.

273. *Lehrbuch*, II, 1925, p. 222; IV, p. 387; and I, pp. 206-207.

274. *Ibid.*, II, 1909, p. 788 and 1925, p. 719.

275. *Ibid.*, III, 1913, p. 95 and 1926, p. 26.

276. *Liberalismus*, I, p. 563-564.

277. *Lehrbuch*, III, 1926, p. 298.

278. *Liberalismus*, I, p. 92 (Pesch cites v. Hertling on this point) and p. 550.

279. *Ibid.*, I, p. 138.

280. *Ibid.*, I, p. 558.

281. *Lehrbuch*, I, p. 381.

282. *Liberalismus*, I, p. 592.

283. *Lehrbuch*, III, 1913, pp. 295-297 and 1926, pp. 233-235. Pesch carefully notes that workers have no legal claim on the basis of the work contract to profit sharing: *Ibid.*, V, p. 632. His argument depends on the nature of the collaborative partnership.

284. *Ibid.*, V, pp. 573-574.

285. "Die Arbeiter und Angestellten sind dazu berufen, gleichberechtigt in Gemeinschaft mit den Unternehmern an der Regelung der Lohn- und Arbeitsbedingungen sowie an der gesamten wirtschaftlichen Entwicklung der produktiven Kräfte mitzuwirken." Cited at *Ibid.*, III, 1926, p. 738.

286. *Ibid.*, IV, p. 385.

287. "Mitverwaltung, Ertragsbeteiligung und Eigentum muss für die Arbeiter ermöglicht werden. . . ." *Ibid.*, IV, pp. 413-414. Pesch refers to the authority of a letter of the American Catholic Bishops' in support of the broadening of the worker's position beyond that of a wage laborer.

288. *Ibid.*, IV, p. 336.

289. *Ibid.*, V, p. 673.

290. *Ibid.*, II, 1925, p. 278.

291. *Ibid.*, III, 1926, p. 739.

292. *Ibid.*, II, 1925, p. 223.

293. *Ibid.*, III, 1913, p. 270 and 1926, p. 205.

294. *Ibid.*, IV, p. 516.

295. *Ibid.*, III, 1913, pp. 83-86 and 1926, pp. 13-17.

296. *Ibid.*, IV, p. 306.

297. *Ibid.*, IV, p. 308.

298. *Ibid.*, IV, p. 333.

299. *Ibid.*, II, p. 225. Also see *Liberalismus*, I, p. 132, fn. 1, where Pesch makes reference to Aquinas in the *Summa Contra Gentiles* (Bk. 3, chap. 130) on this point.

300. *Lehrbuch*, III, 1913, p. 257-260 and 1926, pp. 193-195.

301. *Ibid.*, IV, 315.

302. *Liberalismus*, I, p. 167.

303. *Lehrbuch*, III, 1913, p. 91 and 1926, p. 22.

304. *Ibid.*, IV, p. 384.

305. *Ibid.*, II, 1909, p. 220.

306. "Da es der Staatsgewalt thatsächlich unmöglich bleibt, jeden Einzelnen für jeden Fall gegenüber der wirtschaftlichen Übermacht anderer wirksam zu schützen, so ist es das Naturgemässe, dass diejenigen, welche die gleichen Interessen haben, in der Gesellschaft auch zusammenstehen, um so einerseits ihre Schwäche durch die Vereinigung zu überwinden, andererseits eine solche positive Förderung ihrer speziellen Interesse durch die eigene, vereinigte Kraft zu erwirken, wie sie dieselbe unmittelbar von der Gesamtheit der ganzen staatlichen Gesellschaft nicht erwarten können." *Liberalismus*, I, p. 563.

307. Ruhnau, *Der Katholizismus in der sozialen Bewährung*, p. 167.

308. Bowen, *German Theories of the Corporative State*, pp. 52-53.

309. *Liberalismus*, I, p. 158. Also see *Lehrbuch*, II, 1909, p. 209.

310. *Liberalismus*, I, p. 559.

311. *Lehrbuch*, II, 1925, p. 265.

Chapter 4
Oswald v. Nell-Breuning: Work as Organic Activity

Collaborator with Pesch's successor Gustav Gundlach,[1] theological specialist who was commissioned to write much of the encyclical *Quadragesimo Anno* for Pope Pius XI, professor of social and economic ethics at the Hochschule Sankt Georgen and at the Goethe University, both in Frankfurt am Main—Oswald v. Nell-Breuning, in works published over the decades from 1932 to 1987,[2] has repeatedly examined work activity as it takes place within social structures. On the basis of this understanding, he has consistently argued for a type of workers' participation known as *Mitbestimmung* (codetermination). I shall present Nell-Breuning's theory of work, by examining in turn: his conception of the working person, the necessity for work, the social structure which supports work activity, its external requisite and, finally, the service to which persons direct this activity. Subsequently I shall examine his arguments for *Mitbestimmung* as they are established in his theory of work activity.

The Centrality of the Working Person

Nell-Breuning asserts that his anthropology differs from that of the earlier economists. Where they, especially the neoclassical economists, presented the person primarily as one who exchanged, for Nell-Breuning the person actively contributes to the entire economic process.[3] The human person is the agent and goal of the entirety of economic life.[4] Nell-Breuning, in fact, defines the economy in terms of human work; it is human activity by which persons realize themselves.[5] A crucial, positive dimension to the need for work which is often emphasized by Nell-Breuning is encapsulated in the scholastic axiom that every agent achieves its fulness in its own action (*omne agens agendo perficitur*).[6]

Human need includes the specific requirement for the activity of work. To exert themselves, to achieve something, to demonstrate their usefulness to themselves and others—for persons work activity serves each of these ends.[7]

Work is the privilege of beings who are intelligent and free.[8] In and through the workplace, persons can attain the fulfilling of their individual capacities for understanding and for responsible decision.[9] Since the entire person partakes in work,[10] human persons complete themselves by bringing each of their powers to fullness and thereby achieving personal individuality through work activity.[11] Indeed, in perfecting these powers the person comes to a harmonious whole.[12] Still further, through the exercise of their capacities, persons in society contribute to the well-being of each other.[13]

In the view of Nell-Breuning, persons reveal their individual subjecthood in work activity. As a person, the individual is to be treated as a subject, not solely as an object.[14] The more working persons realize, through their own activity, their position as subjects in society, the more they will demand that the order and shape of society reflect their roles as responsible subjects.[15] The loss of work expresses the loss of the subject role in society. The unemployed person is forced into the status of a mere object.[16]

To the uniqueness and independence of the human being must be assigned a dignity because each is not merely an individual, but more, a self-empowered person.[17] The exercise of his intellectual and volitional capacities in work activity manifests and secures this dignity.[18] In the economic sphere, Nell-Breuning locates the basic value of human dignity pre-eminently in work.[19] Because of their intrinsic dignity, persons are subjects not only apart from the economy, each within him- or herself, but also within it, because their efforts are intimately bound to themselves.[20] The freedom which each person is capable of actualizing through his or her work activity is the basis for this dignity.[21]

Integral to the freedom of working persons is their responsibility; thus they, as workers, engage themselves in a task not only to execute it but also to exercise their capacity for decision in the course of the process and in the shaping of the work situation.[22] Free persons can contribute toward the shaping of the conditions under which they work with each other;[23] and their individual responsiblity allows their fellow workers to call them to account. Because of this responsibility and because of their capacity to understand, their work must fit into a framework of meaning. In part this implies that workers must be able to claim as their own particular ends the tasks which they undertake in and for society.[24]

The Necessity and Limit of Vital Activity

Nell-Breuning recognizes the necessity that persons work for the sake of their survival. He emphasizes two important points concerning this: first, not only is each person forced to work for survival and security, but also, this is our common and mutual lot.[25] Second, the minimum required for existence is not merely physiologically constrained but is a culturally defined limit.[26] Further, the highest levels of our culture remain linked with the embodiment of the human worker.[27] Although the needs of persons are complex in their cultural breadth, the wage offered in the usual pressures of supply and demand in the "labor market" frequently does not meet physiological needs, let alone the more complex ones which would enable a person to partake in the cultural and political life of the community.[28] Although directed toward the maintaining and extension of life, work activity undertaken in order to meet the range of our needs can itself be threatening because it may expose the person to risks which threaten health or life. In both cases the particular modern situation of labor has introduced merely contingent limits to work activity.

The entire person is necessarily limited both in body and spirit. Human organs, incomparably more swiftly than machine parts, grow weary, grow old.[29] An inadequacy to many tasks and a limitedness in them characterizes the human person.[30] The personal achievement of a harmonious, whole self suggests that some limits must border and establish the pattern of work activity; persons may strive for too much or may pursue too great a development in one or another area to the detriment of the remaining personal spheres. The goal of wholeness governs the determination of which tasks are beyond one's limits. Whether one can pursue a particular result depends on the totality of duties and obligations.[31] Such an evaluation can take place for the individual worker and also within society, for the very limitation of persons indicates their sociality.[32] This principle of the determination of limits will gain in significance when I examine below the positive operation of subsidiarity. Given his/her limitations, the person achieves a harmonious, whole self in collaboration with others in society.

In performing their tasks, the various individuals must not disappear into such a group, even while they partake in activities which depend on that group and support it. Each individual's freedom must be shielded.[33] Just as the person must not exceed his or her organic limits, so too the social organism is limited vis-à-vis the organic processes of the individual. The negative dimension of the principle of solidarity guides a group or a society to secure a space for the individual's free and thus responsible action.[34]

When one considers the activity of free and equal persons within the complexities of historical, political and economic processes, the question of authority arises.[35] How is authority to be conceived so that it both maintains the stocks of sustenance required for work and ensures security in it as requisites for personal freedom? How can authority be conceived so that it does this yet still directs social processes away from chaos and toward the achievement of the common good? Nell-Breuning provides a nuanced, complex answer.

Without question, authority must be maintained in social processes, in particular, in the direction of joint work activities. More importantly, the proper exercise of this authority depends on an organic growth to the completion of decisions, a process developing from the lowest up to the highest levels.[36] At the lower levels, within the plant, the authority of the supervisors directs the efforts of their subordinates. At the higher levels, this authority takes on quite a different character, for here in the "labor market" the power of capital and the power of the united workers stand in opposition.[37] Should authority indeed be achieved on an organic pattern, then the members participating in the decision achieve a proper unity.[38] Should this authority not be the result of a continuous, participatory development from level to level, then its exercise is alien to the working persons.

It is important to recognize that free, working persons do not object to their dependence in the work situation but, rather, reject a dependence which is shaped by an authority foreign to the intelligence and responsibility of the workers.[39] Since such an authority would treat persons not as subjects but, rather, as objects, it presents itself as alien to them. In the case where the working person is treated only as an object, then the exercise of authority, since the personal participation of the workers does not inform it, runs the danger of affecting them as arbitrary or capricious choice.[40] The workers perceive such a distant authority as exploitative.[41]

Limited, working persons require an organic, participatory process of decision-making lest those who wield power exercise authority without attending to the needs and limits of persons in society. To secure this process of decision and the proper use of authority, the social structure must be shaped intentionally. Work activity is the basis for economic life and economic life requires mutual shaping by working persons.[42] In the company, one finds a living organism which could dominate its workers unless its legal constitution exactly reflects its life needs, explicitly including those of the persons which constitute it.[43] If such a shaping of the firm through recognized, legal structures does not obtain, then the means toward the economic project—that is, the means to secure human needs—become equivalent to the end.[44] Then economic activity loses shape, loses direction and grows without limit.[45] The loss of a comprehensive, actual objective (*Sachziel*) for the economic life of society represents a failure in the shaping of society's work life.[46] The members of society must design the exercise of authority in order to make possible the treatment of working persons as co-creators of the culture in which they participate.[47]

The Organic Structure of Work Activity

Each human person is oriented in his or her entire self—body and spirit—toward society.[48] The human person is a social being (*ens sociale*) who both needs life and work with others and is capable of them in order to bring him- or herself to completion.[49] Reciprocally, society is directed to the benefit of the individuals who constitute it.[50] Further, this social nature informs the work activity of each person.[51] In order to face one's own limits as well as the demands of external nature, the individual, private efforts to meet these personal needs, require the support of a society.[52] In this manner, working persons provide each other the opportunities for work activity.[53] In addition, work activity submits the naturally given society of persons to reconstruction through the intentions and actions of working persons.[54] Further still, labor has value in exchange for other goods only within a social framework,[55] where persons determine the value of effort according to its contribution in supporting and shaping the macroeconomic life process.[56]

The activity which may be called social action is not equivalent to the action of social structures.[57] These structures are not unities of material substance but of social organization; their activites proceed not immediately and necessarily from themselves but from the free activity of their members.[58] Social action denotes the activity of persons within society,[59] especially their work. Because persons as such are ethically responsible, the existential situation common to all persons leads to a shared responsibility.[60] The meeting of personal needs through work activity takes place in a solidary structure of mutual responsibility.[61] The principle of solidarity, as a basic principle of social living, guides the actual, social interchange that aims to realize this reciprocal responsibility.[62] Rather than ignoring the particular person by leaving him or her in isolation from others, especially within economic processes, the principle of solidarity means that, "all social living, so of course also every economy, proceeds from the individual and therefore also is, in the final analysis, for the sake of the individual once again."[63] As a result, these particular obligations are incumbent on society: the assignment of the proper work task to each, the insuring of the just sharing in production, and the securing of the development of each person.[64] Thus, solidarity must guide social action so that society realizes a proper, organic unity; such unity is a prerequisite for the just distribution of work product and the just remuneration of work effort.[65] Insofar as solidarity is a dynamic principle, it is a principle of development in society, the development of the common good. This common good is structurally organized and dynamically organizing.[66]

Work is social not only because persons coordinate their actions but also because through their work persons create and shape society itself. Work activity creates and re-creates culture.[67] Work and the processes of economics are cultural activities;[68] indeed they form the foundation upon which the higher activities of culture can take place.[69] Whoever works, then, cooperates in the creation and shaping of the entirety of human culture.[70] Work activity manifests a dynamism both in its traversing of a course toward provision of a good or service and in its transformation of the sociocultural matrix within which it achieves its goal. The processes of this social activity are both technical and qualitative. The latter distinguish the the economy as a locus of cultural activities. These activities make use of the technical processes.[71] Social interaction, exactly as the interaction of persons, requires values for regulating the processes which can achieve the betterment of the lives of those engaged there.[72] Nell-Breuning believes that those cultural values are particularly critical which secure the opportunity and dignity of work activity in a society.[73]

Work activity brings workers not only the possession of more things, objects for consumption and use, but also, and more importantly, enables workers themselves to become more, insofar as their work expresses and actualizes their participation in the corporate process.[74] However, the worker's inability to assume the role of a subject characterizes the actual wage relationship in which work activity takes place.[75] The course of history indicates that the freedom of the working person may become still more manifest in this changing situation of work effort as codetermination is implemented within the wage relationship and then new structures supplant, in turn, this pattern of association in work.[76] Work ought to function as a factor in the organization of culture and society. To this end those who engage in this activity must themselves be able to determine the pattern, the ends, the benefit and the position of their own activity.[77]

Both the plant and the firm, in the view of Nell-Breuning, are each constituted by a structure of associated persons. Not things but persons compose the firm.[78] The society of the firm consists not in a natural community but, rather, in a bondedness, the purposive and objective basis of which calls attention to the centrality of the work activity in this community.[79] This social relation based on the purpose of the firm implies a reciprocity between the firm and the workers: each is to benefit the other.[80] In particular, the firm must prevent the waste of human capacities and also secure opportunites for their development.[81]

Nell-Breuning conceives of the firm as the dynamic interaction of three poles.[82] Investors entrust their funds in the hope of an adequate return on their investment. Workers entrust themselves through a period of their work activity in hope of a lasting opportunity for work.[83] The entrepreneur gains his legitimization from both sides and is responsible to both.[84] The entrepreneur or manager directs the economic activity of the combination which he achieves as a result of the work contributions of some persons and the capital contributions of others. A recurrent misunderstanding of the nature of the firm[85] ignores this organizing and coordinating role of the entrepreneur, a role whose functions rank it as productive alongside those of capital and labor.[86] It is clear that the firm cannot endure unless it is directed by a unified and decisive entrepreneurial will, yet this will Nell-Breuning does not conceive as representative of the varied interests of capital and labor. Rather, the entrepreneur is obliged to seek the good of the firm and the common good.[87] Nell-Breuning distinguishes himself from the Marxist view of opposed and conflicting interests; the collaborative activity of persons associated in the firms recognizes the differences of interests, yet allows for these to be subordinated to their common interest.[88]

Upon this understanding of the firm, within which persons cooperate for the purpose of production and service, Nell-Breuning recommends a restructuring of economic units to correlate better with this special social nature of work. Nell-Breuning offers the occupational order (*berufsständische Ordnung*) as the concept for this social restructuring. In his work on *Quadragesimo Anno*, he conceived of the occupational order not solely for individual firms but, rather, on a macro-level.[89] Furthermore, the structures of the occupational order are not solely technical constructs which engage in abstract economic activity but, in addition, are associations of persons who mediate between the various economic units and the larger social structure. The social order suggested by Nell-Breuning attempts to avoid a dangerous mingling of private concerns with the public sphere and, at the same time, to shun the contextless separation of the public and private dimensions of a person's activity.[90] Despite the misuse of the concept by several fascist regimes in order to support policies of centralization, quite in opposition to the principle of subsidiarity upon which Nell-Breuning bases his concept, the various groups within the *Ordnung* strive, in fact, to further the life of society apart from the state, though under its aegis.[91] Not a department or branch of the state, not merely an economic entity, the occupational order organizes and provides a social structuring.[92]

Human work takes place upon a social armature which enables it to produce goods and provide services; through these twin economic functions, work re-works its own social structure. The solidarism of Heinrich Pesch provides the basis for Nell-Breuning's suggested restructuring; his occupational order traces its lineage to the organization of society which Pesch designated as a social system of work (*soziales Arbeitssystem*).[93] The social structure constitutes itself through personal work activity and also sustains itself by means of it. Through their occupations working persons actualize their individual responsibility for the community of which they are members.[94] Each person's occupational engagement unites him or her into a community of the firm and links the workers with the society as a whole.[95] The intention of the occupational order is to supplant the opposition of employers and employees through the structural articulation of the solidarity which is founded upon work activity itself.[96]

The responsibility of working persons in solidarity with one another as it takes effect through their work will reshape all levels of society. In the contemporary debates, Nell-Breuning grants that he concentrates his interpretation of *Mitbestimmung* so that it applies most immediately to the the plant and the firm. Nevertheless, he asserts that he conceives of these two levels, as had Pesch earlier, as part of a larger social scheme, which moves from the activity of the individual worker to an international, social system of work.[97]

In his later work, by using performance association (*Leistungsgemeinschaft*) as a substitute, Nell-Breuning sought to avoid some misunderstandings about the term occupational order (*berufsständische Ordnung*), which arose as a result of an alleged similarity between this *Ordnung* and the attempts to revive the guild tradition.[98] He emphasizes, however, in the interpretation of this phrase, that the focus is not on the achievement which results from the group's performance but, rather, on their collaboration itself. Central to the concept of the occupational order is the joining of both employers and employees in each work-society.[99]

Through the construct of this occupational society, Nell-Breuning moves from the oppositions and conflicts of society divided along class lines, to a cooperative activity which organically incorporates functional differences, without letting society deteriorate into a structureless amalgam. Not an alliance grounded in temporary feelings or passing demands only, as would be the case in a community of interests (*Interessengemeinschaft*), but rather, actual, executed work establishes this performance association (*Leistungsgemeinschaft*).[100] In the occupational order or performance association, Nell-Breuning presents his organic, social system of work activity; the sociality, solidarity, cultural transformation and participation characteristic of work activity inform this system.

The External Requisite for Organic Activity

When the person works, he must bring into his service natural resources, animals, tools and machines. Work requires nature, requires the interaction with the external, in order to complete itself. It also requires tools and machines in order to become more fruitful.[101] Human rationality, demonstrated in work activity by the reasoned pursuit of an end, takes up all dimensions of the embodied spirit which is the person.[102] By being taken into work activity, all these facets partake in human dignity. Uniquely, the person directs the efforts of machines and animals.[103] Uniquely, the person utilizes machines and animals to come to the fulfillment of his potentials.[104] Work implies an outpouring of the human spirit and of the responsible consciousness of the person[105] into resources and instruments. Thus, tools and machines become extensions of the person, because upon these externals, human deliberation and decision have been impressed. Machines relieve persons only of burdensome efforts, not of the exercise of their capacities for knowledge and decision.[106]

Repeatedly, Nell-Breuning refers with approval to Pope Pius XI's conceptual analysis of economic activity which views capital as *res*.[107] In this view capital is external to the activities which guide and accomplish the production of goods and the provision of services, yet is required for the achievement of these ends. Capital thus functions as a tool for attaining a maximal fruitfulness beyond the present. In order that persons may surpass the initial limits to their work activity, capital makes possible the extension of work into the future.

The technical instruments used in the work process have affected persons in their social interaction in multiple ways. While these productive, technical means have altered the coordinated activity of persons within the framework of the division of labor,[108] workers have become separated from these means of production. In the period when tools began developing into systems of machines, when workers at regular intervals and under clear discipline gathered in workplaces[109] and when the division of labor became determinative of social patterns, then it was that the danger of the subordination of the worker to the machine took shape in the place of work. As workers became separated from the means of production, they were rendered servants of those instruments. As the division of labor demonstrated its efficiency in production, the concentration of the means of production intensified. This concentration into larger technological units led to the elimination of the responsibility of the worker in even a small segment of the work process.[110]

The separation of workers from the technical means of production involves: (1) the ownership of the means by persons other than the actual workers; (2) the restriction to the managers of knowledge concerning the purposes and capacities of the means used in the work process; and (3) the elimination of the workers' contribution to decisions about the use, coordination, improvement and limits of these technical means in the work process and workplace. Nell-Breuning concentrates his analysis on this last point.[111]

Not in themselves but in the sociocultural framework[112] in which they stand over against working persons, the inorganic, instrumental means have given rise to a situation marked by the lifelessness of working persons in the course of and as a result of their labor.[113] Rather than being master of the machine and the system of machines, the person is rendered its servant or is treated as the equivalent of the machine.[114] The machine begins to serve as the guide for the kind, pace and procedure of the work to be done.[115] Thus the work process to which these instruments are critical becomes foreign to the worker.[116]

The problematic relationship of the worker to the system of machines presents a pivotal question. Nell-Breuning accepts the separation of the worker from the means of work achievement at least for the present, given the current structure of property relationships. He rejects an alternative where persons are simply recognized as inferior to machines, and he discards the romantic alternative where working persons decline to use machines at all.

What ought to be the alliance of working persons to machine technology, given the gap between them? As was indicated earlier, in the work system where persons are taken into processes as subordinate to instruments, they too become objects. Insofar as persons are thereby mechanized, the work system in which their active efforts are enframed stands alien to them. Alternatively, the working persons may subordinate the instruments to themselves. The work of cooperating persons in the firm will then become even more important relative to machines. The time has arrived in which, just as does a group of people united in a state, so the corporation as a living organism merits a constitution.[117] The present sociocultural framework for technology will be unable to equate economic life with technical decisions,[118] once the understanding reigns that the human bond formed through work activity calls out for a qualitative constitution of the system of work. Such an establishment would also encompass and govern the system's utilization of tools.

The organic alternative can be actualized, if the particular life and activity *(Eigenleben)* of the firm is recognized. Nell-Breuning understands the particular life of the firm as constituted by the union of persons who stand behind the several kinds of work effort which jointly create the effectiveness of the firm. In view of the actual heightening of the technical demands for coordination and cooperation, this union of persons ought to be constituted as a body. This does not remove the separation of workers from the instruments of production,[119] rather it offers a mode which integrates those instruments into an organic pattern of work.

Service: The Reflexivity of Work Activity

Work activity completes itself in service. How Nell-Breuning understands this component of work I shall now examine in terms of: functional relations, work activity both as production and service, subsidiarity and, finally, politics as an expression of social responsibility.

The social activity to aid society itself and the persons in it—both in their growth and at their limits—takes place as a set of interacting functions. Each occupational group (*Berufsstand*) or performance association assumes a particular function in society. This set of functional, organic structures secures not only the stability but, still more, the vital activity of society.[120] The classless society (*klassenfreie Gesellschaft*) organizes itself according to the various goods and services contributed by the different groups toward the common good.[121] Far from constituting a loose assemblage of social activities, these functions together provide actual service for society itself and for the sake of its members.[122]

Work activity entails a set of social tasks; this is to say, the cooperation of society furnishes not a tone of fellowship merely but, rather, a cohesive variety of assignments whose completion society requires for its further life. Persons undertake these multiple tasks for their mutual benefit, that is, for the benefit of other persons and of the society which maintains them. They expend work activity upon the achievement and maintenance of a harmonious life together.

Further still, Nell-Breuning declares that his functional approach reveals a better understanding of the firm as an association of persons. That is, it represents an advance on Heinrich Pesch's understanding of the firm as based upon property.[123] The firm itself achieves its ends by means of the coordination of the functions of management, capital and dependent labor. No longer is one to ask, "How are property in the means of production and the workforce of proletarians who are destitute of these means to be set in relation to each other, but rather how are persons as bearers of different functions to be united with each other in an efficient unity?"[124] An advance on the West German program of codetermination offers a practical way to bring the persons together by way of their activity conceived as functional contributions to the firm.

The highest value resulting from human activity is the value which derives from service. Work, insofar as it is a meaningful activity, satisfies the working person and provides benefit to others.[125] In the performance of service, work most clearly shows its social character.[126]

Production is not a value in itself; rather, the product fabricated in work activity must meet human needs. On the contrary, when productivity becomes the chief value, not only are human needs unmet; but, further, persons themselves are deprived of their humanity.[127] Lest that be the case, work must complete itself in the service of persons, of each person and of the whole person.[128] Nell-Breuning does conceive the productive work activity of society as a kind of service. Because the work activity entails a goal-direction toward provision for the community beyond the area for the exchange and alteration of materials, the production of goods reaches into the economic sphere as a dimension of human culture.[129] In the object created, and more clearly in the aid provided, what is produced lends an ethical value to work insofar as it serves the needs of the community.[130] Indeed, the partial contribution to the completion of a good or service characteristic of a single task within the division of labor also counts as a service to the neighbor.[131]

Productive work operates only as a single function within the activity of society; it has value only within the economic framework which society has established so that such productivity might be to its benefit. Firms are the primary agents for undertaking this function in the modern economy. As was seen earlier, the firm cannot become an entity in itself, a structure of juridical and technical sub-components.[132] Rather, the firm is an association of persons which stands, in turn, as a member of the larger community. The firm's purpose, then, ends not with production but with provision to the community, that is, service. In addition, the firm's purpose does not end with the maximization of profit to secure a high return on capital investment. The return on capital, "must rather serve the personal well-being of the persons active in the individual firm or in the economy as a whole and thus recede and subordinate itself in the case of a conflict of goals or interests."[133] Capital and its functional significance are hardly denied in Nell-Breuning's system; rather, that activity which seeks a return he assesses in terms of service.[134]

Nell-Breuning is careful to note that the fruit of work activity in the modern economy more often takes the form not of produced objects but of services provided.[135] The provision of service to others and to society stands in these cases as both the result of work activity and its explicit intention. Profit expressed in terms of money is a socially determined and readily comprehensible measure of this service. Workers have a claim to their full share of their own product, understood, in view of the range of service activities, not as of share in objects produced but, rather, as a full share in the profit measured in terms of money.[136]

On the strength of their work activity, workers acquire a claim to service. In terms both of productive and service activities, the members of a society have a claim to the service represented in the social "product," a claim based on the work performed in that society.

The listing of functional segments of society, which Nell-Breuning offers, reveals the pre-eminence of the services: the educational system, the health system, research and higher education, literature and publishing, broadcasting, the press, art, the free professions and, as one sphere among the others, "the economy with the production of goods, the system of credit, and the system of transportation."[137] The preponderance lies with the so-called "non-productive" activities which are better denominated "service" activities because of their significance in supporting the entire society in its economic and higher cultural life.[138] The economic realm as such contributes to the developing life of society. Rather than existing as an independent sphere, it serves the whole.[139]

Of special interest are the services (*Dienstleistungen*) offered by the professions. These service functions require on the part of their practitioners the mastery of particular knowledge and a special responsibility.[140] The work contributed by these professionals forms not merely service in some general sense but specifically service to persons in their intimate and manifold needs.

Nell-Breuning regards the service component of work as related to the basic principles of social ethics, because it expresses the beneficent side of subsidiarity.[141] Subsidiarity as helpful assistance (*hilfreicher Beistand*) offers support to those who cannot help themselves past a certain degree. To interfere where the person would otherwise have the chance to act and thereby develop, would be an injury.[142] Where the working person is capable of exercising his or her own powers, there the community must refrain from interference.[143] Where individuals face their limits of inability, there ought the community provide aid, as service to those who need. Furthermore, where the persons are capable of acting but where their powers are insufficient, there the community ought to offer only that support which is sufficient to fill the deficiency in its members' capacities.[144] The service to those at the limits applies not only to needy individuals but also to smaller, needy groups which can be served in their limitations by a larger community.[145]

The common good emerges from the organic life of the community. The organic life of the community itself requires the provision of assistance to society's members so that the achievement of their ends will return to the common good.[146] Similarly, the growth and limits which condition the life of any organism form key targets for service activity; that is, not only must working persons face and shape their own limits, but also society itself must confront and configure its own limits. Through work activity persons shape and maintain the values of culture.[147] The values of the culture emerge pre-eminently from the efforts carefully to attend both to the society's growth and to the limits of this growth.[148] The execution of this care in service completes the order required in society.

Since work is an activity in freedom, the working person assumes a responsibility.[149] He or she assumes responsibility for the world of work, which is by no means a narrow field but, rather, encompasses the physical, personal and institutional environment in which work activity takes place.[150] Through their individual work activity, the members of society assume this responsibility vis-à-vis those affected by their efforts.[151] Their directing this freedom towards service constitutes their work activity most fully.[152] The worker as such is liable—especially as a collaborator in a working group—for the provision of sustenance for all circles of society.[153] The work group mediates the responsibility for society in two directions: (a) the individual's responsibility for society and (b) a reciprocal obligation.[154] Reciprocally, the whole of society must take responsibility for the persons who constitute it. The organic social order allows to individual persons and secures for them particular positions from which they may exercise their freedom. Likewise, the groups within society are particular instances of the service of the individual by society; indeed, the product or service provided to the members of society is the basis for the delimitation of a group, its unification and organization.[155]

A central task for the constituting of society itself as the work-world in the wider sense (namely, as the framework for the functions undertaken in work) is the political activity of persons. Such political activity includes a wide range: where questions of power and its legitimation are being decided, there political activity must be at work. Nell-Breuning recognizes the political dimension of work activity within the economic sphere. Given Nell-Breuning's understanding of the organic integration of society, of work activity and of the functionality of the work undertaken, one can see that the participatory interaction of the members within each of these groups contributes to the life of the entire society. In this sense, participation within the firm furthers the provision of service. The contemporary transformation of work has reduced the need for transitive labor, that is, labor which produces or directly alters material objects, thus freeing persons to engage in activities previously restricted to the few: service activity accomplished on behalf of the community.[156] Indeed, Nell-Breuning holds that the development of a free citizenry and of democratic principles in the political sphere both points ahead toward a democratic organization of work and finds its completion in the field of work.[157] Democratic participation represents that form of political activity which most effectively responds to the demands of personal and social limits and which most aptly respects the organic structure of society.

A continuity stretches from the immediate ambit of the working person to the highest form of social life—government itself. The organic understanding of society implies that a reciprocal relation exists between center and parts. Thus, guided by the principle of subsidiarity, as was presented above, assistance constitutes a significant movement from the center toward the parts. Guided by the principle of solidarity, responsible and full use of capacities denotes the reciprocal contribution of the parts to the whole. The organic structure of work activity in society makes possible a participatory democracy in the economic sphere, which, in turn, supports genuine democracy in the realm of government and politics.[158]

Codetermination

The West German federal laws touching codetermination structure the relationship between employers and employees. Nell-Breuning has often articulated and argued for this complex system of structures for participation in the economic sphere. The structures of this relationship offer a middle option between workers' self-management[159] and the more familiar collective bargaining.[160] A tripartite organization characterizes the typical German firm: the assembly of shareholders (*Hauptversammlung*), the supervisory board (*Aufsichtsrat*) and the board of management (*Vorstand*). A small group of directors, appointed by the supervisory board constitutes this managing board, which attends to the ongoing business of the firm. West German codetermination does not touch all of these levels, yet it does include another structure, the works council (*Betriebsrat*) which pertains to the realities of the plant as distinct from the legal constructs of the firm.

The nineteenth[161] and early twentieth[162] century tradition of codetermination in the workplace and in the firm has been articulated since the establishment of the Federal Republic in several laws. Of these the most important are: the Act for Codetermination in the Iron, Steel and Coal Industries (*Montanmitbestimmungsgesetz*, 1951); the Works Constitution Act (*Betriebsverfassungsgesetz* 1952); the revised Works Constitution Act (*Betriebsverfassungsgesetz* 1972); and the General Act Concerning Codetermination (*Mitbestimmungsgesetz* 1976). The laws of 1952 and 1972 chiefly concern the establishment of works councils and the specification of their rights and powers. Such councils handle matters concerning grievances, information rights, job evaluation, employment policies, schedules and layoffs.[163]

Those laws which introduce codetermination into the corporate structure do so along two general approaches. In some instances (1951 and 1976) the law prescribes the appointment of a member of the managing board, the labor director (*Arbeitsdirektor*), to represent employee interests. In addition, often even when there is no labor director,[164] some members of the supervisory board—the proportion varies from one-third to one-half (the latter is the so-called "parity" codetermination) depending on the relevant law and the size and legal type of the firm affected—represent the interests of the employees. In addition to this variety, again depending on the size of the firm and its legal type, codetermination may affect particular companies only at the level of the plant and not at all in their corporate structure.[165]

When Nell-Breuning, early in his career, as a participant in the *Königswinterer Kreis* research group, studied socio-economic structures and considered possible social policies with students of Sombart,[166] he learned to construct from the epochal, capitalist phenomenon an idealized abstraction of the functional relationship of capital and labor.[167] Influenced by his own reading of Pesch and by regular contact with Pesch's student and successor Gustav Gundlach, Nell-Breuning's earliest formulation for the reconstruction of this functional relationship was the occupational order, especially in his writing connected with the encyclical *Reconstructing the Social Order* (*Quadragesimo Anno*). Later, he reinterpreted this as the performance association. Further still, the contemporary variant for which Nell-Breuning has argued at great length is codetermination (*Mitbestimmung*).[168] Nell-Breuning acknowledges his dependence on Pesch and yet clearly sets himself further along that line of Catholic social teaching. He believes, however, that two older writers (Pesch and H. Roesler—who delineated a social system of work) saw firms as established only upon ownership in things.[169] Moving away from the physicalist basis of one strand of the natural law tradition, Nell-Breuning focuses instead on persons and their activities. With this shift in emphasis, he then argues for a system of codetermination wherein working persons can together direct their joint activity and where persons achieve a predominance over objects and ownership. Nell-Breuning so firmly upholds the dignity of the working person and the priority of persons to things in the immediate workplace, in the firm and in the economy, that, although he has supported codetermination, he remains by no means satisfied even with his own previous proposals. Speaking of the technical composition of the supervisory board, he criticizes that legal variant which relegates one group in the firm to a subordinate role and then continues with a criticism of his own previous approach which was based on the supervisory board which represents the owners of wealth.[170] He looks beyond the present legal structure to a system based on labor activity (*laboristische Ordnung*)[171] where the organization of the workplace and corporation will reflect the pre-eminent value of work. In his 1983 commentary on the encyclical *Laborem Exercens*, Nell-Breuning favorably examines the socioeconomic model of laborism and continues to affirm that workers must assume the roles of leadership and initiative in their work.[172]

Codetermination and the Working Person

Nell-Breuning justifies codetermination by observing that it assists in the unfolding of the human personality.[173] That aspect of *Mitbestimmung* which focuses on the person must be emphasized. Persons ought to be guaranteed freedom and responsibility in their work so that both these capabilities may develop in them.[174] Codetermination secures for workers room for the workers' own decision-making and responsibility, especially at the immediate levels of workplace and shop floor.[175] Just as it calls the directors of the firm and its supervisors in the shop to responsibility, so *Mitbestimmung* enables the simultaneous and equal responsibility on the part of all the employees.[176] At the level of the firm, economic participation implies that the workers will exercise a prudence concerning the demands of the future[177] and thereby grow more responsible for the use of their finite capacities over time. Workers' participation also protects human dignity as it is capable of being actualized in personal freedom.[178] The parity which shapes effective, economic codetermination at the level of the firm secures the participation of the working persons in decision-making.[179] Parity guarantees the protection of the the worker from being coerced into an object role in the firm.[180] The particular force of *Mitbestimmung* is aimed directly at this point: the elimination of heteronomous control over the workers as objects so that they may become at least subjects equal to others in the firm.[181] *Mitbestimmung* enables the person to move from an object to a subject role in the workplace and firm.[182] Through workers' participation, the person can partake in the formation of social will. On the economic level, the set of social decisions necessary for the formation of capital investments constitutes a key instance of this.[183] In its very structure work activity is dynamic because it continually transforms the workers and because their work participates in the formation of ongoing social decisions. Codetermination makes possible the participation of these subjects concerning investment decisions;[184] by this means, workers can achieve economic subjecthood in the historical perspective, that is, over time, not just in the immediacy of consumption decisions. Codetermination shields finite persons from the danger of a dominating authority, while fostering the development of these persons and their growth to organic completeness through the exercise of responsibility and its consequent growth.

Codetermination and the Limits to Work Activity

Nell-Breuning argues for codetermination in terms of the need to work, with particular attention to the limits which are imposed by authority upon the person's work activity and with regard to the necessary shaping of a culture by limited, working persons.

Although codetermination is insufficient to guarantee their position of employment,[185] it does offer workers a greater say in efforts to secure their work.[186] Codetermination requires a reciprocal responsibility:[187] on the one hand, the workers must attend to the need for technical changes and for high productivity; on the other, the managers must answer to the workers for the security of their jobs. In the face of the demands and dangers of work itself, codetermination can contribute to the securing of protection for the worker at the workplace itself. It can provide a means whereby procedures are regulated in order to reduce the risk of accidents.[188] Further, codetermination could lead to the elimination of the cyclic tendencies in extreme, liberal capitalism which prevent full employment.[189] *Mitbestimmung*, then, makes it possible that the required rationalization of work may advance with the utmost attention to personal circumstances.[190]

Both the plant and the firm are larger units in which persons join for common tasks.[191] The constructs in which the instruments of machines and finance interact must not inhibit the actual, active collaboration of persons.[192] The combination of persons at the level of the plant or firm, especially in view of the decisions required for the achievement of corporate ends, raises questions about the nature and use of authority and responsibility. The separation of the worker from the means of production widens the distance of the individual or groups of individuals from the points of decision necessary for the achievement of common goals through a corporate structure. The socialization of the means of production, producers' cooperatives, partnerships and redistribution of wealth all miss a key aspect of the problem of separation: these solutions touch the question of property but do not approach the articulation and control of authority and responsibility within the systems of the plant and firm. Not in order to avoid the problems of property, nor to subvert progress in redistribution but, rather, in order to support solutions attempted in these other spheres, *Mitbestimmung* addresses itself precisely to the problematic of decision and control concerning systems which threaten—insofar as they are disjoined from contact with and control by workers—to ignore the needs and limits of those persons.

Mitbestimmung recognizes the actual composition of the firm: finite, working persons united toward a particular set of goals. *Mitbestimmung* shapes the legal constitution of the firm to correspond to this reality.[193] At the level of the plant, codetermination makes the directorial power constitutional; at the level of the firm it constitutes this power.[194] The emphasis here lies not on the planning of the technical processes involving labor but on the organization of those persons who compose the firm, "The constitution of the firm makes the association of persons a framed structure within the regulated and legal system.[195] The constitution of the social reality of the firm within a legal order represents a step toward recognizing workers as potential subjects in their individual relationships with the firm;[196] for a legal relationship, as a technical arrangement of resources would not, allows for and structures the personal interaction of the members of the firm.

When the employees partake in the decision process, this entails a participation over time in the several, integrated steps leading to the full decision.[197] In this way, finite persons join together to shape a culture out of materials which outlast the individuals who contribute to the process. Embodied and limited persons require for their full development a shaped social field for their activity. This must not merely be the de facto organization but one which reflects a formation by the intellects and wills of persons.[198] The firm, exactly as a social structure, that is, as a living organism, must be brought into an organic form.[199] *Mitbestimmung* offers significant advantages to working persons through the shaping of the community of those who work in the plant; it gives to personal development a space situated in a cooperative project which is comprehensible and appropriable in terms of one's experience.[200] Indeed, *Mitbestimmung* provides an organic configuration of the whole of society, because, just as in the model of the incorporated performance associations (*leistungsgemeinschaftliche Körperschaften*), so now the constituted firms offer a mediating factor between the spheres of private life and official (*staatlich*) life.[201] The procedures and organizations of *Mitbestimmung* integrate the varied social bodies and eliminate the discontinuities between them. In this way embodied and limited persons face institutions with which they are integrally connected and can also utilize the range of their capacities at all levels of society.

Nell-Breuning presents arguments for *Mitbestimmung* upon the basis of his understanding of work.[202] Under the principle of solidarity, work is an activity directed toward the common good. Secondly, work fashions the society and its culture. This formation of culture is most fully social insofar as the workers are incorporated into the processes which determine social decisions.

The reasons for *Mitbestimmung* proceed not from the nature of the wage contract but from reasons of public good.[203] Since the activities of a complete system of codetermination take place both within and outside the firm,[204] the continuity of these activities (just as the corporations in the occupational order mediated the economic and political realms) offers the potential for one such benefit to the public. The commonweal would be advanced because this participatory, structural arrangement could facilitate the political realm's supervision over powerful firms.[205] The structures of codetermination would take into consideration both the interests of the particular firm in which these structures function and the general interests of society.[206]

The secrets kept by management about the actual condition of the company have often caused mistrust to color the association of persons who constitute the firm; codetermination, on the contrary, would help maintain trust, since knowledge, as well as responsibility, would be shared.[207] Trust sustains the firm: investors must entrust the management with their funds, while workers must place their trust in management's ability in the workplace and in the corporate world to insure that their employment remains secure.[208] Codetermination would, then, contribute to the stability of the firm as an association of persons. Codetermination's substitution of cooperation for class conflict provides yet another benefit: the strain of strikes is diminished and their number reduced.[209] Indeed, *Mitbestimmung* makes contemporary still another characteristic of the occupational order because it attempts to defuse the employer-employee conflict whose epicenter lies in the so-called "labor market."[210]

Mitbestimmung actualizes the participation of working persons in the social processes which create culture. Clearly codetermination is distinct from the socialization of the means of production; indeed it is not even a means toward that end but, rather, a superior alternative.[211] When arguing by an analogy to the political realm, Nell-Breuning, in his appeal for the application of democratic principles in economic units, explicitly avoids using the term economic democracy (*Wirtschaftsdemokratie*), since it carries too many historical connections with socialist movements early in this century.[212] He, nevertheless, describes the advance from subjection to subjecthood, from subordination to participation as a progress for the person from the state of economic vassal (*Wirtschaftsuntertan*) to that of economic citizen (*Wirtschaftsbürger*).[213] Nell-Breuning puts forward the idea that the introduction of this means for participation in the larger firms, in such a way that the participants representing both capital and dependent work share the object and subject roles, would, in turn, enable democratic processes to develop through the whole of society.[214]

Grounded upon the principle of solidarity, *Mitbestimmung* provides a structure for participation in the workplace, factory and firm.[215] Codetermination secures the position from which workers speak, so that their voices may have effect upon decisions.[216] Individual persons, insofar as actions proceed immediately from themselves, legitimate those actions in and through their own free decisions. If, however, they perform these actions at the behest of an authority, they must legitimate this authority. Such legitimation proceeds only from the working persons themselves.[217]

Codetermination and Technology

Codetermination, by incorporating the responsibility of each worker into the direction of the work system, precludes the workers' subservience to the machine.[218] Codetermination enables working persons to forestall or correct any situation wherein they are treated as subordinate to present technical instruments—be these machines or accounting procedures—or valued solely as substitutes for machines which are still too costly.[219] An organic pattern of work allows the workers together to be master of the technological system.

Codetermination and Service

When Nell-Breuning elucidates the organic reflexivity of work in terms of (1) function, (2) service and (3) subsidiarity, he also uses these as a further support in his arguments for codetermination. The service to persons achieved in assuring their places in the work community and in society allows them to be responsible citizens in both these areas.[220] In turn, the functioning through participation on the part of citizens secures the dynamic order of society. The integrating of persons into particular functions shapes the society, serves the society and serves the persons so integrated.

Codetermination, in its provision of structures for the responsible interaction of all participants in work, enables the workers to recognize the service inherent in their work contributions. That is, the workers, on the basis of their participation in codetermination, come to see their own work activity not as part of an abstract system of exchange but, rather, in terms of the service provided to themselves on the basis of their work within the firm and, similarly, to other persons of the society through the products and services provided by the firm.[221]

Mitbestimmung supports a better regulation of economic processes, a regulation which is based on human needs and interests and which reaches its completion through the interaction of the several levels of economic action. Further, codetermination serves persons because it enables them, through their participation, to come to a fuller knowledge about these processes and their own role in them. For example, codetermination would enable a broader range of workers to learn responsibility concerning the difficult social task of capital formation.[222] Codetermination provides, in addition, structures wherein a close collaboration between working individuals and the other partners in the work of the firm may grow; the working persons of all levels who constitute the firm may in this way come to see themselves as an association of persons. Grounded in this experience, they may then more readily direct their joint efforts so that they make decisions in terms of the firm's prime orientation toward the persons who constitute it. Similarly, should codetermination at the sector and national level be achieved, then society as a whole will more securely attain the goal of serving both the needs of others in society and those of society itself through corporate efforts—all grounded in personal work activity.

Because codetermination allows for the voicing of particular needs, the representation of personal and local requirements, it more effectively fosters the positive dimension of subsidiarity. Because persons can themselves announce their needs, *Mitbestimmung* bolsters efforts to avoid centralized intrusion. Individual workers and groups of workers present and declare their capacities and needs because this structure of participatory decision-making is ongoing, rather than periodic, and interactive, rather than unidirectional. The system sustains both the initiation of novel notions, as yet unimagined at a particular location, and the response on the part of the workers to this initiative.

Nell-Breuning himself recognizes that the *Mitbestimmung* of the Federal Republic of Germany has not developed far enough to actualize the socio-economic structures which were urged in *Quadragesimo Anno*.[223] Still further, he believes that, although at present no practical scheme exists, those who enter the economic sphere through their work activity should attain the superior position in economic organizations with regard to the decisions made and benefits resulting from the processes of production and service.[224] Such a "laboristic" system would best meet the dignity of work activity.

What are the main elements of Nell-Breuning's understanding of work activity? In his system of thought, the intellectual and volitional capacities of persons both manifest and secure their dignity. A particular social whole culturally determines the limits and needs which impinge on these persons. Within the social whole, the powers which direct society must wield their authority so that they remain in touch, on the one hand, with the requirements of persons and, on the other hand, with the capacities for intelligence and responsibility on the part of all persons.

Work activity constantly re-creates the culture to which the structures of society give shape. Organized so as to secure the participation of workers, plants and firms are pivotal structures in society. Machines and machine systems—located within plants and controlled by firms—are the results of work activity. The human will and mind impress themselves upon the material which makes up machines and machine systems. In the face of human limits and needs, such capital helps persons cooperatively to reach a fruitfulness which extends beyond the present. The properly constituted workplace integrates technology into the organic structure which workers compose and activate.

Along several lines of analysis, work activity evinces service. Work creates a return which embraces the provision for the needs of the workers themselves. It requires the coordinated functioning for the common good; it provides for the community, including the culture of those who constitute it. Those activities and professions which maintain, build or assist the organic life of society, especially the proper development of lower levels, manifest service in particular clarity.

Codetermination emerges from Nell-Breuning's understanding of the working person. The workers as equals are responsible for the unfolding of their own selves; they also take up responsibility for decisions within the firm and through this instrumentality forge and fashion their culture. This structure of production and recognition incorporates workers in social decision for the common good. This complex system makes it possible for workers to avoid subordination to the machines and to take on an effective control of the technological system. In addition, codetermination shapes a space where workers' voices can be effective and serviceable in recognizing others and in regulating subsidiarity.

The design of an economic system ought to secure (a) efficiency, (b) the desirable, moral and intellectual qualities in the participants, (c) economic freedom, (d) justice and (e) democratic process.[225] Such a design requires a conception of the work activity of the person at the center of the system, and it requires that such a concept comprehensively explain the work which maintains and advances the system. The concluding chapter will present such a conception, a constructive theory of work activity which I have developed through the analysis of the British and German traditions examined in the four preceding chapters. I situate the theory in terms of the preceding discussion by briefly offering here some criticisms of Nell-Breuning's arguments for codetermination and their implicit theory of work activity.

Perspective on Nell-Breuning and Codetermination

Nell-Breuning's concept of the working person so depends on the debates found in the modern German industrial situation that it appears to identify work with activity done in the course of employment. Such a view would overlook children, the infirm and the elderly—those who either have yet to seek work or those whom work has already formed. Moreover, although Nell-Breuning stresses the importance of "being" over "having" and although he underlines the significance of subjecthood, the import of "doing," that is, of free activity in production and recognition, requires further elucidation.

Nell-Breuning tends to overlook the pressure of human limits because he does not focus on the situation where they impinge most immediately, namely on the shop floor and in the codetermination structure which governs there, the councils (*Räte*). Because he has argued for so many years and with so much effort for codetermination on more comprehensive levels, the emergence of this difficulty is understandable. Nonetheless, it is critically important in a theory of work activity to pay heed to embodied persons in their immediate surroundings. There it is that they first wish their individual voices to be heard. There it is that they begin to learn responsibility. There it is that they see how their own efforts initiate the entire cooperative action of work in society and thereby begin to learn how this collaborative effort can carry them past their own time-bound limits into a common history.

Nell-Breuning, although aware of the other, possible, broader levels of codetermination, concentrates on the plant and company levels. As a result, the danger emerges that the recognitive element in work activity will not assume its proper scope—the whole of society. Consequently, a company egoism over against other firms, a failure to recognize the unemployed as potential co-workers or a blindness to the neighbors of the locale who do not work in a particular plant, may arise as instances of this threat.

Nell-Breuning self-critically understands the need to reconceive the corporate side of codetermination.[226] He recognizes, furthermore, that codetermination on levels above the corporation—concerning, for example, regional and national bodies—would plainly manifest the force of work activity in shaping the entirety of civilization.

Although certainly an advance in the organization of workers' participation, Nell-Breuning's proposals do not move far enough ahead in terms of the degree of participation—save for some references to and general recommendations concerning a scheme of "laborism." *Mitbestimmung* remains at the level of consultation and limited control. It does not reach joint- or self-management.[227] The evolution of the theory of codetermination requires the exploration of the quality of participation; more specifically, it requires that new programs for codetermination meet the assumptions and criteria of democratic process[228] (for example, in terms of the setting of agenda or equality of voice).

An incorporation of the understanding of technology available in the fuller concept of work activity would strengthen the theory behind codetermination. Work activity as innovation, although mentioned by Nell-Breuning,[229] does not receive the attention it deserves. Through their individual work, persons discover. As an instance of this, they discover new procedures for implementing the task at hand more efficiently. The activity of discovery (which proceeds from the range of each person's capacities (including the individual's manual dexterity and intelligence) provides a critical connection between working persons and the devices, tools, machines and systems of these which result from their activity. Personal and collaborative work activity gives rise to technology. To the information about technological systems and devices, working persons have a claim since these proceed from their own activity. Such a claim subverts the otherwise unexamined privilege assigned to alien, technical structures and expert knowledge and counters the subordination which they impose. Once the information about the varied technologies which shape the workplace becomes available to the workers (although this availability does not imply an actual, technical knowledge on the part of each worker), they then possess the prerequisite for their own informed and responsible decision-making. These decisions can then lead to a technology better adapted to the desires and limits of working persons.[230]

Finally, a fuller analysis of service activity as a constitutive element of work would complete Nell-Breuning's concept of work activity. Without question, Nell-Breuning understands the subjecthood of the free person, but he does not specify the orientation of this freedom in terms of the limits and needs of others. Further research might also explore the connection between the training in participation which occurs in the workplace and active participation in the civic world. Still further, an exploration made of the fulfilling of service activity in political action could lead to the elimination of the nervous gap between federal and union politics, on the one side, and work activity in the plant and company, on the other. I shall now present such a exploration, which derives from the thought of the economic writers I have hitherto examined and which can respond to the challenges placed by advances in technology and by contemporary demands for participatory structures.

1. Gundlach was a student and collaborator of Pesch from 1924 in Berlin while, at the same time, he studied with and wrote his dissertation under W. Sombart, Schmoller's successor. See his "Lebenslauf" in his dissertation, *Zur Soziologie der katholischen Ideenwelt und des Jesuitenordens* (Berlin: Emil Ebering, 1927), p. 120.

2. Nell-Breuning's first publication appeared in 1924. His dissertation on the morality of the stock market was published in 1928. His first work which deals explicitly with work activity dates from 1932.

3. *Wirtschaft und Gesellschaft Heute*, 3 Bände (Freiburg: Herder, 1956-1960), Bd. 1, pp. 171-172.

4. *Ibid.*, Bd. 1, p. 208.

5. "Wirtschaft ist das menschliche Tun, bei dem der mit Vernunft und Selbstbestimmung begabte Mensch sich der vernunftlosen Schöpfung, der in ihr liegenden Möglichkeiten und in ihr waltenden Gesetze als Mittelsystem bedient, um sich selbst als Geist-Leben-Wesen zu verwirklichen." See *Gerechtigkeit und Freiheit: Grundzüge katholischer Soziallehre* (Wien: Europaverlag, 1980), p. 143.

6. Oswald v. Nell-Breuning und Hermann Sacher, Hrsg., *Wörterbuch der Politik*, Heft 3, *Zur sozialen Frage* (Freiburg: Herder, 1949), Sp. 119; and Oswald v. Nell-Breuning, *Baugesetze der Gesellschaft: Gegenseitige Verantwortung—Hilfreicher Beistand* (Freiburg im Breisgau: Herder, 1968), p. 82.

7. ". . . Haben wir Menschen das Bedürfnis, etwas zu tun, nicht nur irgendwie unsere Kräfte zu regen, sondern etwas zu leisten, das uns als achtbare Glieder der menschlichen Gesellschaft erweist und damit zugleich unser eigene Selbstachtung bestätigt; wir haben das Bedürfnis etwas Vernünftiges, Rechtschaffenes zu vollbringen, etwas, das die Anspannung unserer Kräfte erfordert, sie aber auch lohnt. . . ." See "Arbeitslosigkeit: Unser aller Recht auf Arbeit," in *Kirche in der Gesellschaft: Der katholische Beitrag*, hrsg. von Jürgen Wichmann, Geschichte und Staat, Bd. 220-222 (München: Gunter Olzog Verlag, 1978), p. 53.

8. Nell-Breuning und Sacher, Hrsg., *Wörterbuch der Politik*, Heft 3, Sp. 113.

9. ". . . Die wertvollen Menschen wünschen Verantwortung zu tragen und wünschen als Persönlichkeiten in der Verantwortung zu wachsen und zu reifen." See *Wirtschaft und Gesellschaft Heute*, Bd. 3, p. 246.

10. *Arbeit vor Kapital: Kommentar zur Enzyklika "Laborem Exercens"* (Wien: Europaverlag, 1983), p. 75.

11. *Ibid.*, p. 75. See also "Arbeitslosigkeit," p. 56 where Nell-Breuning presents work as actualizing the possibilities in oneself and in the world.

12. The capability of enjoying leisure, one element in the full development of the person, depends on the pattern of that individual's conduct in the workplace. "Arbeit und Musse," *Stimmen der Zeit* 160 (1957): 96-97.

13. *Wirtschaft und Gesellschaft Heute*, Bd. 1, p. 14.

14. Nell-Breuning und Sacher, Hrsg., *Wörterbuch der Politik*, Heft 3, Sp. 126. Note also Nell-Breuning's agreement with Werner Sombart: "Wirtschaftssubjekte haben Interessen; Wirtschaftsobjekt ist derjenige, der keine Interessen hat oder richtiger: dessen Interessen nicht zählen." See *Streit um Mitbestimmung* (Frankfurt am Main: Europäische Verlagsanstalt, 1968), p. 28.

15. ". . . Je stärker sie [die Menschen] gerade als arbeitende Menschen sich dieser ihrer Subjektstellung bewusst werden, je eindringlicher und nachdrücklicher wie sie anhalten, diese ihre Subjektstellung auszufüllen, um so entschiedener werden sie verlangen, diese ihre Subjektstellung in der Ordnung und Gestaltung von Gesellschaft und Wirtschaft voll und ganz anerkannt zu sehen, nicht aber als blosse Objekte des gesellschaftlichen und wirtschaftlichen Lebensprozesses behandelt und verschlissen zu werden." *Wirtschaft und Gesellschaft Heute*, Bd. 1, p. 175.

16. *Ibid.*, Bd. 1, p. 64.

17. "Dem Menschen eignet der Vorzug der Geschlossenheit, der Eigenständigkeit und Unmittelbarkeit des vernünftigen, selbstmächtigen Wesens. Der Mensch ist also nicht bloss Individuum schlechthin, sondern selbstmächtiges Individuum, Person." *Ibid.*, Bd. I, p. 20.

18. "Menschliche Arbeit ist Betätigung menschlicher (geistiger und/oder leiblicher) Anlagen und Kräfte in der Richtung auf ein ernstgenommenes Ziel, das erreicht, oder Ergebnis, das verwirklicht werden soll. Es ist der Mensch, der dieses Ziel setzt oder dieses Ergebnis anstrebt und es ernst nimmt, d.h. den ernsthaften Willen hat, es zu erreichen bzw. zu verwirklichen, und der, weil es ihm wirklich ernst ist, sich darum bemüht, seine Kräfte anstrengt. . . . Diese ihr vom menschlichen Willen gegebene Zielgerichtetheit ist es, die die menschliche Arbeit von allem anderen, worauf wir die Bezeichnung "Arbeit" anwenden, unterscheidet und ihr den Adel der Menschenwürde verleiht." See "Arbeitslosigkeit," pp. 55-56.

19. *Wirtschaft und Gesellschaft Heute*, Bd. 1, p. 51.

20. *Wirtschaft und Gesellschaft Heute*, Bd. 2, p. 128.

21. As a Christian, Nell-Breuning establishes this dignity in that each person is created in the image and likeness of God (see *Baugesetze der Gesellschaft*, p. 80) and in that (see *Wirtschaft und Gesellschaft Heute*, Bd. 1, p. 184) each person is both created a child of God and called to community with the Trinity. These bases for human dignity, although not denied, do not receive as great an emphasis as does the grounding of dignity in personal freedom. This latter does not, of course, necessarily oppose the explicitly theological position.

22. *Wirtschaft und Gesellschaft Heute*, Bd. 2, p. 113.

23. *Mitbestimmung: Wer mit Wem?* (Herder: Freiburg, 1969), p. 135.

24. "Kirche und Arbeitswelt," *Stimmen der Zeit* 203 (Dez., 1985): 801.

25. Oswald v. Nell-Breuning and Hermann Sacher, Hrsg., *Wörterbuch der Politik* Heft 1: *Zur christlichen Gesellschaftslehre* (Freiburg: Herder, 1947), Sp. 120; and Oswald v. Nell-Breuning, *Baugesetze der Gesellschaft*, p. 17.

26. Nell-Breuning and Hermann Sacher, Hrsg., *Wörterbuch der Politik*, Heft 3, Sp. 147; and Oswald v. Nell-Breuning, *Gerechtigkeit und Freiheit*, p. 228. One such culturally defined need is that for stability of location—a need felt both by the person and especially on the part of a family of a worker. See "Kommentar zum 3. Kapitel des 2. Teils 'Gaudium et Spes,'" in *Lexikon für Theologie und Kirche*, 2. Auflage, *Das Zweite Vatikanische Konzil*, Teil III, p. 496.

27. "Ziel der Wirtschaft und darum Sachziel alles wirtschaftlichen Handelns ist das menschliche Kulturleben, nicht unter allen erdenklichen Rücksichten, sondern nur unter einer, nämlich insofern es zufolge der Geist-Leib-Natur des Menschen an stoffliche Voraussetzungen gebunden ist und bis in seine höchsten Stufe hinein an das Stoffliche gebunden bleibt. Mit Bedacht bezeichnen wir als Sachziel der Wirtschaft das menschliche Kulturleben selbst." *Wirtschaft und Gesellschaft Heute*, Bd. 1, p. 173.

28. *Ibid.*, Bd. 1, p. 243.

29. *Baugesetze der Gesellschaft*, p. 82.

30. *Ibid.*, p. 24.

31. "Ob ich etwas 'kann' oder 'nicht kann', bestimmt sich nicht danach, ob meine Kräfte, wenn ich nichts anderes zu tun und keine andere Verpflichtungen hätte, gerade eben noch aus reichen würden, um diese eine Leistung zustande zu bringen; es bestimmt sich vielmehr danach, ob es sich sinnvoll und verantwortbar in die Gesamtheit meiner Aufgaben und Pflichten einordnen lässt oder ob ich dadurch an anderem, das wichtiger, wertvoller, oder dringlicher wäre, gehindert würde." See *Ibid.*, p. 84.

32. "Die Ergänzungsbedürftigkeit des Menschen ist zwar gewiss Ausdruck seiner geschöpflichen Begrenztheit, nichtsdestoweniger aber viel mehr ein Reichtum als eine Armut: die scheinbare Schwäche ist in Wahrheit seine Stärke. Der Mensch besitzt die Fähigkeit, Kultursachgebiete aufzubauen, die ihrem Wesen nach sich nicht aus dem einzelnen herausspinnen lassen, sondern nur als Wertbereiche zwischen sozial verbundenen Individuen denkbar sind." *Wirtschaft und Gesellschaft Heute*, Bd. 1, p. 20. Also see *Ibid.*, p. 42.

33. Nell-Breuning und Sacher, Hrsg., *Wörterbuch der Politik*, Heft 1, Sp. 69-70.

34. *Baugesetze der Gesellschaft*, p. 79.

35. "Auch 'Macht und ökonomisches Gesetz' schliessen einander nicht in dem Sinne aus, als ob Wirtschaft machtfrei wäre oder auch nur machtfrei sein könnte. . . ." *Wirtschaft und Gesellschaft Heute*, Bd. 1, p. 45.

36. Nell-Breuning und Sacher, Hrsg., *Wörterbuch der Politik*, Heft 3, Sp. 129; and *Mitbestimmung*, Reihe Theorie und Praxis der Gesellschaften (Frankfurt am Main: Europäische Verlagsanstalt, 1968), pp. 18-19.

37. "Arbeit und Mitbestimmung," in *Mitbestimmung: Referate und Diskussion auf der Tagung katholischer Sozialwissenschaftler von 17. bis 19. Februar 1968 in Mönchengladbach*, hrsg. von Anton Rauscher (Köln: J.P. Bachem, 1968), p. 29. Also see *Wirtschaft und Gesellschaft Heute*, Bd. 1, p. 45.

38. Nell-Breuning und Sacher, Hrsg., *Wörterbuch der Politik*, Heft 3, Sp. 110.

39. *Mitbestimmung*, p. 15.

40. *Mitbestimmung: Wer mit Wem?*, pp. 222-223.

41. *Mitbestimmung*, p. 15.

42. "Wirtschaft ist eine Gestaltungsaufgabe." *Wirtschaft und Gesellschaft Heute*, Bd. 1, p. 11.

43. *Aktuelle Fragen der Gesellschaftspolitik* (Köln: Verlag J.P. Bachem, 1970), p. 249.

44. *Wirtschaft und Gesellschaft Heute*, Bd. 1, p. 275.

45. "Verneinung eines Sachziels der Wirtschaft macht den Begriff der wirtschaftlichen Leistung unvollziehbar oder muss ihn mit Befolgung des Rationalprinzips gleichsetzen und damit uferlos machen." *Ibid.*, Bd. 1, p. 46.

46. "Die Kombination von Einzel- und Teilzielen zu einem Gesamtziel und die Wertung dieses Gesamtziels ist an erster Stelle ein Gestaltungsproblem. . . ." *Wirtschaft und Gesellschaft Heute*, Bd. 1, p. 49.

47. Nell-Breuning und Sacher, Hrsg., *Wörterbuch der Politik*, Heft 3, Sp. 105.

48. *Baugesetze der Gesellschaft*, p. 22.

49. *Wirtschaft und Gesellschaft Heute*, Bd. 1, p. 20.

50. "Solidarismus" in *Philosophisches Wörterbuch*, 14. Auflage, hrsg. von Walter Brugger (Freiburg: Herder, 1976), p. 364.

51. *Arbeit vor Kapital*, pp. 75-76.

52. *Baugesetze*, p. 18.

53. *Ibid.*, Bd. 1, p. 168.

54. *Wirtschaft und Gesellschaft Heute*, Bd. 1, p. 22.

55. *Die soziale Enzyklika: Erläuterungen zum Weltrundschreiben Papst Pius' XI. über die gesellschaftliche Ordnung* (Köln: Katholische Tat-Verlag, 1932), p. 115. Nell-Breuning remarks that a scholastic understanding of valuation returns to a degree when Marx, in rejecting Lassalle's *ehernes Lohngesetz*, indicates the significance of the socio-cultural estimation. See *Gerechtigkeit und Freiheit*, p. 250 n.

56. Nell-Breuning und Sacher, Hrsg., *Wörterbuch der Politik*, Heft 1, Sp. 148. Nell-Breuning also indicates the social valuation for work when he refers to the scholastic terms for the processes which create such a valuation in a dynamic, expanding economy: *raritas*, *utilitas*, *communis aestimatio*. See *Gerechtigkeit und Freiheit*, pp. 239-240.

57. *Ibid.*, Heft 1, Sp. 57.

58. *Wirtschaft und Gesellschaft Heute*, Bd. 1, p. 23. With reference to this point, Nell-Breuning recalls the scholastic axiom, *Actiones sunt suppositorum*.

59. "Der Weg von der socia(bi)litas zur activitas socialis ist also dieser: Die metaphysische Sozialanlage des Menschen wird durch die physischen Prinzipien des Wesenswillens und Kurwillens aktualisiert zu den verschiedenen Sozialgebilden, die als solche nicht substantielle Einheiten, sondern Ordnungseinheiten sind. Eben darum eignet den Sozialgebilden der metaphysische Wesenszug der Organisiertheit. Diese Organisiertheit hinwiederum ist aktualisiert in den physischen Organen mit ihrer verschiedenen Organstellung als leitenden und ausführenden Organen usw." *Wirtschaft und Gesellschaft Heute*, Bd. 1, p. 24.

60. "In der menschlichen Gemeinschaft besteht nicht nur die umfassende Gemeinverstrickung, sondern—weil die Menschen sittlich verantwortliche Wesen sind—auch die ganze gleich umfassende Gemeinhaftung." See *Baugesetze der Gesellschaft*, p. 21. Also see *Gerechtigkeit und Freiheit*, p. 47.

61. Nell-Breuning und Sacher, Hrsg., *Wörterbuch der Politik*, Heft 1, Sp. 133.

62. *Baugesetze der Gesellschaft*, pp. 11 and 14. "Das Solidaritätsprinzip besagt die wechselseitige Bezogenheit oder Hinordnung des Gliedes auf das Ganze und des Ganzen auf die Glieder (sogenannte 'Rückbindung') aus der die genau entsprechende wechselseitige Verpflichtung des Füreinander-Einstehens ('Solidarität') zwingend folgt. . . ." For this see *Wirtschaft und Gesellschaft Heute*, Bd. 1, p. 25.

63. ". . . Alles gesellschaftliche Leben, selbstverständlich also auch jede Wirtschaft, geht vom einzelnen aus und ist darum auch letzten Endes wieder um des einzelnen willen da." See *Die soziale Enzyklika*, p. 145. Here Nell-Breuning, expressing his agreement with Leo XIII, demonstrates in the section from which these words are taken the connection between the principles of solidarity and subsidiarity. Also see Oswald von Nell-Breuning *Reorganization of Social Economy: The Social Encyclical Developed and Explained*, edited and translated by Bernard W. Dempsey (New York: Bruce Publishing Company, 1936), p. 206. See further *Baugesetze der Gesellschaft*, p. 74.

64. *Baugesetze der Gesellschaft*, p. 65.

65. *Die soziale Enzyklika*, p. 137. As an example of a social result of the principle of solidarity, Nell-Breuning chooses the membership of workers in society. See *Baugesetze der Gesellschaft*, p. 56.

66. *Baugesetze der Gesellschaft*, p. 32.

67. *Wirtschaft und Gesellschaft Heute*, Bd. 1, p. 4. Also see pp. 2-3 where the cultural goods are presented as possible only within society. The divine command when humans lost paradise involved a command to work for culture. See *Die soziale Enzyklika*, p. 134.

68. Nell-Breuning und Sacher, Hrsg., *Wörterbuch der Politik*, Heft 1, Sp. 149. And also *Wirtschaft und Gesellschaft Heute*, Bd. 1, p. 291.

69. *Die soziale Enzyklika*, p. 138. And also *Wirtschaft und Gesellschaft Heute*, Bd. 1, p. 12.

70. Nell-Breuning und Sacher, Hrsg., *Wörterbuch der Politik*, Heft 3, Sp. 105.

71. *Wirtschaft und Gesellschaft Heute*, Bd. 1, p. 44.

72. *Ibid.*, Bd. 1, pp. 20 and 89.

73. "Alle Kulturwerte können nur durch A[rbeit] geschaffen und erhalten werden; um so wichtiger ist es, die A[rbeit] selbst u[nd] die Lebensbedingungen des abeitenden Menschen . . . kulturwürdig zu gestalten." See "Arbeit" in *Philosophisches Wörterbuch*, 14. Auflage, hrsg. von Walter Brugger (Freiburg: Herder, 1976), p. 25. Also note Nell-Breuning und Sacher, Hrsg., *Wörterbuch der Politik*, Heft 3, Sp. 122.

74. "Being more" as opposed to "having more" are terms which Nell-Breuning draws from Pope Paul VI's address to the International Labor Organization in 1969. See *Aktuelle Fragen*, pp. 251-252.

75. *Wirtschaft und Gesellschaft Heute*, Bd. 1, p. 61.

76. *Mitbestimmung: Wer mit Wem?*, p. 103.

77. *Ibid.*, p. 89.

78. *Ibid.*, p. 75; and *Arbeit vor Kapital*, pp. 98-99.

79. *Mitbestimmung: Wer mit Wem?*, p. 193.

80. *Wirtschaft und Gesellschaft Heute*, Bd. 1, p. 211.

81. *Ibid.*, Bd. 1, p. 212.

82. In the words of Pope Pius XI, *res* represents the capital input, *opera* the contribution of workers, and *intellectus* the entrepreneurial guidance. See *Quadragesimo Anno*, in *Seven Great Encyclicals*, ed. by William J. Gibbons (Glen Rock, New Jersey: Paulist Press, 1963), section 69. For the three-pole model, see *Mitbestimmung: Wer mit Wem?*, p. 117.

83. *Mitbestimmung: Wer mit Wem?*, p. 119.

84. "Arbeit und Mitbestimmung," p. 22.

85. This error equates the external role of the firm as a subject of ownership and the interal and external life of the firm as a social structure. See *Aktuelle Fragen*, p. 240.

86. *Mitbestimmung*, p. 72. Unlike Marx, who failed to recognize the importance of the entrepreneur, Nell-Breuning does recognize this agent's significance in the producing firm. See *Mitbestimmung: Wer mit Wem*?, p. 117.

87. *Mitbestimmung*, pp. 80 and 96.

88. *Ibid.*, pp. 84 and 128.

89. Also in a later work, Nell-Breuning sees the tradition of the *berufsständische Ordnung* as a macroeconomic approach, that is, one directed to economic activity within a range broader than that of the single firm. See *Gerechtigkeit und Freiheit*, p. 230.

90. *Mitbestimmung: Wer mit Wem?*, p. 207. He grants in his retrospective on Catholic social teaching that *Quadragesimo Anno* lacked clarity in explicating the social dimension of the various economic groups. See *Wie sozial ist die Kirche?: Leistung und Versagen der katholischen Soziallehre* (Dusseldorf: Patmos Verlag, 1972), p. 111.

91. For this point see Nell-Breuning und Sacher, Hrsg., *Wörterbuch der Politik*, Heft 1, Sp. 67. Note also that, "Als Gliedverband mit besonderen Aufgaben im Rahmen der Volksgemeinschaft ist der Berufsstand eine jener dem Staate vorgeordneten natürlichen Gemeinschaften, deren Sein und Recht nicht erst von ihm abgeleitet, sondern gleich ursprünglicher Art wie das des Staates selbst ist." See *Ibid.*, Heft 1, Sp. 70.

92. See "Um die '*berufsständische Ordnung*'" *Stimmen der Zeit* 142 (1948): 6-7 and 9. For the independence of the *berufsständische Ordnung* from the state also see *Wirtschaft und Gesellschaft Heute*, Bd. 1, pp. 222 and 235.

93. *Soziallehre der Kirche: Erläuterungen der lehramtlichen Dokumente* (Wien: Europaverlag, 1977), p. 123.

94. *Wirtschaft und Gesellschaft Heute*, Bd. 1, p. 194.

95. ". . . Die Berufsgemeinschaft oder der Berufsstand: der Berufstätigkeit eignet eine doppelte gemeinschaftsbildende Kraft; sie stiftet Gemeinschaft der Berufsgenossen untereinander und verknüpft zugleich durch die berufliche Leistung mit dem Volksganzen. . . . " *Ibid.*, Bd. 1, p. 19. Also see pp. 219 and 278. The term *Stand* is used, as in G. Gundlach's understanding, in place of "class." Nell-Breuning thereby obtains two results: first a de-emphasis of oppositional and conflictual dimensions of class and, at the same time, room for differences in interests; and second, the conceptual integration of a *Stand* into society—each *Stand* has its place in the society. See *Arbeit vor Kapital*, pp. 112-113.

96. *Wirtschaft und Gesellschaft Heute*, Bd. 1, p. 236. Also, for the link between *berufsständische Ordnung* and *klassenfreie Gesellschaft* (N.B. not the Marxist term *klassenlose*), see *Wie sozial ist die Kirche*?, p. 81.

97. "Als Gegenstände der Mitbestimmung pflegen aufgezählt zu werden: personelle, soziale und wirtschaftliche Angelegenheiten; als Ebenen, auf denen Mitbestimmung stattfinden soll, der Arbeitsplatz des einzelnen Arbeiters, der Betrieb, in dem er tätig ist, das Unternehmen, bei dem er Arbeit genommen hat, der Wirtschaftszweig, dem das Unternehmen angehört, schliesslich die Volkswirtschaft und—wenn man ganz vollständig sein will—die Weltwirtschaft." See *Mitbestimmung*, p. 19.

98. An early use of this term as equivalent for *berufsständische Ordnung* appeared shortly after a declaration by Cardinal Frings concerning the consequences of the key principles of *katholische Sozialarbeit*. See *Wirtschaft und Gesellschaft Heute*, Bd. 2, p. 92.

99. Nell-Breuning und Sacher, Hrsg., *Wörterbuch der Politik*, Heft 1, Sp. 65.

100. *Ibid.*, Heft 1, Sp. 66.

101. *Die soziale Enzyklika*, p. 91; and *Wie sozial ist die Kirche*, p. 66.

102. "Rationalisierung und Automatisierung: Segen oder Fluch," *Stimmen der Zeit* 175 (1965): 136.

103. "Arbeit," p. 25.

104. *Baugesetze der Gesellschaft*, pp. 82-83.

105. Nell-Breuning und Sacher, Hrsg., *Wörterbuch der Politik*, Heft 3, Sp. 106.

106. "Rationalisierung und Automatisierung," p. 144.

107. *Quadragesimo Anno*, Sec. 69.

108. *Mitbestimmung*, p. 9.

109. Nell-Breuning und Sacher, Hrsg., *Wörterbuch der Politik*, Heft 3, Sp. 115.

110. *Baugesetze der Gesellschaft*, p. 121.

111. For example, see *Wirtschaft und Gesellschaft Heute*, Bd. 1, pp. 84-85. Also note "Rationalisierung und Automatisierung," p. 146, where Nell-Breuning observes that information technology increases the gap between top managers and most employees. And further, *Wirtschaft und Gesellschaft Heute*, Bd. 2, p. 113 notes that technology, when properly deployed, calls upon the employee's responsible problem-solving, prudent judgment and capacity to form.

112. "Die ganze Problematik der Rationalisierung, die uns Kummer bereitet, ist eine Eigentumlichkeit der Arbeitnehmergesellschaft. . . ." "Rationalisierung und Automatisierung," p. 137. Also see *Ibid.*, p. 142.

113. Nell-Breuning und Sacher, Hrsg., *Wörterbuch der Politik*, Heft 3, Sp. 116.

114. *Ibid.*

115. *Ibid.*, Heft 3, Sp. 119.

116. "So verläuft der Weg von der Arbeit im eigenen Raum oder Heim am eigenen Stoff und mit eigenen Mitteln für den eigenen Bedarf zur Arbeit im fremden Werkraum, am fremden Stoff, mit fremden Arbeitsmitteln, nicht nur für fremden Bedarf, sondern auch auf fremde Rechnung." Nell-Breuning und Sacher, Hrsg., *Wörterbuch der Politik*, Heft 3, Sp. 116.

117. "Mehr und mehr aber entfaltet sich das Eigenleben des Unternehmens; immer gewichtiger wird das Tun der in Betrieb und Unternehmen tätigen Menschen, im Vergleich mit dem die Sachmittel immer ausgeprägter zum blossen Zubehör herabsinken. Damit ist der Zeitpunkt gekommen, um dem Unternehmen als lebenden Organismus—ähnlich wie der staatlich geeinten Volksgemeinschaft—eine Verfassung zu geben. . . ." See *Mitbestimmung: Wer mit Wem?*, p. 60.

118. "Die Prädominanz des *Quale*, die das Unterscheidungsmerkmal der Wirtschaft von der Technik und ihre Überlegenheit über die Technik ausmacht, hat zur Folge, dass Wirtschaft . . . als solche nicht rechenhaft ist." *Wirtschaft und Gesellschaft Heute*, Bd. 1, p. 44.

119. Even in the "laborist" version of governance in the firm, the separation would remain. *Gerechtigkeit und Freiheit*, p. 232.

120. *Wirtschaft und Gesellschaft Heute*, Bd. 1, pp. 255-256.

121. *Arbeit vor Kapital*, p. 115.

122. *Die soziale Enzyklika*, p. 155.

123. *Soziallehre der Kirche*, p. 124.

124. "Wie sind Eigentum an Produktionsmitteln und die Arbeitskraft produktionsmittelentblösster Proletarier zueinander in Beziehung zu setzen, sondern: Wie sind Menschen als Träger unterschiedlicher Funktionen miteinander zu einer effizienten Einheit zu verbinden?" *Ibid.*, p. 142. Here Nell-Breuning refers to Vatican II, Ecumenical Council, *Gaudium et Spes*. See this text in *Documents of Vatican II*, ed. by Walter M. Abbot and Joseph Gallagher (London: Geoffrey Chapman, 1967), sec. 68.

125. *Gerechtigkeit und Freiheit*, p. 222. He also writes, "Kommt dann hinzu, dass der Mensch erlebt, wie er selbst in seiner Arbeit und durch sie menschlich reift und wächst und zu seiner vollmenschlichen Entfaltung gelangt und zugleich als Ergebnis seiner Sachgüter oder Dienstwerte schafft, die dazu beitragen, sein oder seiner Mitmenschen Dasein und Leben zu erhalten, zu bereichern, zu verschönern oder zu veredeln, dann wird ihm die Arbeit trotz aller mit ihr verbundenen Mühe und Last zum Bedürfnis. See "Arbeitslosigkeit," p. 57.

126. Nell-Breuning's definition of work takes into consideration both dimensions of the results of work activity: "Der Mensch arbeitet, wenn er seine geistigen oder körperlichen Kräfte betätigt in der Richtung auf ein ernst genommenes Ziel, das erreicht oder verwirklicht werden soll. Lernen und Beten sind echte A[rbeit], obwohl beide keine Erzeugnisse hervorbringen; andere geistige u[nd] wohl alle körperlichen A[rbeit]-en fuhren zu einem äusserlich wahrnehmbaren Ergebnis, sei es einem Erzeugnis, sei es einer Zustandsänderung." I emphasize that in Nell-Breuning's view the products created maintain or better the circumstances of persons in society. Indeed, the alteration of circumstance contributes to the development of the social and historical situation of persons who jointly shape their culture. See "Arbeit," p. 24.

127. Nell-Breuning und Sacher, Hrsg., *Wörterbuch der Politik*, Heft 3, Sp. 110-111.

128. Here Nell-Breuning comments on the words of Pope Paul VI at his Geneva address to the International Labor Organization on its fiftieth anniversary. See *Soziallehre der Kirche*, p. 182.

129. Nell-Breuning und Sacher, Hrsg., *Wörterbuch der Politik*, Heft 3, Sp. 105.

130. *Ibid.*, Heft 3, Sp. 120. Even work for the sake of earning one's sustenance is a service to the community.

131. *Wirtschaft und Gesellschaft Heute*, Bd. 1, p. 193.

132. In the conception of the *Unternehmen an sich*, "ob die erbrachten Leistungen für irgend jemand von Nutzen sind, d.h. der Bedarfsbefriedigung dienen, ist belanglos." See *Mitbestimmung*, pp. 72-73.

133. The return on capital ". . . muss vielmehr dem persönlichen Wohl der im Einzelunternehmen oder in der Gesamtwirtschaft tätigen Menschen dienen und darum im Falle einer Ziel- oder Interessenkollision zurücktreten und sich unterordnen. . . ." See *Arbeit vor Kapital*, p. 101.

134. The work of the entrepreneur represents a particular form of service to society, indeed to a particular group in society. The entrepreneur is the first servant of his firm. *Mitbestimmung: Wer mit Wem?*, p. 76.

135. *Arbeit vor Kapital*, p. 77.

136. *Gerechtigkeit und Freiheit*, p. 225.

137. ". . . Die Wirtschaft mit der Gütererzeugung . . . mit dem Kreditwesen, mit dem Verkehrswesen." See Nell-Breuning und Sacher, Hrsg., *Wörterbuch der Politik*, Heft 1, Sp. 62.

138. Nell-Breuning warns of the tendency in economics to forget "service" as a dimension of human work activity: "Dagegen hiesse es den ganzen Sinn der menschlichen Arbeit und des menschlichen Kulturlebens auf den Kopf stellen, wollte man—wie es von Kreisen der Wirtschaft manchmal versucht wird—nur die Fertigung von Sachgütern als nützliche Arbeit gelten lassen, alle übrige Arbeit dagegen mit dem Stigma 'unproduktiv' als wertlos oder doch als notwendiges Übel brandmarken." See *Ibid.*, Heft 3, Sp. 117-118.

139. *Ibid.*, Heft 3, Sp. 110.

140. *Ibid.*, Heft 3, Sp. 153.

141. In the earlier section on limits, I referred to the negative side of subsidiarity as a guarantee for the protection of the individual. For the twin aspects, see *Baugesetze der Gesellschaft*, p. 79.

142. "Die Gemeinschaft soll ihrem Gliede hilfreichen Beistand in dem leisten, was der einzelne, der Glied der Gemeinschaft ist, auf sich allein gestellt und aus eigenen Kräften nicht vermag; was dagegen der Einzelmensch selbst leisten kann, entzöge sie ihm ja die Gelegenheit, durch die eigene Tätigkeit tüchtiger, vollkommener zu werden, als er bereits ist; das aber wäre kein hilfreicher Beistand, sondern wäre eine Schädigung." *Ibid.*, p. 83.

143. Nell-Breuning refers to the position of Bishop Ketteler opposing the increasing activity of the centralized state. See *Die soziale Enzyklika*, p. 142.

144. *Baugesetze der Gesellschaft*, p. 94.

145. *Ibid.*, p. 83.

146. *Wirtschaft und Gesellschaft Heute*, Bd. 1, p. 51.

147. Nell-Breuning und Sacher, Hrsg., *Wörterbuch der Politik*, Heft 3, Sp. 120.

148. Nell-Breuning himself warns us away from a static, physiologically based understanding of the metaphor of society as organism, "In Wirklichkeit ist natürlich die Gesellschaft als ein moralisch-juridischer Organismus, als ein Gebilde, das zwar von der Natur gefordert ist, in

dessen Ausgestaltung die Natur aber grösste Freiheit lässt, viel freier und beweglicher als der physiologische Organismus. Darum ist der Analogie völlig genügt, wenn entgegen jenem Leistungsprinzip der liberalen Konkurrenz, wo sich in jedem Augenblick um meinen gesellschaftlichen Standort zu kämpfen habe und jeden Augenblick gewärtigen muss, durch eine nur vorübergehende Minderleistung ihn zu verlieren, eine gewisse Sicherung des gesellschaftlichen Standortes geboten ist. . . ." See *Die soziale Enzyklika*, p. 156. Also see *Reorganization of Social Economy*, p. 221.

149. *Wirtschaft und Gesellschaft Heute*, Bd. 1, p. 195.

150. "Kirche und Arbeitswelt," p. 795.

151. *Wie Sozial ist die Kirche*, p. 69.

152. *Wirtschaft und Gesellschaft Heute*, Bd. 1, p. 196.

153. *Ibid.*, Bd. 2, p. 13.

154. "Sozialethik," in *Philosophisches Wörterbuch*, 14. Auflage, hrsg. von Walter Brugger (Freiburg: Herder, 1976), pp. 366-367.

155. Nell-Breuning und Sacher, Hrsg., *Wörterbuch der Politik*, Heft 1, Sp. 64.

156. "Kirche und Arbeitswelt," p. 802.

157. See, for example, *Mitbestimmung: Wer mit Wem?*, p. 192 and *Mitbestimmung*, p. 113.

158. *Wirtschaft und Gesellschaft Heute*, Bd. 1, p. 235. Also see *Ibid.*, p. 258 for the role subsidiarity plays in keeping the state to its proper functions.

159. On the basis of research in the extensive and long-lasting Yugoslav experience in this field and through an extension of neoclassical analysis (see his *The General Theory of Labor-Managed Market Economies* (Ithaca: Cornell University Press, 1970)), seminal work in this field has been notably accomplished by Jaroslav Vanek. Other developments in self-management have been the Mondragon system in the Basque country, designed through the ideas and impetus of Jose Maria Arizmendi, and the U.S. application and specification of the Mondragon model in the work of David P. Ellerman. Consult the bibliography for references to his writing.

160. Paul Bernstein, *Workplace Democratization: Its Internal Dynamics* (n.p.: Kent State University Press, 1976), p. 48.

161. Franz v. Baader and Heinrich Koch figure as forerunners of these ideas. See Eckhardt Barthel and Joachim Dikau, *Mitbestimmung in der Wirtschaft*, Zur Politik und Zeitgeschichte, No. 43 (Berlin: Colloquium Verlag, 1980), p. 66.

162. See Article 165 of the Weimar Constitution which guarantees the "Gleichberechtigung von Kapital und Arbeit." Also note that the *Räte* law of 1920 prescribed the election of two council members to the board of directors. Quentin Lauer, "Co-management in Germany: Toward Industrial Democracy," *Social Order* I (January, 1951) p. 19.

163. Quite helpful in providing an overview of the law is Joerg Dierolf, *Codetermination: An Introduction to "Industrial Democracy" in West Germany* (Bergisch-Gladbach: Heider Verlag, 1977); here, pp. 11-19.

164. Robert J. Kühne, *Codetermination in Business: Workers' Representatives in the Boardroom* (New York: Praeger, 1980), p. 34.

165. Barthel and Dikau, *Mitbestimmung in der Wirtschaft*, p. 104. Also see Dierolf's summary of this variety, *Codetermination: An Introduction*, p. 25.

166. Sombart was a student of G. Schmoller and A. Wagner, became professor of economics at Berlin and served as last president of the original *Verein für Sozialpolitik.*

167. *Wie Sozial ist die Kirche*, p. 106.

168. *Wirtschaft und Gesellschaft Heute*, Bd. 2, pp. 92 and 128.

169. "In Gesellschaft und Wirtschaft sollte nicht das sachliche Element (Kapital, Vermögen), sondern das personale Element, der Mensch mit dem, was er leistet, mit seiner Arbeit, bestimmend sein. . . . Für die von der christlichen Soziallehre geforderte Ordnung haben schon Hermann Roesler und ihm folgend Heinrich Pesch die Bezeichnung 'soziales Arbeitssystem' vorgeschlagen, tatsächlich haben sie, mit einem 'sozial temperierten Kapitalismus' (heute 'soziale Marktwirtschaft' genannt) zufriedengegeben." See *Soziallehre der Kirche*, pp. 123-124.

170. "Eine Lösung, die a priori den einen Faktor in den Stand setzt, den anderen Faktor zu überfahren—und das ist ja im Grunde genommen die unausgesprochene Rechtfertigung für die Drittelung im Aufsichtsrat nach dem Betriebsverfassungsgesetz—ist grundsätzlich und unbedingt abzulehnen.

"Die Folgerung, die ich daraus ziehe, ist allerdings nicht, den Aufsichtsrat anders zusammenzusetzen, sondern obwohl ich selbst Miturheber oder Miterfinder der Mitbestimmung auf dem Wege über den Aufsichtsrat der Vermögensträgerin bin, erachte ich diese Lösung, gleichviel, ob der Aufsichtsrat gedrittelt oder quasi-paritätisch zusammengesetzt ist, als eine—damals allerdings verführerisch naheliegende und daher sich geradezu aufdrängende—Fehlkonstruktion." See "Arbeit und Mitbestimmung," p. 48. Also see *Mitbestimmung: Wer mit Wem?*, p. 80.

171. *Soziallehre der Kirche*, p. 124.

172. *Arbeit vor Kapital*, p. 91.

173. *Wirtschaft und Gesellschaft Heute*, Bd. 2, p. 112.

174. *Ibid.*, Bd. 1, pp. 216-217.

175. *Baugesetze der Gesellschaft*, p. 124. In addition see "Arbeit und Mitbestimmung, p. 25.

176. *Aktuelle Fragen*, pp. 215-216.

177. *Wirtschaft und Gesellschaft Heute*, Bd. 2, p. 153.

178. *Baugesetze der Gesellschaft*, p. 74. Note also: "Jedem Mann im Betrieb an seinem Arbeitsplatz soviel Selbständigkeit, soviel Entschlussfreiheit, soviel Eigenverantwortung wie nur irgend mit der einheitlichen Ausrichtung des Betriebsgeschehens und seinen geordneten, reibungslosen Ablauf vereinbar: das ist nicht nur menschlich, sondern auch im engeren und engsten Sinne wirtschaftlich das Wesentliche an der ganzen . . . Eröterung über das Mitbestimmungsrecht im Betrieb." *Wirtschaft und Gesellschaft Heute*, Bd. 1, p. 84.

179. Nell-Breuning und Sacher, Hrsg., *Wörterbuch der Politik*, Heft 1, p. 70.

180. *Mitbestimmung*, p. 86.

181. ". . . Die Uberwindung der einseitigen und darum als Fremdherrschaft sich auswirkenden Fremdbestimmung der blossen Objektrolle der Arbeit, [und] ihr Aufstieg zu wenigstens gleichberechtigter Subjektrolle." *Ibid.*, p. 30.

182. *Ibid.*

183. *Wirtschaft und Gesellschaft Heute*, Bd. 1, p. 60; and *Arbeit vor Kapital*, pp. 106-107.

184. *Arbeit vor Kapital*, p. 106.

185. *Wirtschaft und Gesellschaft Heute*, Bd. 2, p. 127.

186. *Ibid.*, p. 109. "Hier entspringt also das Verlangen nach Mitbestimmung dem Bedürfnis nach Sicherheit. Weil der Unternehmer weder rechtlich noch tatsächlich der Belegschaft die Erhaltung ihre Arbeit gewährleistet, deshalb will die Belegschaft selbst für die Erhaltung ihre Arbeitsplätze sorgen." *Ibid.*, pp. 109-110.

187. Nell-Breuning und Sacher, Hrsg., *Wörterbuch der Politik*, Heft 3, p. 136.

188. *Mitbestimmung*, pp. 23 and 25.

189. Nell-Breuning und Sacher, Hrsg., *Wörterbuch der Politik*, Heft 3, p. 134.

190. *Wirtschaft und Gesellschaft Heute*, Bd. 2, pp. 110-111.

191. Nell-Breuning sees the plant as a "*produktionstechnischer*" concept and the company as a "*juristisch-ökonomischer*" one. Both are types of the technical structures in which questions of decision and authority arise. See *Mitbestimmung*, p. 25.

192. The "*Unternehmen an sich*" stands as an example of a conception which does not recognize the association of persons. Such a view allows the firm's activity to be executed solely on the basis of guidance by the measure of its technical achievements. See *Ibid.*, p. 72.

193. *Mitbestimmung: Wer mit Wem?*, p. 83. Also see *Aktuelle Fragen*, p. 234.

194. *Mitbestimmung*, p. 29.

195. "Verfasst werden soll selbstverständlich nicht der Prozess, sondern der Verbund dieser Menschen, soll über die blosse Faktizität zu einem verfassten Gebilde der Rechtsordnung

erhoben werden; das und nichts anderes ist mit Unternehmensverfassung gemeint." *Ibid.*, p. 73. Also see *Aktuelle Fragen*, p. 235.

196. *Mitbestimmung*, p. 79.

197. *Soziallehre der Kirche*, p. 203.

198. *Wirtschaft und Gesellschaft Heute*, Bd. 1, p. 256.

199. *Mitbestimmung: Wer mit Wem?*, p. 180 and *Aktuelle Fragen*, p. 232. Nell-Breuning describes this organic character of the firm in these words, for example: "[ein] leibhaftes Sozialgebilde in der Entfaltung seines Lebens nach innen und aussen, seines inneren Kräftespiels und seiner Interaktion mit der Aussenwelt." For this, see *Aktuelle Fragen*, p. 240.

200. "Einen weiteren und wichtigen Beitrag kann und soll die Sozialpolitik leisten durch eine Gestaltung der menschlichen Lebensgemeinschaften, insbesondere der Betriebsgemeinschaft, die der Entfaltung der Persönlichkeit jeden nur möglichen Raum gibt in gleichzeitigen Einbettung in überschaubare und darum erlebnismässig erfahrbare Gemeinschaften (team-work)." See *Wirtschaft und Gesellschaft Heute*, Bd. 1, p. 17.

201. *Ibid.*, Bd. 1, pp. 13 and 38.

202. ". . . Seine Arbeit soll der Rechtsgrund seiner Mitbestimmung sein." *Mitbestimmung: Wer mit Wem?*, p. 6. Nell-Breuning also states that the bond resulting from work draws persons together more strongly than does a geographic community. *Wirtschaft und Gesellschaft Heute*, Bd. 1, p. 228.

203. *Wirtschaft und Gesellschaft Heute*, Bd. 2, p. 123.

204. See, for example, *Ibid.*, Bd. 1, p. 302.

205. The ". . . Macht der Gross- und Grösstunternehmen bedarf der Kontrolle. Die wirtschaftliche Mitbestimmung institutionalisiert eine wechselseitige Kontrolle, die Arbeit und Kapital in der Unternehmensleitung gegeneinander ausüben, aber darum geht es hier nicht. . . . Der Zusammenhang zwischen wirtschaftlicher Mitbestimmung in Gestalt einer Unternehmensverfassung und Kontrolle der Unternehmensmacht ist ganz anderer. Die dank der Unternehmensverfassung mit ihrer interessendualistischen oder gar interessenpluralistischen Struktur ausgeglicheneren inneren Machtverhältnisse im Unternehmen erleichtern den zur Machtkontrolle berufenen politischen Instanzen ihre Aufgabe." See *Mitbestimmung*, p. 49.

206. Nell-Breuning mentions the interests, ". . . der Allgemeinheit beispielsweise konjunktur- und beschäftigungspolitischen, währungs- und zahlungsbilanzpolitischen usw. Interessen." See *Ibid.*

207. *Wirtschaft und Gesellschaft Heute*, Bd. 2, p. 127.

208. *Mitbestimmung: Wer mit Wem?*, p. 119. Nell-Breuning also maintains that Catholic social teaching too rarely emphasizes the *oberstes moralisches Gesetz* of preserving trust in processes of exchange. See *Wie sozial ist die Kirche*, p. 69.

209. *Mitbestimmung: Wer mit Wem?*, p. 106.

210. *Wirtschaft und Gesellschaft Heute*, Bd. 1, pp. 283 and 286.

211. *Aktuelle Fragen*, p. 217.

212. See the important work by Fritz Naphtali, *Wirtschaftsdemokratie: Ihr Wesen, Weg und Ziel.* Sammlung "Res Novae"—Veröffentlichungen zu Politik, Wirtschaft, Soziologie und Geschichte, Bd. 42 (1928; reprint ed., Frankfurt am Main: Europäische Verlagsanstalt, 1966).

213. *Mitbestimmung*, p. 132.

214. *Mitbestimmung: Wer mit Wem?*, pp. 34-35.

215. *Baugesetze der Gesellschaft*, p. 70.

216. *Mitbestimmung*, p. 25.

217. ". . . Soweit ihr Handeln durch Weisungen der Unternehmensleitung bestimmt werden soll, bedarf diese dazu einer Legitimation, die sie sich nicht selbst erteilen, auch nicht aus Ihrem Eigentumsrecht an den sachlichen Produktionsmitteln, sondern ganz offenbar von den arbeitenden Menschen selbst herleiten kann." See *Ibid.*, p. 137.

218. *Ibid.*, p. 25.

219. *Wirtschaft und Gesellschaft Heute*, Bd. 2, p. 112; and *Aktuelle Fragen*, pp. 213-214.

220. "Die Herstellung der rechten Ordnung der Gesellschaft . . . gliedert denjenigen, der durch seine Arbeitsleistung zum Wirtschaftsvollzuge und damit zu dem Beitrag des Berufsstandes zum Gesamtwohl des Gesellschaftsganzen mitwirkt, als vollwertigen und vollberechtigten Berufsburger zunächst in den engeren Kreis der Berufsgenossen damit aber zugleich auch in das grosse Ganze der Volksgemeinschaft wieder ein; sie sichert ihm seinen Standort in der Gesellschaft, gewährleistet ihm wieder die verlorene Standschaft. So schafft sie wieder eine wahre Volksordnung." See *Die soziale Enzyklika*, p. 166. Also see *Reorganization of Social Economy*, p. 232.

221. Nell-Breuning speaks of this principle as applied to the Berufsstand in Nell-Breuning und Sacher, Hrsg., *Wörterbuch der Gesellschaft*, Heft 1, Sp. 68 and in *Wirtschaft und Gesellschaft Heute*, Bd. 1, p. 242.

222. *Wirtschaft und Gesellschaft Heute*, Bd. 2, p. 154.

223. *Mitbestimmung*, p. 160 and *Arbeit vor Kapital*, pp. 96 and 101.

224. *Mitbestimmung*, p. 7; see also *Mitbestimmung: Wer mit Wem?*, pp. 78-79.

225. Robert A. Dahl, *A Preface to Economic Democracy* (Berkeley: University of California Press, 1985), pp. 84-88. Dahl presents these as five goals for democracy in the economic order.

226. This move will have wide-ranging consequences not only in profoundly questioning the control by management but also in critically examining the allocation of capital income and the ownership of capital. See Jaroslav Vanek, *Self-Management: Economic Liberation of Man* (Harmondsworth: Penguin, 1975), pp. 14-15.

227. J.Y. Tabb and A. Goldfarb, "The Typology of Systems of 'Participation'" in *Self-Governing Socialism*, ed. by B. Horvat, M. Marković and R. Supek (White Plains, New York: International Arts and Sciences Press, 1975), Vol. 2, p. 63.

228. Robert A. Dahl, *A Preface to Economic Democracy* (Berkeley: University of California Press, 1985), pp. 57-59.

229. *Wirtschaft und Gesellschaft Heute*, Bd. I, p. 12.

230. Jaroslav Vanek, "Technology, Self-Management and the Social Order: An Inquiry with a Concrete Dimension," *Socialism in the World* 6, No. 33 (1982): 60.

Chapter 5
The Theory of Work Activity

Emerging out of the criticisms presented in the final section of the preceding chapter, the constructive theory of work activity, which I offer here, opens with a consideration of the working person, followed by an account of the four central components of work activity: limits, sociality, technology and service. Each section of this treatment concentrates on those factors integral to work activity as they delineate an ethic of work. An ethic of work, however, is not a work ethic.[1] A work ethic frequently offers only a set of heteronomous moral structures which indicate the preferred attitude and conduct of laborers within a conventional system of tasks, without a sufficient consideration of the intrinsic significance of work or of the personal and collaborative ends pursued. Work encompasses matter, activity and goal in reciprocal interaction. In the following analysis, the ethics of work is characterized by the description of work as a human action[2] and by the denomination of the goals of this action in terms of the human community and the persons who compose it.

My presentation begins with a treatment of the person who works. In this section, I treat in turn the working person (1) who is engaged in this activity which is a good in itself, (2) who is a social being and (3) who manifests a four-fold freedom.

The Working Person

Work activity is not simply a means. As a central human activity, it is an end in itself.[3] The speech and action (as presented by H. Arendt) which maintain the polis are exceeded by the varied species of work activity, which manifest it as intrinsically human action. Arendt's well-known, triple distinction within work activity (labor, work and action) is useful but only as a description of the three elements each of which is found in every work activity.[4] Across the range of

these three elements, work activity is a good in itself which is characterized by self-determination, self-realization, creativity, rationality and sociality.[5]
As Hegel's theory demonstrated, the individual in society is the source of work activity regarded as both recognition and production. Individual persons are the origin of work—in their needs, choices, professions and activities—within the network of social relations and structures. Living, embodied persons generate an incessant, unstilled abundance of efforts which both create and express. Because of this continuous generation of products, creations, expressions and actions, persons themselves, ever novel, exceed codification. This originality, this capacity for generation, indicates a fundamental substrate of freedom in the person. Furthermore, in this, their expressive and productive activity, persons orient themselves fundamentally to other persons, indeed, to other persons in a social formation.

Through their work activity, persons come to a mutual recognition. The intimacy of the family initiates such a recognition but does not complete it. Within market and labor processes, persons recognize one another through their work activity. In this individual recognitive capacity, working persons acknowledge the present dignity of the others, and also discover in them new needs and further desires. In his/her ability to produce, the working person provides both for present needs and for the goods required in the further unfolding of others in society. Recognition and production each affirm the inherent capacity of the other person for freedom and contribute to the further actualization of this liberty.

Work does not mean the efforts in labor by an individual; nor does the ethics of work imply simply a morality of individual action at the place of work. Such interpretations rely on an understanding of work as merely productive. Through work the persons achieve their own selves only on the strength of society's mediation. It is through their work activity that persons draw from and respond to the society within which they complete their individual selfhood.[6] The expressive, reflexive and reciprocal character of recognition implies that work activity is essentially social. This social factor, linked with the dynamism of work, makes it possible for work to pertain not only to individual morality but further to those social structures which actualize ethics.

Work activity as an economic input requires a transferability and consequent anonymity of work acts. The conception of work activity, in so far as it is ethical, must encompass this transferability without eliminating the importance of embodied persons and their free wills in shaping and regulating the economic system. Hence I emphasize the creative action of working persons who both create themselves within the social network and contribute to the creation of society itself. While acknowledging the importance of productivity on the part of the working person,

I also maintain that the worker is a moral agent who exercises freedom and a concomitant responsibility.[7] The freedom of the working person pertains to each of the four key elements of work activity and actualizes itself in coming to responsibility in society. I now turn to each of these four elements.

Desire, which first seeks the satisfaction of minimal needs and, in providing for these, comes to pursue still higher satisfactions, expresses the freedom of work activity in the face of human limits. The satisfaction of any particular set of needs and desires is not the mark of free activity; rather, the establishment of an order—within the framework of which the individual and society satisfy their conjoint needs—signifies the achievement of freedom.[8] The satisfaction of needs within this harmonious pattern results from work activity.[9] Work activity represents a freedom which, confronted by limitations and impelled by human desire, proceeds to a still greater actualization which is ultimately the "equilibrium of civilization."[10] In the face of limits, work activity provides both freedom and liberation for the person and society.[11]

Sociality permeates the freedom of the working person. Hence, the unity and balance of pattern to be achieved in society finds its analogy in the ordering of the needs and desires of the individual.[12] Actualizating the freedom of work activity requires the recognition of each other and of one another's needs. One recognizes the other with that person's needs and desires. A responsibility toward the other in that individual's need expresses the freedom of the working person.[13] The other person does not constrain freedom; in calling the person to responsibility, the other makes possible the exercise of freedom.[14] The recognitive operation of work activity establishes liberty in society.

The technological side of work activity instantiates the freedom of work activity in still another way. Technology does not comprise merely a conglomeration of instruments and systems; it is integral to human activity and thus expresses the freedom of this activity. The concept of work activity as "the presentation of ideas through material forms"[15] makes clear that technology derives from human work activity. Here the intelligent exercise of discovery "frees man from servitude to a determined, always identical nature. . . ."[16] The contributions which technology has introduced throughout history in order to shorten the time of work have had the effect of requiring a lesser expenditure of energy; and further, the technological world has enabled persons to realize reflexively their capacity for greater, free activity.[17] The intelligent interaction of workers and tools both engenders skills and strengthens the exercise of their personal freedom.[18]

Service activity completes the exercise of freedom in work. As a working person contributes to the formation of a world where more persons can exercise their liberty, he or she becomes freer.[19] The work of persons as artificers frees them from the incessant demands which require labor for simple sustenance, because they create durable objects which facilitate the provision of that sustenance. Likewise, the work of the citizen frees the laborer and the artificer for the service of establishing a society where freedom may be achieved for all.[20] The freedom of work activity is not an individual license but a liberty exercised along with that of other persons. In this sense, freedom is organic. (1) It works in the face of human limits in shaping a life for persons; (2) it achieves its fullness through interrelation with others in society; (3) it extends itself into the world as machines and tools, which remain closely bound up with this free activity. (4) It completes itself in shaping a whole where all members of society may participate in its life. In the succeeding sections I shall explore each of these elements of freedom.

Work Activity and Personal Limits

The classical economists initiated their inquiry with regard to the actual needs and activities of working persons. Paradoxically, this focus on human need often led them to grant a flawed theoretical attention to the neediness of human persons. Death threatens; the necessities of life mark the inescapable margin. Persons must expend labor, itself painful, in order to meet the pains of living and the threat of death.[21] Because they desired to pursue their inquiry as a comprehensive theory and because they wished to offer remedy from the enormity of evils affecting the working classes, the classical economists concentrated their analysis on the conditions for economic growth, that is, for the development of the productive and exchange capacity of the entire nation. In order to treat these conditions theoretically, they set themselves at one remove from the actual situation. So much did economic theory concern itself with growth, that, while it focused on the production and exchange of surpluses, it underplayed the centrality of subsistence. The attention to theory, conceived on the model of the unchanging laws of physics, and the interest in the national wealth, set economic study on a trajectory which distanced the theoretical consideration from actual and embodied, multiple and localized, historical persons who work. Classical economics betrayed a propensity to overlook the evaluation of those socio-economic benefits which are conveyed to individual persons.

The classical economists recognized the link of the particular working person to the natural world but then applied the laws of this nature to the aggregate of personal activities. Rather than focusing on need alone, Paul Schrecker calls attention to the other side of the dual status of work activity, namely, the free exercise and historical deposit of this personal action.[22] In emphasizing this free, personal side, I present now a theoretical account of the human limits intrinsic to work activity. I examine in turn (1) the generation of history in the midst of limits, (2) the stimulus toward development which these limits provide and (3) the engendering of more intricate social desires which can lead to the formation of a proper shape out of these manifold needs.

Work activity is the primary human means for staving off death, or, more precisely, for rendering one's mortality in a particular pattern, since one can only delay death.[23] Indeed, the tie which work fashions between the person and the world offers yet another avenue along which the wearing and weakening natural forces bear down on the person.[24]

Whereas, in an obvious sense, time imposes limits on work activity, in another and profounder sense, it is work activity itself which, in its shaping of the world, makes time. Work activity postpones the effects of natural, unmet demands. In working, persons accord themselves a delay.[25] In overcoming the otherness of the natural world in order to create that from which one lives, the working person comes to have time.[26] Time then is not a mere receptacle for actions, but is itself created and re-shaped in a history of efforts.[27] Although limits are integral to human work activity because they constantly confront this effort, this effort continually changes these limits. Human work, of course, inserts itself into the circuit of the recurrence and satisfaction of need; it also creates expectations and objects which endure and thus reshape the world within which persons work. Still further, personal work sets up the original and the novel: new births, the outcomes of which are unpredictable and rich. In sum, whether within time's circuit, or in the movement along the line of history, or in the reorientation of the world and history onto a new plane, work activity confronts and reshapes temporal limits.

Work activity does not confront human limits simply as a means for their removal or for the satisfaction of certain requirements. The satisfaction found in work and the integration of this activity with the remainder of a person's life—both these contribute to the shaping of limits, for example, those imposed by age and health.[28] Persons satisfied in their work are frequently healthier; the dissatisfied, more often ill. Human needs represent a lack which must be satisfied. They also constitute an opportunity which stimulates both toward action and, consequently, toward the development of potentials. The existence of the person as a part of nature stimulates the development of the individual's capacities through work. Matter and spirit together make up the human being. These elements are uniquely conjoined and intermingled in the individual. The consubstantiality of matter and spirit within the human is the basis for the interpenetration of person and nature. Work activity mediates this complex interaction.[29] Each person engages him- or herself in activity in order to "live from" his or her needs.[30] Those needs feed his or her life as they stimulate him/her to move from privation to enjoyment.[31] Human work activity, the very possibility for meeting needs, itself becomes a requirement, an advanced, personal need.

In the satisfaction even of simple needs, work activity brings about more complex and more social needs, including those connected with the emergence of surplus labor, of free time and of advanced universality.[32] The fact that an individual's work is socially embedded reveals two further limits to his or her activity. Working persons cannot create *ex nihilo* but necessarily utilize the efforts of those who have preceded themselves. In addition, although workers—through both their activity and its results—attain a kind of existence beyond their own individual and collective limits, their efforts and products, in turn, exceed their own control.[33]

The rising from the subordination to nature's demands and the shaping of these new needs into a whole constitute the ethical element in work activity with regard to human limits. Through their individual efforts, persons determine the order and proportion among their needs and thereby act ethically. In the interaction with nature, persons achieve a form, a shape which is ethical.[34] This order must take into account the needs of the whole society and those needs, obtainable only through society, which individual persons require. Singly and jointly, persons desire (1) consumer goods to secure their lives, (2) useful and perduring objects through which they shape their world and (3) processes which devolve from their own activity whereby their further actions and expressive interchange remain secured. Work activity is ethical when it achieves a proper order amidst these multiple, desired goods.

The Sociality of Work Activity

The theoretical treatment of the sociality in work examines its multifaceted development (1) from cooperation in the face of limits, (2) through the desire for society as a constitutive element in work, (3) through the function both of recognition and expression, (4) to the ordering of a society in solidarity and free participation. I now treat each of these in turn.

Work activity is a principal achievement and expression of solidarity in the face of the limits which restrict society externally and internally. The institution of the wage contract, in which an apparently discrete exchange of money for work takes place solely between a laborer and an employer, makes it difficult to see the intrinsic sociality of work. The difficulty increases because some "unproductive" workers, that is, those whose efforts do not yield a material product, appear to contribute nothing to the total social product, even though their efforts make it possible that collectively working persons may survive and shape their needs. Nevertheless, through work activity persons cooperatively confront the limits which hedge them about. Furthermore, because many of the most important dependencies and vulnerabilities which confront persons emerge from a particular social matrix,[35] persons ought to undertake responsible action through their work activity for the shaping of the needs and desires which emerge in their particular society.

Work activity comprises not only the efforts to give a particular form to the limits which confront persons in society but also those exertions involved in the building up of society. In fact, work continually transforms the very idea of the community within which one works. Human work fabricates and organizes civilization in its entirety.[36] In the construction and maintenance of this entire range of civilization, individual working persons, although primarily active in a single province of civilization, partake in the entirety of the range of activities in their cosmos.[37] Their particular efforts cannot logically be simply economic,[38] for each work activity assumes a goal for which, in turn, a system of ends provides the framework.[39] This system implies the existence of other persons whose motivations and purposes constitute the organization of a civilization as one which is open and evolving.[40] Now I turn to the second element of the sociality of work, the desire for society.

Upon the field where continual collaboration creates, shapes and maintains all provinces of civilization through work activity, the significant, critical element for each particular person is work activity itself. It is through this that persons orient themselves toward each other in society. All results of productive or service activity, objects or actions, separate themselves from their originators and are handed over to others. Work aims at results; others take these up.[41] Work entails an orientation toward society. H. de Man articulates this orientation toward others in more concrete categories:

> For persons whose intellectual faculties are enlisted in the task of production, it entails . . . an element of need to do work that shall be qualitatively good—and the effort to do work of good quality is the purest expression of the spirit of workmanship, is a form of community feeling. But even when this motive is entirely excluded, in the worker's mind the aim of productivity remains inseparable from the

> aim of social utility. The conviction that their work has social value underlies all the demands of the organized workers, from the demand for a just wage upwards.[42]

Persons evaluate their work from the vantage of community standards and also expect that it be socially useful.

Social need is the basis for the evaluation of the quality and utility of the results of work activity. Yet need does not represent the final category of human want, instead desire does, a desire which work activity actualizes. This desire fulfills itself in relationship with the other.[43] Human need finally requires the other person. Persons undertake collaborative work, in part, in order to engage in relationships with others. This desire does not seek individuals merely in order to establish particular, useful or affective relationships; but, rather, reflexively and internally it searches for the appropriate arrangements of the members of society by which they can attain structured relationships with others through their work activity. Individual persons need and desire the other, communities and society. Society meets both its own profoundest need and its desire in achieving an internal order for the sake of persons. This order is dynamic since it is based on work activity. In attaining this order, society itself becomes ethical.

Sociality is one dimension where the ethical character of work unfolds. I now threat the third range of sociality: recognitive expression. The recognitive dimension of work activity implies that persons signal or symbolize themselves through the results of their particular work.[44] While persons disclose themselves in their individual work activities, the work which they undertake contributes to the creation and maintenance of a world. Alone, a person cannot bring forth a world; hence, to commence working is to initiate the social division of labor.[45] Persons attend to the expression of the others in their individual work because each recognizes the other as a collaborator in the common work. One can understand the expression of another in that person's work and can grasp the significance of it because the mutual recognition of persons provides a ground for this comprehension. In order to create a world, working persons must coordinate this division of labor; the recognition inherent in work activity represents the capacity for this cooperative organizing.

Community does not result from the merely contingent proximity of persons. Recognition constitutes community. Social life derives both from the "likeness of consciences" and from the division of labor.[46] The latter element corresponds to the productive side of work activity, the former to the recognitive. The recognitive element, that is, the "likeness of consciences" among those who acknowlege one another as collaborators, proceeds from and attains expression in work activity. The usual understanding of workers—as persons who are deployed in and subjected to a preordained system—presents the dependence of the worker.[47] In contrast to that view, the incorporation of the personal exercise of recognition into the theory of work activity, alongside productive enterprise, makes it possible for persons to secure freedom in their work.

Work activity conjoins both productive and recognitive elements; yet beyond establishing communal life in general, it further provides the basis for exchange. Exchange depends in a particular manner on signaling and symbols, which working persons interpret on the basis of their mutual recognition. Certainly exchange cannot be viewed as a mere extension of production, but neither does it occur apart from production. Exchange represents one result of work activity and reveals action (as it is understood in Arendt's theory).[48] A closer examination of the link between language and work will make clearer the broad extent of work activity in the constitution of society.

Theorists generally draw too great a distinction between work and language. M. Riedel acknowledges the close interdependence of society, work and language, yet distinguishes between the latter two by declaring that work requires artificial implements while speech does not.[49] Neither declaration is always true. E. Levinas attempts a distinction which only underlines the quality of expression found in work:

> [Language] . . . is a first action over and above labor, an action without action, *even though* speech involves the efforts of labor, *even though*, as incarnate thought, it inserts us into the world, with the risks and hazards of all action. At each instant it exceeds this labor by the generosity of the offer it forthwith makes of this very labor.[50]

Yet this generosity at most differs in degree not quality, for work activity itself, in entering the social sphere where others can make use of it and its products[51] and where they interpret it in ways which exceed the control of the working person, is itself generous in this orientation toward others. In fact, Levinas himself describes the self-effacement of the worker and his abandonment of the work result, but he still distinguishes the activity of expression from productive work when he compares personally expressive *action* with a produced *object* while failing properly to compare *expression* with work *activity*.[52]

Work activity is social through both its productive and recognitive elements. Just as in speech the sound conjoined with meaning yields expressions of a person understood within a society; so too in work activity, production and recognition together constitute expressive work activity. In work activity there emerges human meaning and creative expression which take on a form other than that found in speech or other such commonly recognized means of expression. Persons can achieve freedom (the exercise of which Arendt wishes to secure at the level of "action") in the public sphere, only if they also attain it in the "private" sphere of work and the economy.[53] One step to the attainment of this active freedom is to perceive the recognitive and expressive dynamics which are integral to work activity. In this way, theory need not accord a privilege to speech, a prerogative which then becomes calcified in a separation from and a denigration of productive activity. One can discern the close link between speech and work upon realizing the twin elements—productive and recognitive—within work activity.

The recognitive side of work activity makes possible expression among persons so that they can coordinate their system of work, that is, so that they may collaborate effectively. This coordination and collaboration represents the ordering of work activity in society. The ethical force of sociality derives from an aesthetic of order among cooperating workers.[54] Persons in society work so as to achieve a triple order, each within him- or herself, among the members of a society and, finally, between the personal and social orders. Work is an activity which is an element of some larger whole or which contributes to the uniting of parts into a whole.[55] Earlier I made use of H. de Man's insight that the spirit of workmanship represents one indication of the social orientation of work activity. This instinct for workmanship is another link between an aesthetic sense and a direction toward the social, for this spirit of craft derives from the bringing of a whole into existence.[56]

At this point I turn to the treatment of the fourth facet of sociality insofar as it is ethical: the ordering of society in solidarity and free participation.

The social whole represents an order across the entire range of civilization as it maintains a balance among the varied human desires[57] and through the history created by human activity.[58] In addition to taking both the economic and the immediate into account, the aesthetic of order must organize all civilized activity within a history. This aesthetic of order seeks to establish a social whole which is ethical, for this organization guarantees the freedom of the members of the community. The members attain their individual freedom only within a system where each individual's freedom coheres with that of each of the other members.[59] The principle of solidarity specifies this understanding of the ethics manifested in the sociality of work activity.

By solidarity I mean that principle which guides the organization of society so that it assures production and recognition both for that society and for the persons who constitute it.[60] Freedom in solidarity requires that the society not be a mechanical organization nor a system modelled on physiological processes but, rather, an organic whole, the life of which proceeds from its ethical character.[61] The character of this life can be ethical because it is free, specifically because each member of the group can contribute to its common purpose.[62] The common purpose of the group is to guarantee the productivity and recognition of persons and society. Ethical standards are those which maintain the fundamental conditions of this solidarity.[63] In order to attain this common purpose, reflection on and communication about the interaction of the society's members assume a greater significance here than that found in a society where working persons exist isolated one from another. Persons work and act properly in a society which directs its activities according to solidarity.

Solidarity as a principle for free human action leads to a more complete understanding of the parameters of social interaction than that found in either of the two major political-economic models. Whereas socialism emphasizes duties and liberalism secures individual rights, the interrelation of rights and duties in a system of solidary participation presents a model of social organization which is more adequate to the recognitive and productive sides of work activity. The guarantee of production in society requires that the benefits of this production reach those who work; the guarantee of recognition requires their participation in group and social decisions.[64] Prior to the industrial era, work activity had taken place within a communal system of status which then was dissolved in favor of the individualized and anonymous provision of labor for the sake of Promethean achievements in technology and production. A reform of the present system, a revamping which depends upon a reconstruction of the theory of work, cannot recover the former communal model of sociality. The contemporary demand is for the construction of a pattern of sociality which is not externally imposed—by either tradition or centralized authority—on the workers. A sociality based on participation can avoid the twin difficulties of a romance for the past or a sacrifice to an externally imposed scheme.

A solidary system established upon participation would answer the growing discontent with concentrations in industry and would meet the growing need for public, decentralized control of those goods immediately needed by persons, such as food and health services.[65] If a solidary system of participation is established, then recognition and expression interrelate producer, product and user. The alternative is the autonomous life of production, exchange and consumption, each isolated in its respective region. In a solidary system, workers, each with an individual voice, can distinguish themselves from and be superior to the labor power which each provides. The persons who make use of goods and services are not merely collocations of needs, and the finality of production is not merely utility. Production, in a constant challenge and response to the recognitive side of work activity, enables workers to provide for the dynamic order of society.[66] Such a society is just, for it not only allocates persons their proper shares of produced goods and services but also accords recognition to each: it acknowledges the uniqueness and pays heed to the voice of each.[67] Where the order of society guarantees the exercise of freedom in solidarity, there social justice prevails.

Technology and Work Activity

Persons are social not only as mediated by the productive and recognitive processes of work activity. They are also social in that they work with tools and machines, which are shared and traditional, toward establishing a common world. As I shall now present, technology is an element intrinsic to work activity (1) because it provides the capacity for liberation in the face of limits (especially through the reshaping of time), (2) because it mediates between working persons and the world of their work, (3) because it implies a society of collaborators and (4) because it concretizes human and social discovery. The technological devices, structures and processes do not constitute a neutral ground for activity and decision. Rather, the liberation which technology makes possible, its extension of personal capacities into the world, its implication of a shared world and its incorporation of the capacity for revealing a world pregnant with novel potentials for persons in society—all these indicate the ethical significance of this component of work activity.

Working persons design tools and machines to avoid or diminish the effects of the limits upon their work activity. Here lies the first of the four ethical elements of technology. The simple tool when put to use substitutes its wear for the wearying of the person. The tool takes the task at hand into its own time. Technology, the comprehensive and reflexive development of tools into machines and systems, deploys human curiosity and knowledge, the source of which often lies in work, in order to deflect the alacrity of decay and the imminence of demise. In this way, the tools and machines promise a liberation; but too often they offer a new bondage. Then, freed from the old toil and trouble, yet, owing to the restriction of access to the knowledge which has been built into machines, a new subservience emerges.[68] The effort expended in research and invention becomes idolatrous in so far as proper order is lost; that is, the priority of attention directed toward working persons disappears. Instead, in their flight from human limits, the technologies of production become their own ends and not only ignore human persons but spurn them as loci of weakness and limitation. Production mounts to the destruction of recognition.

A more complete understanding of technology vis-à-vis human limits treats the avoidance of limits together with the creating of a space for freedom. Through technology, the person is freed from the needs of the moment so that he or she can plan and order a history. Machines and tools set persons free from self-concern so that they can move towards each other.[69] Technology then not only puts time to efficient use, it also reshapes and reconstitutes time. Simply put, machines make possible time for rest and spare time on a different pattern and of a different quality than exist in work realms which are unaffected by technology. Understood more deeply, technology reconstitutes time with the result that under its liberating force time may become an ordered history and may support an ordered society.[70]

The second ethical element in technology lies in the particular mediation between persons and their world. The instruments of work activity—tools and machines—like the symbols of language, form a middle (that is, both external and internal), permanent ground upon which working persons may control natural processes. That the technological constitutes a world external to persons is evident. Yet, the activity of persons also constructs an internal bond between themselves and technology, both at its inception and in its continued use.

The utilization of tools engages working persons in their own particular capacities—sensate, voluntary and intellectual. Beyond the relief of some amount of pain, the recourse to technology brings about an enjoyment.[71] The use of tools and machines also implies a control and deployment reflecting the deliberate choice on the part of the working person.[72] Furthermore, technology is the embodiment of the achievements of the intellect; and the use of technology sometimes requires and often leads to the continued use of the person's intellectual capacities.[73] The invention and installation of technology derives from human skills which are simultaneously embodied, voluntary and intellectual. The resulting technology is a deposit not only of the willed direction of a particular force along a certain path but also of the insight, analysis and synthesis of human intelligence. Technology in many socio-economic arrangements may exist over against human persons. It proceeds, nevertheless, from their own capacities; and the exercise of the abilities of these persons in the utilization of implements can increase their own potential as well as develop these tools and machines.[74] The consubstantiality of spirit and matter constitute the human person;[75] intelligence, will and sensation interpenetrate one another in the work activity of the person. Technology, the deposit and active extension of working persons, deepens, sets in motion and guides this interpenetration between intelligence and the material.

The third ethical feature of technology is that it implies a collaborative society. The consubstantiality which exists in the person and which, in turn, analogously marks the technology that an individual creates or operates, also prevails at the level of society. The objective techniques required by particular tools and machines exist only within a particular society. Society and technology reciprocally affect one another.[76] Particular techniques and tools delimit the potential of a given society, which, in turn, gives form to these processes and machines according to its purposes.

The multiple objects and activities of technology indicate once again the social character of human work activity.[77] A particular set of apparatuses (for example, a workshop) does not stand ready solely for the work of a solitary individual. The technology of a workplace points to the public world and to its surrounding nature. The workshop of a city, its "roads, streets, bridges and buildings," connotes the larger world, including nature, on which it has imposed a definite orientation.[78]

Technology indicates its corresponding society. The society requires that this technology instantiate the possibilities of efficient production and mutual recognition. To this end, all members of society ought to partake in the control and direction of technology. Technology results from the collective efforts of their physical labor and intelligence. So too, both in their choices in the course of the utilization of machines for production and in their free exercise of will concerning the development and application of technology, the members of society will secure the realization of the recognitive side of their work.

In the area of discovery lies the fourth ethical characteristic of technology. Work is an activity which leads to discoveries, and technology is the deposit of these discoveries.[79] Work activity is not simply the completion of a task, for it includes the search for and the discovery of means.[80] Even the use of tools already developed entails discovery:

> The hand that bears itself toward the hammer or the nail that is to be driven in, is not simply the efficient cause of this end, while the end will be the final cause of the movement. For in the movement of the hand it is always to some measure a matter of seeking and catching hold of the good, with all the contingencies this involves. . . . Groping is not a technically imperfect action, but the condition for all technique.[81]

The intent of this passage from Levinas is not, of course, that the hand alone is significant in the discovery intrinsic to work activity; the hand represents the entire body and, thus, the entire person who engages in work activity marked by discovery.[82] Creation surpasses production because it discovers and generates a pattern of time as well as a world. The work of a particular individual or group is not merely instrumental; nor do the intentions and ends of those workers circumscribe it. The utilization of technology and the discovery which proceeds from work activity create a new world, the forces of which subsume the goals of the workers as it opens out into unanticipated possibilities which others can actualize.[83]

The discoveries brought to light through work activity eventuate in the construction of devices which contain and manifest the knowledge and learning that had led to their creation. This deposited knowledge is universal because in the machine it stands ready to be engaged by anyone. Knowledge transmitted by society led to the development of technology; the particular concretizations of this knowledge become available to all in society for use. These same tools are indeed the repositories of the potential not solely of the persons who utilize them but of the entire society.[84] Whether these potentials are utilized for abstract ends, solely for particular groups, for technology itself or, rather, for meeting the needs of persons, constitutes the question of the service element in work activity and the question of how this element provides ethical direction for the economic activity based on work.

Service and Work Activity

Working persons in society can apply technology to the confrontation with human limits in two opposed ways. The production of goods which meet abstractly conceived needs can utilize technology. In this scheme, the habits of abstraction subserve a situation where production becomes its own end. Alternatively, the provision of the desired means to meet human needs can make use of technology. In order to emphasize the importance of this alternative and to make it more feasible, the recovery of the full concept of work activity brings to light its service element. Service is that element of work activity which deploys the potential of society so as to assist persons in their need. Service means the cultivation of the earth and the generation of objects for the benefit of persons. Service cherishes and protects persons.

The service element in work activity includes (1) a contemplative and dynamic recognition of the limits of others, (2) an extension or unfolding of society, (3) the exercise of freedom in responsibility and (4) the establishment of a reflective, communal life. Each of these facets of service I shall examine in turn.

In service, persons regard each other and discover their capacities and limits. This is the first ethical element of service. This contemplative stance in work activity comprises two elements: recognition and discovery. When these two elements are understood, service can replace a deficient notion of contemplation which disjoins it and sets it above work activity.

The attention to the other person initiates the confrontation with human limits, avoids the misdirection of work activity toward abstract goals and provides an ethical basis for mutual assistance.[85] Contemplation directed toward the other focuses work activity in accordance with an awareness of the limits and potentials of that individual. Abstract processes do not then bind work and exchange. To contemplate the other is to recognize that particular person: to remark the difference, the location, the need and the claim of that particular person.[86]

To contemplate the other person leads to service. One regards the face of the other; therein by synechdoche, one sees the entire person in that individual's embodiment and, consequently, in his or her need.[87] The society of working persons in their contemplation of one another undertakes the meeting of needs, the reshaping of limits and the preparation for death. A caring service emerges when these concerns are recognized. Some of the needs are, "created, shaped, or sustained by current social arrangements."[88] One form of service is collaboration to reduce or reshape such contingent needs and the limits. Implicit in the attention to and focus upon the actual person is the awareness that the other is similar to oneself and that all are capable of engaging in (or actually do engage in) such a contemplative regard for the other. This sense of equality prevents this care from becoming a dominating intrusion. Persons, in their activities together, direct themselves to the service of one another, each in his or her respective needs.

The contemplation which is the initial stage of service is dynamic in this way: it discovers the other and discovers what this other expresses. The emphasis here lies on bringing the person into focus and on uncovering the less conspicuous needs. The natality of action[89] I ascribe to service activity; for it reveals the new. The capacity for discovery inherent in work activity here aims at the other members of society. Persons come to the edge, limit, margin of their particular capacities, especially of their individual abilities to work. On the basis of self-awareness and by means of the contemplation of others, one can discover the needs of others, especially of those others who have until now remained unseen and unheard in the usual actions of society. Service discovers the new: new needs to be met and other persons to be newly integrated as participants in society. One's collaborators—those accepted and those freshly discovered—become the final object of work activity.

Service also includes a second ethical element: the extension of society. Work activity conduces to the unfolding and completion of society. Through work activity needs are met and more needs develop. Work creates wealth, the purpose of which is to secure new activity.[90] Only other persons can originate this activity. The end of work is not merely provision for needs but a betterment of human existence itself.[91]

In a person's work activity each individual is recognized and reciprocally he/she manifests a certain recognition toward others. Since the person is a historical being, each individual grows or declines. An individual's growth depends in part on the shape of that person's society; the development of that society, in turn, depends on the individual's growth. Personal growth in recognition proceeds insofar as an individual extends recognition to others and, indeed, to ever more members of society. In this, the need for other persons becomes more pronounced. As work continues, persons see the others in society more clearly and come to understand their needs more fully. Work brings about discoveries not only in the realm of objects and techniques but also in the sphere of persons and their interactions. Through the discovering activity of cooperative work, persons recognize more persons as members of society and comprehend a broader range of their needs.

Work activity moves beyond the completion of a particular object or service. Work does not exhaust itself in laboring to meet present need, not even the most advanced needs of the entire society. Embodied as personal activity and established as a social action, work pursues not just the present goal but the larger purpose of bringing society to its fullness.[92] Sociality is intrinsic to work activity and persons bring society to its fullness by assuring that step by step all members of society come to participate in it and by guaranteeing provision for their needs, including the higher and more complex ones. Persons who make a history strive to perfect the shape and structure of society.[93] Through work activity persons deepen their initial sociality; service represents the most significant pattern of a profounder sociality.[94]

The third ethical component of service is the exercise of a responsible freedom. More than the particular result of their efforts, in work activity what matters is the free activity of persons in their working. Service pursues, above all, the guarantee of this central need, namely, the securing of the freedom of each person. In so doing, persons actualize their own individual freedom.[95]

Ideally, the freedom of the working person stimulates the freedom in another.[96] This constitutes the work of service. The exercise of freedom by individual persons entails their responsibility for those affected by their actions. The sociality of work activity implies that their work affects many directly and indirectly. Yet a central need of the other is the necessity to have his or her freedom secured. Because of their own freedom, working persons assume a responsibility individually and collectively to guarantee the freedom of others.[97]

Work activity comprises three elements. The first two—the production of consumption goods and the making of lasting objects (the devices of technology and the articles of the human dwelling)—make possible and find their final sense in the third part—the action of service. The securing of the necessities of existence, the creation of a world and society where persons recognize one another and where their expression has meaning—both these are integral to and find completion in emancipation, that is, in the securing of free participation for ever more members of society.

The service side of work activity makes possible the establishment of relationships among free persons by creating a world and a society fit for action and speech. It is here where the fourth element of the ethical significance of service lies. Service work is reproductive,[98] because it maintains the society and its world over time and because it contributes to the development of this human world. Even production itself can occur on the basis of service, for it requires the continuity of social structures.[99] Yet to care for and nurture and sustain the human world requires clarity about the reflective character of work activity. Service as the maintenance of society and its world requires the mutual reflection of persons about their work activity. Service points toward a communal interaction which springs from work activity itself.

This communal interaction refers to the political component of work activity which the theoretical literature has so often ignored. To ignore the political element would be to neglect the element of power which extends not solely over materials and processes but, further, over persons. This would disregard the conflicts which emerge in the work situation. It would be an error in another direction to focus only on these conflicts and the misuse of power over persons. The maintenance of a world and society where persons freely collaborate requires a political cooperation that enables them to direct the use of power and to attempt to resolve conflicts. The contemporary operation of systems of employment, publicly pruned for the most part of political elements, makes it difficult for many to perceive this interactive, service side of their own activity. Yet cosmetic approaches which introduce more leisure after or more variety during the hours of employment will not alleviate the modern, wide-spread dissatisfaction with work.[100] The transformation of work in theory and in practice must acknowledge the political forces active in the workplace.[101]

In order to complete work as an ethical activity, working persons must undertake to shape their society so that it excludes evil uses for the social product and so that persons may be free and act virtuously.[102] The discovery and contemplation of the other leads to the recognition of many others and develops into the demand for a system to secure this recognition.[103] The attention directed toward others also reveals their weaknesses and limits. This attention, coupled with the person's responsibility, evolves into the political activity to organize and to create structures of coordinated action.[104] The work of service pursues the establishment and maintenance of the polis in the face of the transience of human expression and activity.[105] Service is a human work which makes it possible for persons to continue not only as consumers and craftspeople and also as free collaborators and citizens.

In the preceding presentation, the emphasis on meeting human needs has implied a stress on the use value of goods as opposed to their abstract, exchange value. Similarly, I have hoped to rectify the theoretical bias in favor of production by offering a schema in which service can assume a significant position alongside production. At the same time, the ethical interest has focused not on universal norms governing conduct but, rather, on the developing capacities of persons in society, which itself can advance in knowledge, organization and common values.

All work activity encompasses, in varying degrees, production for the maintenance of life, the creation of durable devices and objects which furnish the world, and action which continues and fosters communal life. Civilization, as it is borne by this production, creation and public action, stands in an intimate reciprocity with work activity and personal freedom.[106] In the course of the development of civilization, different types of work activity have been the predominant force in society's advance. Agriculture, commerce, and industry have been three highly significant stages of this progress. All three operate at each step, but each in turn has been the focus for the development of civilization. As this unfolding has taken place, an increasingly larger portion of the working population has been able to participate in political processes.

In considering the four key elements of work activity, there has been a similar development. Work activity for millenia concerned itself with providing sustenance. With the onset of the industrial era, it was quite clearly the case that the role of workers in society was crucially important. For several decades now—as a consequence of the sweeping changes introduced by internal combustion, electrical power, medical technology and nuclear science—the ambiguous function of technology in determining the contours of the workplace and, consequently, shaping the life of persons, has assumed a central position in the public forum. In the inchoate, current discussion of service activities, the study of workers has returned to a theme which Smith himself considered. Now, however, in a world of leisure activities, reduced work weeks, structural unemployment and the deterioration in basic services, the question of service is moving to a pivotal position in analyses of work activity. As a greater proportion of the population has gained the possibility of participating in the shaping of society through the action of politics, so it is the occasion for understanding work activity as the basis for their service of one another and to grasp political activity as a privileged form of this service of the common good. The substantive actualization of the formal possibility for political action can take place through the service of participation. Such a society will be ethical, as working persons will move from passivity to action, will prize human interchange, will strive to alleviate suffering and will endeavor to secure freedom.

1. Fritz Giese, *Philosophie der Arbeit* (Halle a. S.: Carl Marhold, 1932), p. 244.

2. "An der Spitze der Wertordnung aller Zwecke und Tätigkeiten steht also kein fertiges Werk (érgon), sondern das Am-Werke-Sein (enérgeia), der zu Leben gehörige Vollzug der Praxis, dessen Zweck wiederum das 'gute' d. h. tugendhafte und glückliche Leben (eû zên) ist." Manfred Riedel, "Arbeit" in *Handbuch philosophischer Grundbegriffe*, hrsg. von H. Krings, H.M. Baumgartner und C. Wild (München: Kosel, 1973), Bd. 1, 129.

3. "The 'work of man' is no end because the means to achieve it—the virtues or *aretai*—are not qualities which may or may not be actualized, but are themselves actualities." Here Hannah Arendt understands "end" as one which is extrinsic to the person. See *The Human Condition* (Chicago: University of Chicago Press, 1968), p. 207.

4. James T. Knauer, "Rethinking Arendt's 'Vita Activa': Toward a Theory of Democratic Praxis," *Praxis International* 5, No. 2 (July, 1985): 187-188.

5. Mihailo Marković, *From Affluence to Praxis: Philosophy and Social Criticism* (Ann Arbor: The University of Michigan Press, 1974), pp. 65-66.

6. Dietmar Mieth, "Solidarity and the Right to Work," *Concilium* 160 (1982): 63.

7. David P. Ellerman, "Capitalism and Worker's Self-Management," in *Self-Management: Economic Liberation of Man*, ed. by Jaroslav Vanek (Harmondsworth: Penguin, 1975), p. 155.

8. Paul Schrecker, *Work and History: An Essay on the Structure of Civilization* (Princeton: Princeton University Press, 1948), p. 264.

9. "Since freedom consists in the opportunity for conjoint satisfaction of the several needs and desires of civilized man, and since the coordination and reciprocal adaptation of the norms articulating this satisfaction cannot be accomplished except through work, all freedom is essentially freedom of work." *Ibid.*, p. 259.

10. *Ibid.*, p. 268.

11. "Die Arbeit ist Ausdruck der Befreiung des Individuums und der Gesamtheit." Giese, *Philosophie der Arbeit*, p. 298.

12. Schrecker, *Work and History*, p. 300.

13. "Before the hunger of men responsibility is measured only objectively; it is irrecusable. The face opens the primordial discourse whose first word is obligation. . . ." Emmanuel Levinas, *Totality and Infinity: An Essay in Exteriority*, translated by Alphonso Lingis, Duquesne Studies, Philosophical Series, 24 (Pittsburgh: Duquesne University Press, 1969), p. 201.

14. *Ibid.*, p. 197.

15. For "die Darstellung von Ideen durch materiellen Formen," see Viktor von Weizsäcker, "Zum Begriffe Arbeit," in *Synopsis: Festgabe für Alfred Weber* (Heidelberg: Verlag Lambert Schneider, 1948), p. 724.

16. Adriano Tilgher, *Work: What It Has Meant to Men Through the Ages*, translated by Dorothy Canfield Fisher (New York: Harcourt Brace and Company, 1930; reprint ed., Harcourt Brace Jovanovich, 1977), p. 98.

17. Karl Rahner, "Theological Remarks on the Problem of Leisure," in *Theological Investigations*, Vol. IV, translated by Kevin Smith (Baltimore: Helicon Press, 1966), p. 385.

18. The special position of the skilled worker indicates the close association between technology and freedom. This person, skilled in his use of technological implements, is more difficult to control than the unskilled worker, less subject to workplace discipline. The more complex the task workers undertake with the machines, the more difficult it is to determine what is to be expected from them. David Schweickart, *Capitalism or Workers' Control?: An Ethical and Economic Appraisal* (New York: Praeger, 1980), p. 165.

19. Tilgher, *Work*, p. 134.

20. This represents a development of ideas from Arendt, *Human Condition*, p. 236.

21. Leo Strauss, *Natural Right and History* (Chicago: University of Chicago Press, 1953), p. 250.

22. ". . . All events in the economic province . . . are events of nature at the same time as they are events of history . . . . To contend that, considered as events of nature and considered as events of history, acts of economic work exhibit the identical structure would be tantamount to considering the fundamental principle of economy as a natural law . . . ." Schrecker, *Work and History*, p. 169.

23. "The real 'economic' problem is indeed created by scarcity: not the 'economic' scarcity of factors of production like natural resources, labor and capital, or the products and services produced for sale to others; but the *existential* scarcity of the penultimate resources of life, time and energy, a scarcity created by our finitude and mortality. And there is indeed in human nature a problem of allocation of resources, not merely between what goods to produce and to consume but how to allocate our entire life-time and life-energy, the totality of our being between the various dimensions of our existence. . . ." Walter A. Weisskopf, *Alienation and Economics* (New York: E.P. Dutton and Co., 1971), p. 187.

24. Tilgher, *Work*, p. 146.

25. Levinas, *Totality and Infinity*, p. 165.

26. *Ibid.*, p. 117.

27. M.D. Chenu, *The Theology of Work: An Exploration*, translated by Lilian Soiron (Chicago: Henry Regnery Co., 1966), p. 69.

28. Ronald Mason, *Participatory and Workplace Democracy: Theoretical Development in Critique of Liberalism* (Carbondale, Illinois: Southern Illinois University Press, 1982), p. 129.

29. Chenu, *Theology of Work*, p. 31.

30. Levinas, *Totality and Infinity*, p. 111. Also see p. 114 where the human "thrives on" and "is happy from" his needs.

31. "Need is also the time of labor: a relation with an other yielding its alterity. To be cold, hungry, thirsty, naked, to seek shelter—all these dependencies with regard to the world, having become needs, save the instinctive being from anonymous menaces and constitute a being independent of the world, a veritable subject capable of ensuring the satisfaction of its needs, which are recognized as material, that is, as admitting of satisfaction." Levinas, *Totality and Infinity*, p. 116.

32. Heller, *Theory of Need in Marx*, pp. 90-91.

33. Levinas, *Totality and Infinity*, p. 176; and Tilgher, *Work*, p. 137.

34. "[The worker] . . . realizes form when he creates time, and opposes constancy with alteration, the eternal unity of his ego with the diversity of the world; he gives form to matter when he proceeds to annul time, affirms persistence within change, and subjects the diversity of the world to the unity of his ego." Friedrich Schiller, *On the Aesthetic Education of Man: In a Series of Letters*, translated and introduced by Reginald Snell (Frederick Ungar Publishing Co., 1954), p. 63.

35. Robert E. Goodin, *Protecting the Vulnerable: A Reanalysis of Our Social Responsibilities* (Chicago: University of Chicago Press, 1985), p. 191.

36. Schrecker, *Work and History*, p. 197.

37. *Ibid.*, p. 258.

38. ". . . No act of work is completely determined by the economic norm alone. The latter exclusively determines a formal notion of expenditure and effort, and not the element entering this proportion, nor the means of actualizing it." *Ibid.*, p. 74.

39. Max Scheler clarifies this point: ". . . Dass der Imperativ 'du sollst arbeiten' schlechthin ohne nähere bedingende Angaben, kein sittliches Gesetz sein kann. Wie alles mittelbare Tun . . . ist die sittliche Qualität der Arbeit abhängig von Zwecken, und zwar von den Zwecksystemen der Organisation und des Rechtes." See "Arbeit und Ethik," in *Gesammelte Werke*, hrsg. von Maria Scheler und Manfred S. Frings (Bern und München: Francke Verlag, 1971), Bd. 1, p. 186.

40. *Ibid.*, p. 170.

41. For this understanding see Levinas, *Totality and Infinity*, p. 227: ". . . Inscription in a foreign will is produced through the mediation of the work, which separates itself from its author, his intentions, and his possession, and which another will lays hold of. The labor which brings being into our possession ipso facto relinquishes it, is in the very sovereignty of its powers unceremoniously delivered over to the Other." Although Levinas' insight utilizes the image of the produced object, it does not depend on it.

42. Henri DeMan, *Joy in Work*, trans. by Eden and Cedar Paul (London: George Allen & Unwin, 1929), pp. 140-141.

43. Levinas, *Totality and Infinity*, p. 179.

44. *Ibid.*, p. 176.

45. Karl Jaspers, *Von der Logik: Philosophische Logik*, Erster Band (München: R. Piper & Co. Verlag, 1947), p. 331.

46. Emile Durkheim, *The Division of Labor in Society*, translated by George Simpson (Glencoe, Illinois: The Free Press, 1933), p. 226.

47. Scheler, "Arbeit und Ethik," p. 176.

48. H. Arendt distinguishes between labor, work and action as three types of human activity. Surprisingly, despite her repeated efforts to maintain the conceptual boundaries between these, she writes that: ". . . While the public realm as exchange market corresponds most adequately to the activity of fabrication, exchange itself already belongs in the field of action and is by no means a mere prolongation of production. . . ." See *Human Condition*, p. 209.

49. Riedel, "Arbeit" in *Handbuch philosophischer Grundbegriffe*, hrsg. von H. Krings, H.M. Baumgartner und C. Wild, Bd. 1, p. 138.

50. Levinas, *Totality and Infinity*, p. 174; emphases added. He adds later, "I speak a word that, in the measure that it welcomes the Other as Other, offers or sacrifices to him a product of labor, and consequently *does not play above economy*." *Ibid.*, p. 231; emphases added.

51. Weizsäcker notes, "Die Selbstverborgenheit der Arbeit, die darin besteht, dass man mit dem Produkt verschiedenes anfangen kann, gehört also unabtrennbar zu ihrem Begriffe." See "Zum Begriffe Arbeit" in *Synopsis*, p. 735.

52. Levinas, *Totality and Infinity*, p. 227. This same failure to integrate the activity of work into his theory may also be found on p. 297 where he speaks of the "abyss" between language and labor.

53. Mason, *Participatory and Workplace Democracy*, p. 238.

54. "Though need may drive Man into society, and Reason implant social principles in him, Beauty alone can confer on him social character. Taste alone brings harmony into society, because it establishes harmony in the individual." Schiller, *On the Aesthetic Education*, p. 138.

55. Richard Burke, "'Work' and 'Play'," *Ethics* 82 (October, 1971-1972): 44.

56. *Ibid.*, p. 46.

57. Schrecker, *Work and Human History*, p. 133.

58. *Ibid.*, p. 246.

59. F. Schiller describes a person who encounters another within this community: ". . . In the sphere of cultivated society, in the aesthetic state, he need appear to him only as shape, confront him only as an object of free play. To grant freedom by means of freedom is the fundamental law of this kingdom." *On the Aesthetic Education*, p. 137.

60. "The material impulse and the formal are equally earnest in their demands . . . in its action the first is directed towards the maintenance of life, the second towards the preservation of dignity . . . ." Schiller, *On the Aesthetic Education*, p. 78.

61. Franz Kempel, *Göttliches Sittengesetz und Neuzeitliches Erwerbsleben: Eine Wirtschaftslehre in sittlich-organischer Auffassung der* gesellschaftlichen Erwerbsverhältnisse (Mainz: Verlag von Franz Kirchheim, 1901), pp. 151-153.

62. For the idea of a group united through its common purpose, see Ernest Barker, "Introduction," In Otto Friedrich v. Gierke, *Natural Law and the Theory of Society: 1500-1800*, translated by Ernest Barker (Cambridge: Cambridge University Press, 1958), p. lxxviii.

63. ". . . In general, the characteristic of moral rules is that they enumerate the fundamental conditions of social solidarity." Durkheim, *Division of Labor*, p. 398.

64. H. de Man describes the connection between solidarity and participation: "No eagerness for work, until there is a working community; no working community, until there is a community of wills and interests; no community of wills until all have an equal say in the matter; and no community of interests, until all have a right to share in the fruit of production." See *Joy in Work*, p. 74.

65. H.C. Binswanger et al. address this issue when speaking about the centralization of production: "Die Kehrseite besteht oftmals darin, dass die abseits der Zentren befindlichen Konsumenten, Einwohner, Bürger ihre Bedürfnisse und Vorstellungen nicht mehr so wirksam geltend machen konnen. Bei ihnen entsteht als Folge der Zentralisierung ein neuer Bedarf an dezentralisierter Versorgung und Mitentscheidung, den sie nun aber oftmals selbst decken müssen." See Arbeit ohne Umweltzerstörung: *Strategien einer neuen Wirtschaftspolitik* (Frankfurt am Main: S. Fischer, 1983), pp. 173-174.

66. Jean Baudrillard, *The Mirror of Production*, translated and introduced by Mark Poster (St. Louis: Telos Press, 1975), pp. 102-103.

67. Levinas, *Totality and Infinity*, p. 298.

68. Max Horkheimer describes a peril of the contemporary age: "Human toil and research and invention is a response to the challenge of necessity. The pattern becomes absurd only when people make toil, research and invention into idols." See *Eclipse of Reason* (New York: Continuum, 1974), p. 153.

69. F. Giese writes of a liberation toward society and into history: "Die Technik befreit den Menschen vom Raum und Zeit. Ihre Sinngebung bedeutet damit eine Befreiung der zufälligen Tageserscheinung aus der Vergessenheit punktuell historischen Existenz und die Lösung des Menschen als Einzelperson von der Ortsfixierung seines Ichs." *Philosophie der Arbeit*, pp. 286-287.

70. Adriano Tilgher, in commenting on a scene from Goethe's *Faust*, provides an interesting indication of the connection between technology, the establishment of a society and the opening into a history. ". . . In his old age, he [Faust] worked at the draining of a swamp, where some day a new nation was to live in fruitful toil. As he contemplates the future results of his work, a beatific joy comes to him and he cries out to the moment as it passes, 'Ah stay! Thou art so

fair!' He dies and Mephistopheles is about to catch his soul, but loses the wager because at that moment Faust was enjoying a pleasure that lay not in the present but in the future." *Work*, p. 139.

71. "The handling and utilization of tools, the recourse to all the instrumental gear of life, whether to fabricate other tools or to render things accessible, concludes in enjoyment." Levinas, *Totality and Infinity*, p. 133.

72. "The hand comprehends the thing not because it touches it on all sides at the same time (it does not touch it throughout), but because it is no longer a sense-organ, pure enjoyment, pure sensibility, but is mastery, domination, disposition—which do not belong to the realm of sensibility." *Ibid.*, p. 161.

73. M. Horkheimer observes, "The human intellect, which has biological and social organs, is not an absolute entity, isolated and independent. It has been declared to be so only as a result of the social division of labor, in order to justify the latter on the basis of man's natural constitution." *Eclipse of Reason*, p. 54.

74. "La parole donne la maitrise intérieure de la pensée, la technique celle de la nature. A l'enrichissement de l'esprit humain par la parole correspond l'enrichissement de la réalité humaine par la technique." Henri Arvon, *La Philosophie du Travail* (Paris: Presses Universitaires de France, 1961), p. 44.

75. Chenu, *Theology of Work*, p. 22.

76. *Ibid.*, pp. 87-88.

77. "As this destining, the coming to presence of technology gives man entry into something which, of himself, he can neither invent nor in any way make. For there is no such thing as a man who exists singly and solely on his own." Martin Heidegger, "The Question Concerning Technology," in *Basic Writings*, edited and introduced by David Farrell Krell (New York: Harper and Row, 1977), p. 313.

78. Martin Heidegger, *Being and Time*, translated by John Macquarrie and Edward Robinson (New York: Harper and Row, 1962), p. 100. Also see p. 105.

79. F. Giese draws attention to the significance of machine technology beyond its utilitarian benefit and points to a significant human capacity which work deposits in this technology, namely the ability to discover. *Philosophie der Arbeit*, p. 285.

80. Scheler, "Arbeit und Ethik," p. 170.

81. Levinas, *Totality and Infinity*, p. 167.

82. "The body as possibility of a hand—and its whole corporeity can be substituted for the hand—exists in the virtuality of this movement betaking itself toward the tool." *Ibid.*

83. For a suggestion toward this idea, see M. Heidegger, "The Question Concerning Technology," p. 302: "In enframing that unconcealment comes to pass in conformity with which the work of modern technology reveals the real as standing reserve. This work is therefore neither only a human activity nor a mere means within such activity."

84. Sohn-Rethel, *Intellectual and Manual Labour*, p. 177.

85. ". . . Going beyond death is produced not in the universality of thought but in a pluralist relation, in the goodness of being for the other, in justice." Levinas, *Totality and Infinity*, pp. 301-302.

86. *Ibid.*, p. 75.

87. "I analyze the inter-human relationship as if . . . his face, the expressive in the Other (and the whole human body is in this sense more or less face), were what ordains me to serve him." Emmanuel Levinas, "Ethics and Infinity," *Cross Currents* (Summer, 1984): 195.

88. Goodin, *Protecting the Vulnerable*, p. xi.

89. Arendt, *Human Condition*, pp. 177-178.

90. Tilgher, *Work* p. 137.

91. M.D. Chenu, "Work," in *Sacramentum Mundi*, edited by Karl Rahner et al. (New York: Herder and Herder, 1970), p. 371.

92. "Neben das 'arbeiten' tritt also als ein Neues eine Pflicht zu energischen Mitschaffen an der rationellen Vervollkommnung des gegebenen Staates, der als das Medium erscheint, durch welchen die von allen Bürgern als objektiv gültig für alle Bürger erkannten Zwecke auf alle Bürger in den Form eines systematischen Ganzen zurückwirken." Scheler, "Arbeit und Ethik," p. 181. Also note that A. Tilgher asserts that the Hebrews understood work as conducing to completion and unity. *Work*, pp. 13-14.

93. For this perspective on work activity, see Giannino Piana, "Human Work: Blessing and/or Curse?" *Concilium* 160 (1982): 69.

94. "Man, who is by nature social, becomes more social, not only through exchange and voluntary contracts, but through the interweaving of his activities into a social fabric. . . . The realm of 'the neighbor' emerges with new and unforeseen dimensions." M.D. Chenu, "Work," in *Sacramentum Mundi*, p. 372.

95. "To be free is to build a world in which one could be free." Levinas, *Totality and Infinity*, p. 165.

96. ". . . There is also the possibility of a kind of solicitude which does not so much leap in for the Other as leap ahead of him . . . in his existentiell potentiality-for-Being, not in order to take away his 'care' but rather to give it back to him authentically as such for the first time. This kind of solicitude pertains essentially to authentic care—that is, to the existence of the Other, not to a 'what' with which he is concerned; it helps the Other to become transparent to himself in his care and to become free for it." Heidegger, *Being and Time*, pp. 158-159.

97. Here I have made use of R. Goodin's First and Second Principles of Individual Responsibility. See *Protecting the Vulnerable*, pp. 118 and 139.

98. Jürgen Moltmann, "The Right to Meaningful Work," in *On Human Dignity: Political Theology and Ethics* translated and introduced by M. Douglas Meeks (Philadelphia: Fortress Press, 1984), p. 55.

99. Martin Hollis and Edward J. Nell, *Rational Economic Man: Philosophical Critique of Neo-classical Economics* (Cambridge: Cambridge University Press, 1975), p. 251.

100. Arvon, *Philosophie du Travail*, p. 108.

101. R. Mason has taken the approach of reading the political back into the workplace. Given the resistance of actual workers to the political, indeed especially since they all too often understand it as an intrusion upon the sphere wherein they gain security, I contend that the more cogent approach is to examine how political activity emerges from work itself and how this activity is actually a species of the larger category of service.

102. M. Scheler states a critical measure for evaluating whether work is ethical: "Arbeiten für die Gesellschaft ist nur dann ein gutes Arbeiten, wenn zuvor festgestellt ist, dass diese Gesellschaft aus guten Menschen besteht." See "Arbeit und Ethik," p. 185. By this Scheler means that the working person has a responsibility to make certain that the results of his efforts are used only to good ends. See *Ibid*.

103. "The metaphysical relation of the I with the Other moves into the form of the We, aspires to a State; institutions, laws, which are the source of universality." Levinas, *Totality and Infinity*, p. 300.

104. Here I have made some use of R. Goodin's Principle of Group responsibility. *Protecting the Vulnerable*, p. 136.

105. Arendt, *Human Condition*, pp. 197-198.

106. Schrecker, *Work and History*, p. 269.

Sources Consulted

Primary Sources: Books

Delbrück, H., Schmoller, G. und Wagner, Ad. *Über die Stumm'sche Herrenhaus-Rede gegen die Kathedersozialisten—Schreiben an den Geh. Justizrath Professor D. Dr. Hinschius*. Berlin: Verlag von Georg Stilke, 1897.

Hegel, Georg Wilhelm Friedrich. *Hegel and the Human Spirit: A Translation of the Jena Lectures on the "Philosophy of Spirit" (1805-1806) with Commentary*. Translation and commentary by Leo Rauch. Detroit: Wayne State University Press, 1983.

________. *Jenäer Systementwürfe I, Gesammelte Werke*, Hrsg. von der Rheinisch-Westfalischen Akademie der Wissenschaften, Bd. 6 hrsg. von Klaus Düsing und Heinz Kimmerle. Hamburg: Felix Meiner Verlag, 1975.

________. *Jenäer Systementwürfe III, Gesammelte Werke*, Hrsg. von der Rheinisch-Westfälischen Akademie der Wissenschaften, Bd. 8 hrsg. von Rolf-Peter Horstmann unter Mitarbeit von Johann Heinrich Trede. Hamburg: Felix Meiner Verlag, 1976.

________. *Phenomenology of Spirit*. Translated by A.V. Miller with analysis and foreword by J.N. Findlay. Oxford: Oxford University Press, 1977.

________. *Philosophy of Mind*. Part Three of the *Encyclopaedia of the Philosophical Sciences (1830)*. Translated by William Wallace, together with the Zusätze in Boumann's text (1845) translated by A.V. Miller. Oxford: Clarendon Press, 1971.

________. *Philosophy of Right*. Trans. and ed. by T.M. Knox. Oxford: Oxford University Press, 1952.

________. *System of Ethical Life (1802/1803)* and *First Philosophy of Spirit, Part III of the System of Speculative Philosophy 1803/1804*. Edited and translated by H.S. Harris and T.M. Knox. Albany: SUNY Press, 1979.

________. *Werke* in 20 Bände. Theorie Werkausgabe hrsg. von Eva Moldenhauer und Karl Markus Michel. Frankfurt am Main: Suhrkamp Verlag, 1970-1971.

Keynes, J.M., ed. *Official Papers by Alfred Marshall.* Published for the Royal Economic Society. London: Macmillan and Co., 1926.

Marshall, Alfred and Marshall, Mary Paley. *The Economics of Industry.* 2nd ed. London: Macmillan and Co., 1881.

Marshall, Alfred. *The Early Writings of Alfred Marshall, 1867-1890.* 2 vols. Edited and introduced by J.K. Whitaker. London: Macmillan for the Royal Society, 1975.

________. *Elements of Industry*, being the First Volume of *Elements of Economics*. 2nd ed. London: Macmillan, 1898.

________. *Principles of Economics*. 2 vols. Ninth (Variorum) Edition with annotations by C.W. Guillebaud. Vol. 1: Text. London: Macmillan, 1961.

Marx, Karl and Engels, Frederick. *Collected Works*. Vol. 3: 1843-1844. New York: International Publishers, 1975. Vol. 5: 1845-1847. Published 1976. Vol. 9: 1849. Published 1977.

Marx, Karl. *Capital: A Critique of Political Economy*. Volume I: The Process of Capitalist Production. Translated from the third German edition by Samuel Moore and Edward Aveling. Ed. by Frederick Engels with changes from the fourth German editiion. New York: International Publishers, 1967.

________. *Capital: A Critique of Political Economy*. Vol. I. Introduced by Ernest Mandel and translated by Ben Fowkes. Appendix translated by Rodney Livingstone. Harmondsworth: Penguin, 1976.

________. *Capital: A Critique of Political Economy*, Vol. II. Introduced by Ernest Mandel and translated by David Fernbach. Harmondsworth: Penguin, 1978.

________. *Capital: A Critique of Political Economy*, Volume III. Translated by David Fernbach. Harmondsworth: Penguin, 1981.

________. *A Contributiion to the Critique of Political Economy*. Translated by S.W. Ryazanskaya; edited by Maurice Dobb. New York: International Publishers, 1970.

________. *Grundrisse der Kritik der politischen Ökonomie (Rohentwurf)* 1857-1858. Anhang 1850-1851. Frankfurt: Europäische Verlagsanstalt, 1967.

________. *The Grundrisse*. Edited and translated by David McLellan. New York: Harper and Row, 1971.

________. *Grundrisse: Foundations of the Critique of Political Economy*. Translated with a foreword by Martin Nicolaus. New York: Vintage Books, 1973.

________. *Theories of Surplus Value*. Parts I-III. Moscow: Progress Publishers, 1963-1971.

________. *Wage-Labour and Capital* and *Values, Price and Profit*. Introduction by Frederick Engels. New York: International Publishers, 1933 and 1976.

Nell-Breuning, Oswald v. *Aktuelle Fragen der Gesellschaftspolitik*. Köln: Verlag J.P. Bachem, 1970.

________. *Arbeit vor Kapital: Kommentar zur Enzyklika "Laborem exercens."* Wien: Europaverlag, 1983.

________. *Baugesetze der Gesellschaft: Gegenseitige Verantwortung—Hilfreicher Beistand.* Freiburg im Breisgau: Herder, 1968.

________. *Gerechtigkeit und Freiheit: Grundzüge katholischer Soziallehre.* Wien: Europaverlag, 1980.

________. Deutsche Bearbeitung von *Kirche—Arbeit—Kapital*, verfasst von Liederik deWitte. Übersetzt v. Georg Hermanowski. Limburg: Lahn-Verlag, 1964.

________. *Der Mensch in der heutigen Wirtschaftsgesellschaft.* München: Günter Olzog Verlag, 1975.

________. *Mitbestimmung*. Reihe Theorie und Praxis der Gewerkschaften. Frankfurt am Main: Europäische Verlagsanstalt, 1968.

________. *Mitbestimmung: Wer mit Wem*? Herder: Freiburg, 1969.

________. *Reorganization of Social Economy: The Social Encyclical Developed and Explained.* Edited and translated by Bernard W. Dempsey. New York: Bruce Publishing Company, 1936.

________. *Die Soziale Enzyklika: Erläuterungen zum Weltrundschreiben Papst Pius' XI. über die gesellschaftliche Ordnung*. Köln: Katholische Tat-Verlag, 1932.

________. *Soziale Sicherheit? Zu Grundfragen der Sozialordnung aus christlicher Verantwortung*. Freiburg: Herder, 1979.

________. *Soziallehre der Kirche: Erläuterungen der lehramtlichen Dokumente.* Wien: Europaverlag, 1977.

________. *Soziallehre der Kirche: Erläuterungen der lehramtlichen Dokumente.* Zweite erweiterte und ergänzte Auflage. Wien: Europaverlag, 1978.

________. *Streit um Mitbestimmung*. Frankfurt am Main: Europäische Verlagsanstalt, 1968.

________. *Unsere Verantwortung: Für eine solidarische Gesellschaft*. Freiburg im Breisgau: Herder Verlag, 1987.

________. *Wie sozial ist die Kirche*?: *Leistung und Versagen der katholischen Soziallehre.* Düsseldorf: Patmos Verlag, 1972.

_______. *Wirtschaft und Gesellschaft Heute.* 3 Bände. Freiburg: Herder, 1956-1960

Nell-Breuning, Oswald v. and Sacher, Hermann, Hrsg. *Wörterbuch der Politik.* Heft 1: *Zur christlichen Gesellschaftslehre.* Freiburg: Herder, 1947.

_______ and _______. *Wörterbuch der Politik.* Heft 3: *Zur sozialen Frage.* Herder: Freiburg, 1949.

_______ and _______. *Wörterbuch der Politik.* Heft 4: *Zur Wirtschaftsordnung.* Freiburg: Herder, 1949.

Pesch, Heinrich. *Lehrbuch der Nationalökonomie.* 1. Bd.: *Grundlegung.* Freiburg im Breisgau: Herdersche Buchhandlung, 1905.

_______. *Lehrbuch der Nationalokomie.* 2. Bd.: *Allgemeine Volkswirtschaftslehre I.* Freiburg im Breisgau: Herdersche Buchhandlung, 1909. 4. und 5. Auflage, 1925.

_______. *Lehrbuch der Nationalökonomie.* 3. Bd.: *Allgemeine Volkswirtschaftslehre, II—Die aktiven Ursachen im Volkswirtschaftlichen Lebensprozesse.* Freiburg im Breisgau: Herdsche Verlagshandlung, 1913. 2.-4. Auflage, 1926.

_______. *Lehrbuch der Nationalökonomie.* 4. Bd.: *Allgemeine Volkswirtschaftslehre III, 1 und 2.* Erste und Zweite Auflage. Freiburg im Breisgau: Herder, 1922.

_______. *Lehrbuch der Nationalökonomie.* 5. Bd.: *Allgemeine Volkswirtschaftslehre III, 3-5.* Erste und Zweite Auflage. Freiburg im Breisgau: Herder, 1923.

_______. *Liberalismus, Socialismus und christliche Gesellschaftsordnung.* 3 Theile in 2 Bänden. 2. Auflage. Freiburg im Breisgau: Herder'sche Verlagshandlung, 1901.

Pigou, A.C., ed. *Memorials of Alfred Marshall.* London: Macmillan and Co., 1925.

Ricardo, David. *The Principles of Political Economy and Taxation.* Introduction by Donald Winch. London: Dent, 1973.

_______. *The Works and Correspondence of David Ricardo.* 11 vols. Edited by Piero Sraffa with the collaboration of M.H. Dobb. Cambridge: University Press, 1951.

Schmoller, Gustav. *Grundriss der allgemeinen Volkswirtschaftslehre*, Erster Teil. Leipzig: Duncker and Humblot, 1908.

_______. *Grundriss der allgemeinen Volkswirtschaftslehre*, Zweiter Teil, ergänzt und vermehrt. München: Verlag von Duncker und Humblot, 1920.

_______. *The Mercantile System and its Historical Significance.* New York: Macmillan, 1897; reprint ed., New York: Augustus M. Kelley Publishers, 1967.

_______. *Die soziale Frage: Klassenbildung, Arbeiterfrage, Klassenkampf.* München und Leipzig: Verlag von Duncker und Humblot, 1918.

________. *Über das Maschinenzeitalter in seinem Zusammenhang mit dem Volkswohlstand und der sozialen Verfassung der Volkswirtschaft.* Vortrag gehalten in der Hauptversammlung des Vereines deutscher Ingenieure zu München am 30. Juni, 1903. Berlin: Julius Springer Verlag, 1903.

________. *Über einige Grundfragen der Socialpolitik und der Volkswirtschaftslehre.* Leipzig: Verlag von Duncker und Humblot, 1898.

________. *Zur Litteraturgeschichte der Staats- und Sozialwissenschaften.* Burt Franklin: Bibliography and Reference Series, No. 169. Leipzig: 1888; reprint ed., New York: Burt Franklin, 1968.

________. *Zur Social- und Gewerbepolitik der Gegenwart: Reden und Aufsätze.* Leipzig: Verlag von Duncker and Humblot, 1890.

Smith, Adam. *An Inquiry into the Nature and Causes of the Wealth of Nations.* W.B. Todd, Textual Editor. Vols. 2:1 and 2:2 of *The Glasgow Edition of the Works and Correspondence of Adam Smith.* Edited by R.H. Campbell and A.S. Skinner. Oxford: Clarendon Press, 1976.

________. *An Inquiry into the Nature and Causes of the Wealth of Nations.* Ed. by E. Cannan. New York: Modern Library, 1937.

________. *The Theory of Moral Sentiments.* Vol. 1 of *The Glasgow Edition of the Works and Correspondence of Adam Smith.* Ed. by D.D. Raphael and A.L. Macfie. Oxford: Clarendon Press, 1976.

________. *The Wealth of Nations*, Books 1-3. Ed. by Andrew Skinner. Harmondsworth: Penguin, 1974.

Primary Sources: Articles

Nell-Breuning, Oswald v. "Arbeit." In *Philosophisches Wörterbuch*, 14. Auflage, pp. 24-25. Hrsg. von Walter Brugger. Freiburg: Herder, 1976.

________. "Arbeit und Mitbestimmung." In *Mitbestimmung: Referate und Diskussion auf der Tagung katholischer Sozialwissenschaftler von 17. bis 19. Februar 1968 in Mönchengladbach.* Hrsg. von Anton Rauscher. Köln: J.P. Bachem, 1968.

________. "Arbeit und Musse." *Stimmen der Zeit* 160 (1957): 93-99.

________. "Arbeitslosigkeit: Unser aller Recht auf Arbeit." In *Kirche in der Gesellschaft: Der katholische Beitrag*, pp. 53-66. Geschichte und Staat, Bde. 220-222. Hrsg. von Jürgen Wichmann. München: Günter Olzog Verlag, 1978. (Same as "Das Recht auf Arbeit." *Stimmen der Zeit* 196 (1978): 523-532.)

_______. "The Drafting of "Quadragesimo Anno." In *Readings in Moral Theology, No. 5: Official Catholic Social Teaching*, pp. 60-68. Ed. by Charles E. Curran and Richard A. McCormick. New York: Paulist Press, 1986.

_______. "35-Stunden Woche." *Stimmen der Zeit* 202 (April, 1984): 217-218.

_______. "Hat die katholische Soziallehre ihre Identität verloren?" *Internationale Katholische Zeitschrift* 10 (Mär.-Ap., 1981): 107-121.

_______. "Die Kirche als Arbeitgeber." In *Menschenrechte in der Kirche*, pp. 70-90. Hrsg. von Michaela Pilters und Knut Walf. Düsseldorf: Patmos Verlag, 1980.

_______. "Kirche und Arbeitslosigkeit." *Das Neue Reich* (12. März, 1932): 457-459.

_______. "Kirche und Arbeitswelt." *Stimmen der Zeit* 203 (Dez., 1985): 795-802.

_______. "Kommentar zum 3. Kapitel des 2. Teils 'Gaudium et Spes'." In *Lexikon für Theologie und Kirche*, 2. Aufl. *Das Zweite Vatikanische Konzil*, Teil 3, pp. 487-532.

_______. "Der Königswinterer Kreis und sein Anteil an *Quadragesimo* Anno." In *Soziale Verantwortung: Festschrift für Goetz Briefs*, pp. 571-585. Hrsg. von J. Broermann und Ph. Herder-Dornreich. Berlin: Duncker und Humblot, 1968.

_______. "Marxismus—zu leicht genommen." *Stimmen der Zeit* 203 (Feb., 1985): 87-91.

_______. "Das Menschenbild: Unterschiede und Gemeinsamkeiten der evangelischen und der katholischen Soziallehre." In *Spannungsfelder der evangelischen Soziallehre: Aufgaben und Fragen vom Dienst der Kirche an der heutigen Gesellschaft*. Hrsg. von Friedrich Karrenberg und Wolfgang Schweitzer. Bd. 7 der Studien zur Evangelischen Sozialtheologie und Sozialethik. Hamburg: Furche Verlag, 1960.

_______. "Menschliche Arbeit." *Orientierung* 45 (30. Sept., 1981): 195.

_______. "Die optische Täuschung in der Rentendiskussion." *Stimmen der Zeit* 204 (1986): 737-743.

_______. "The Peschian Interest Theory." *Social Order* 1 (Apr., 1951): 177-180.

_______. "Preis." In *Lexikon für Theologie und Kirche*, 2. Auflage, Bd. 8, Sp. 719-721. Freiburg im Breisgau: Herder, 1963.

_______. "Rationalisierung und Automatisierung—Segen oder Fluch." *Stimmen der Zeit* 175 (1965): 134-146.

_______. "Solidarismus." In *Philosophisches Wörterbuch*, 14. Aufl., pp. 364-365. Hrsg. von Walter Brugger. Freiburg: Herder, 1976.

_______. "Sozialethik." In *Philosophisches Wörterbuch*, 14. Aufl., pp. 366-367. Hrsg. von Walter Brugger. Freiburg: Herder, 1976.

_______. "Soziallehre der Kirche im Ideologieverdacht." *Theologie und Philosophie* 58 (1983): 88-99.

_______. "Um die 'Berufsständische Ordnung'." *Stimmen der Zeit* 142 (1948): 6-19.

_______. "Um den berufsständische Gedanken: Zum Enzyklika *Quadragesimo anno* von 15. Mai, 1931." *Stimmen der Zeit* 122 (1932): 36-52.

_______. "Der Wahrheitsanspruch der katholischen Soziallehre." In *Testimonium Veritati: Philosophische und theologische Studien zu kirchlichen Fragen der Gegenwart.* Hrsg. von Hans Wolter. Frankfurt am Main: Joseph Knecht, 1971.

_______. "Die Wirtschaft im Urteil des 2. Vatikanischen Konzils (Constitutio pastoralis in mundo huius temporis, Pars II, Caput III)." *Theologie und Philosophie* 41 (1966): 188-210.

_______. "Wirtschaft und Recht." In *Wirtschafts- und Sozialpolitik in der berufsständischen Ordnung.* pp. 48-63. Erste Soziale Woche des Volksvereins für das katholische Deutschland. Hrsg. von Joseph van der Velden. Köln: Verlag J.P. Bachem, 1933.

Pesch, Heinrich. ["Selbstdarstellung"]. In *Volkswirtschaftslehre der Gegenwart*, Bd. 1. Hrsg. von Felix Meiner. Leipzig: Felix Meiner Verlag, 1924.

Schmoller, Gustav. "Die Arbeiterfrage." *Preussische Jahrbücher* 14 (1864) 393-424 and 523-547; 15 (1865): 32-63.

_______. "Idea of Justice in Political Economy." *Annals of the American Academy of Political and Social Science* 4 (Mar., 1894): 697-737.

Tucker, Robert C. *The Marx-Engels Reader*, 2nd ed. New York: W.W.Norton and Co., 1978.

Secondary Sources: Books

Achinger, Hans, Preller, Ludwig and Wallraff, Hermann Josef, Hrsg. *Normen der Gesellschaft: Festgabe für Oswald von Nell-Breuning, S.J. zu seinem 75. Geburtstag.* Mannheim: Pesch Haus Verlag, 1965.

Althusser, Louis. *For Marx.* Translated by Ben Brewster. London: Verso, 1979.

Anderson, Nels. *Man's Work and Leisure.* Monographs and Theoretical Studies in Sociology and Anthropology in Honor of Nels Anderson, no. 4. Leiden: E.J. Brill, 1974.

Appleby, Joyce Oldham. *Economic Thought and Ideology in Seventeenth Century Thought.* Princeton: Princeton University Press, 1978.

Arendt, Hannah. *The Human Condition.* Chicago: University of Chicago Press, 1958.

_______. *Vita Activa oder vom taetigen Leben*. Stuttgart: W. Kohlhammer Verlag, 1960.

Arvon, Henri. *La Philosophie du Travail*. Paris: Presses Universitaires de France, 1961.

Avineri, Shlomo. *Hegel's Theory of the Modern State*. Cambridge: Cambridge University Presss, 1972.

_______. *The Social and Political Thought of Karl Marx*. Cambridge: Cambridge University Press, 1968.

Barber, William J. *A History of Economic Thought*. Harmondsworth: Penguin, 1967.

Barth, Karl. *Church Dogmatics* Vol. III, *The Doctrine of Creation*, Pt. 4. Edinburgh, T. and T. Clark, 1961.

Barthel, Eckhardt and Dikau, Joachim. *Mitbestimmung in der Wirtschaft*. Zur Politik und Zeitgeschichte, Nr. 43. Berlin: Colloquium Verlag, 1980.

Baudrillard, Jean. *The Mirror of Production*. Translated with "Introduction" by Mark Poster. St. Louis: Telos Press, 1975.

Baum, Gregory, ed. *Work and Religion*. *Concilium*, Vol. 131, No. 1. New York: Seabury Press, 1980.

Bendix, Reinhold. *Work and Authority in Industry: Ideologies of Management in the Course of Industrialization*. Berkeley and Los Angeles: University of California Press, 1956.

Bernstein, Paul. *Workplace Democratization: Its Internal Dynamics*. n.p.: Kent State University Press, 1976.

Bernstein, Richard J. *Praxis and Action: Contemporary Philosophies of Human Activity*. Philadelphia: University of Pennsylvania Press, 1971.

_______. *The Restructuring of Social and Political Theory*. Philadelphia: University of Pennsylvania Press, 1976.

Binswanger, H. C.; Frisch, H.; and Nutzinger, H. G. u. a. *Arbeit ohne Umweltzerstörung: Strategien einer neuen Wirtschaftspolitik*. Frankfurt am Main: S. Fischer, 1983.

Blanchot, Maurice. *The Space of Literature*. Translated, with an Introduction, by Ann Smock. Lincoln: University of Nebraska Press, 1982.

Blaug, Mark. *Economic Theory in Retrospect*. 3rd ed. Cambridge: Cambridge University Press, 1978.

_______. *Ricardian Economics: A Historical Study*. New Haven: Yale University Press, 1958.

Bluestone, Barry and Harrison, Bennett. *The Deindustrialization of America: Plant Closings, Community Abandonment, and the Dismantling of Basic Industry.* New York: Basic Books, Harper Colophon, 1982.

Böhme, Kurt. *Solidarismus und Liberalismus: Eine Kritik des Solidarismus von Heinrich Pesch S.J. vom liberalen Standpunkt.* Inaugural-Dissertation zur Erlangung der Doktorwürde der Hohen Wirtschafts- und Sozialwissenschaftlichen Fakultät der Universität Frankfurt am Main. Kirchhain N.-L.: Zahn und Baendel, 1931.

Bordollo, Otto. *Der Begriff "Produktiv" bei Adam Smith und bei Friedrich List.* Inaugural Dissertation zur Erlangung der Doktorwürde der hohen philosophischen Fakultät der Friedrich-Alexanders-Universität zu Erlangen. Borna-Leipzig: Buchdrückerei Robert Noske, 1909.

Boulding, Kenneth Ewart. *Adam Smith as an Institutional Economist.* The Frank E. Seidman Distinguished Award in Political Economy. Memphis, Tennessee: P.K. Seidman Foundation, 1976.

Bowen, Ralph H. *German Theories of the Corporative State: With Special Reference to the Period 1870-1919.* New York: Whittlesey House, McGraw Hill Book Company, 1947.

Bowley, Marian. *Studies in the History of Economic Theory Before 1870.* London: Macmillan, 1973.

Boyer, John W. *Political Radicalism in Late Imperial Vienna: Origins of the Christian Social Movement 1848-1897.* Chicago: University of Chicago Press, 1981.

Bradley, Keith and Gelb, Alan. *Worker Capitalism: The New Industrial Relations.* Cambridge, Mass.: The MIT Press, 1983.

Braverman, Harry. *Labor and Monopoly Capital: The Degradation of Work in the Twentieth Century.* New York: The Monthly Review Press, 1974.

Brenkert, George G. *Marx's Ethics of Freedom.* London: Routledge and Kegan Paul, 1983.

Brinkmann, Carl. *Gustav Schmoller und die Volkswirtschaftslehre.* Stuttgart: Verlag W. Kohlhammer, 1937.

Brunner, Otto; Conze, Werner; and Kosseleck, Reinhart, Hrsg. *Geschichtliche Grundbegriffe: Historisches Lexikon zur politisch-sozialen Sprache in Deutschland.* Stuttgart: E. Klett, 1972- .

Büchele, Herwig. *Christsein im gesellschaftlichen System: Sozialethische Reflexionen über den Zusammenhang von Glaube und sozioökonomischen Strukturen.* Wien: Europaverlag, 1976.

Bundesminister für Arbeit und Sozialordnung. *Mitbestimmung: Mitbestimmungsgesetz, Anmerkungen zum Mitbestimmungsurteil des Bundesvefassungsgericht, Montanmitbestimmung, Betriebsverfassung.* Bonn: Referat Presse und Information, 1983.

Bydekarken, Hennig. *Die Interpretation der Theorie David Ricardos als geschlossenes nichtarbeitswertaxiomatisches Gleichgewichtssystem*. Volkswirtschaftliche Schriften, Heft 85. Berlin: Duncker und Humblot, 1965.

Canetti, Elias. *Crowds and Power*. Translated by Carol Stewart. London: V. Gollanz, 1962; reprint ed., New York: Farrar Straus Giroux, 1984.

Cannan, Edwin. *A History of the Theories of Production and Distribution in English Political Economy from 1776-1848*. 3rd ed. London: King and Son, 1917; reprint ed., London and New York: Staples Press, 1953.

Caravale, Giovanni A., and Tosato, Domenico A. *Ricardo and the Theory of Value, Distribution and Growth*. London: Routledge and Kegan Paul, 1980.

Carnoy, Martin, and Shearer, Derek. *Economic Democracy: The Challenge of the 1980s*. White Plains, New York: M.E. Sharpe, Inc., 1980.

Chamley, Paul. *Economie, Politique et Philosophie chez Steuart et Hegel*. Annales de la Faculté de Droit et des Sciences Politiques et Economiques de Strasbourg, XII. Paris: Librairie Dalloz, 1963.

Chenu, M.D. *The Theology of Work: An Exploration*. Translated by Lilian Soiron. Chicago: Henry Regnery Co., 1966.

Christ, Harry. *Die Verhaltensweisen der wirtschaftenden Menschen bei Adam Smith*. Inauguraldissertation zur Erlangung des Doktorgrades, Universität Köln. Köln: Photostelle der Universität, 1960.

Clark, Dennis. *Work and the Human Spirit*. New York: Sheed and Ward, 1967.

Clark, J.M., et al. *Adam Smith, 1776-1926: Lectures to Commemorate the Sesquicentennial of the Publication of the "Wealth of Nations."* Chicago: University of Chicago Press, 1928.

Clayre, Alasdair. *The Political Economy of Co-operation and Participation: A Third Sector*. Oxford: Oxford University Press, 1980.

________. *Work and Play: Ideas and Experience of Work and Leisure*. London: Weidenfeld and Nicolson, 1974.

Cohen, G.A. *Karl Marx's Theory of History: A Defence*. Princeton: Princeton University Press, 1978.

Cronin, John and Flannery, Harry. *The Church and the Workingman*. Twentieth Century Encyclopedia of Catholicism, Vol. 10. New York: Hawthorn Books Publishers, 1965.

Cropsey, Joseph. *Polity and Economy: An Interpretation of the Principles of Adam Smith*. The Hague: Martinus Nijhoff, 1957.

Curran, Charles E. *American Catholic Social Ethics: Twentieth Century Approaches*. Notre Dame: University of Notre Dame Press, 1982.

Dahl, Robert A. *A Preface to Economic Democracy*. Berkeley: University of California Press, 1985.

Dankert, Clyde E. *Adam Smith: Man of Letters and Economist*. Hickesville, New York: Exposition Press, 1974.

de Man, Henri. *Joy in Work*. Translated from the German by Eden and Cedar Paul. London: George Allen and Unwin, 1929.

Deane, Phyllis. *The Evolution of Economic Ideas*. Cambridge: Cambridge University Press, 1978.

Deutscher Bundestag und Deutscher Bundesrat. *Verhandlungen* (Protokolle, Drucksachen, Register). Mikrofiche-Edition. München: C.H. Beck, K.G. Saur.

Dierolf, Joerg. *Codetermination: An Introduction to "Industrial Democracy" in West Germany*. Bergisch-Gladbach: Heider-Verlag, 1977.

Dmitriev, V.K. *Economic Essays on Value, Competition and Utility*. Trans. by D. Fry. Edited and introduced by D.M. Nuti. Cambridge: Cambridge University Press, 1974.

Dobb, Maurice. *Political Economy and Capitalism: Some Essays in Economic Tradition*. New York: International Publishers, 1945; reprint ed., Westport, Conn.: Greenwood Press, Publishers, 1972.

________. *Theories of Value and Distribution Since Adam Smith: Ideology and Economic Theory*. Cambridge: Cambridge University Press, 1973.

Donahue, John W. *Work and Education: The Role of Technical Culture in Some Distinctive Theories of Humanism*. Chicago: Loyola University Press, 1959.

Dumont, Louis. *From Mandeville to Marx: The Genesis and Triumph of Economic Ideology*. Chicago: University of Chicago Press, 1977.

Dunkmann, Karl. *Soziologie der Arbeit*. Handbuch der Arbeitswissenschaft, Bde. VIII/IX, Teil 1. Halle a. S.: Carl Marhold Verlagsbuchhandlung, 1933.

Dupré, Louis. *Marx's Social Critique of Culture*. New Haven: Yale University Press, 1983.

________. *The Philosophical Foundations of Marxism*. New York: Harcourt, Brace and World, 1966.

Durkheim, Emile. *The Division of Labor in Society*. Trans. by George Simpson. Glencoe, Illinois: The Free Press, 1933.

Eisermann, Gottfried. *Die Grundlagen des Historismus in der deutschen Nationalökonomie*. Stuttgart: Ferdinand Enke Verlag, 1956.

Ellerman, David P. *Management Planning with Labor as a Fixed Cost: The Mondragon Annual Business Plan Manual*. Somerville, Mass.: Industrial Cooperative Association, 1984.

________. *The Socialization of Entrepreneurship: The Empresarial Division of the Caja Laboral Popular*. Somerville, Mass.: Industrial Cooperative Association, 1982.

________. *The Union as the Legitimate Opposition in an Industrial Democracy*. Somerville, Mass.: Industrial Cooperative Association, 1979.

________. *What is a Worker Cooperative*? Somerville, Mass.: Industrial Cooperative Association, 1984.

________. *Worker Ownership: Economic Democracy or Worker Capitalism*? Somerville, Mass.: Industrial Cooperative Association, 1985

Findlay, J.N. *Hegel: A Re-examination*. New York: Humanities Press, 1958.

Flessati, Dominic. *The Mondragon Experiment*. BBC: Horizon Films, 1980.

Flubacher, Joseph F. *The Concept of Ethics in the History of Economics*. New York: Vantage, 1950.

Föppl, Christine. *Arbeitslosigkeit Ist Heilbar: Ein Aufruf zum gemeinsamen Handeln*. München: Kösel Verlag, 1983.

Foley, Vernard. *The Social Physics of Adam Smith*. West Lafayette, Indiana: Purdue University Press, 1976.

Foucault, Michel. *The History of Sexuality*. Vol. 1: *An Introduction*. Trans. by Robert Hurley. New York: Random House, 1978.

________. *The Order of Things: An Archaeology of the Human Sciences*. New York: Random House, 1970.

Franciscono, Marcel. *Walter Gropius and the Creation of the Bauhaus in Weimar: The Ideals and the Artistic Theories of its Founding Years*. Urbana, Illinois: University of Illinois Press, 1971.

Freeman, Richard B. and Medoff, James L. *What Do Unions Do*? New York: Basic Books, 1984.

Fremantle, Anne, ed. *The Papal Encyclicals in Their Historical Context*. New York: G.P. Putnam's Sons, 1956.

Freyer, Hans. *Die Bewertung der Wirtschaft im philosophischen Denken des 19. Jahrhunderts.* Abhandlungen der sächs. staatl. Forschungsinstitute zu Leipzig—Forschungsinstitut für Psychologie Nr. 6. Leipzig: W. Engelmann, 1921; reprint ed., Hildesheim: Georg Olms Verlagsbuchhandlung, 1966.

Friedrichs, Arno. *Klassische Philosophie und Wirtschaftswissenschaft: Untersuchungen zur Geschichte der deutschen Geistesleben im neunzehnten Jahrhundert.* Gotha: Friedrich Andreas Perthes, 1913.

Fulton, Robert Brank. *Adam Smith Speaks to Our Times: A Study of His Ethical Ideas.* Boston: The Christopher Publishing House, 1963.

Furlong, James. *Labor in the Boardroom: The Peaceful Revolution.* Princeton, New Jersey: Dow Jones Books, 1977.

Furniss, Edgar S. *The Position of the Laborer in a System of Nationalism: A Study in the Labor Theories of the Later English Mercantilists.* Boston: Houghton Mifflin, 1920.

Galbraith, John Kenneth. *The New Industrial State.* 2nd ed. Harmondsworth: Penguin, 1972.

Gibbons, William J., ed. *Seven Great Encyclicals.* Glen Rock, New Jersey: Paulist Press, 1963.

Gierke, Otto Friedrich v. *Natural Law and the Theory of Society: 1500-1800.* With a lecture on the ideas of natural law and humanity by Ernst Troeltsch. Translated and introduced by Ernest Barker. Cambridge: The University Press, 1958.

Giese, Fritz. *Philosophie der Arbeit.* Bd. 10, *Handbuch der Arbeit.* Halle a. S.: Carl Marhold, 1932.

Gilson, Etienne, ed. *The Church Speaks to the Modern World: the Social Teaching of Leo XIII.* Garden City, New York: Image Books, 1954.

Glahe, Fred R., ed. *Adam Smith and "The Wealth of Nations": 1776-1976 Bicentennial Essays.* Boulder, Colorado: Colorado Associated University Press, 1978.

Görres-Gesellschaft. *Staatslexikon*, 1-5 Bände. 5. ed. Freiburg im Breisgau: Herder, 1932.

Goodin, Robert E. *Protecting the Vulnerable: A Reanalysis of Our Social Responsibilities.* Chicago: University of Chicago Press, 1985.

Gootzeit, Michael J. *David Ricardo.* New York: Columbia University Press, 1975.

Gould, Carol C. *Marx's Social Ontology: Individual and Community in Marx's Theory of Social Reality.* Cambridge: MIT Press, 1978

Gremillion, Joseph, ed. *The Gospel of Peace and Justice: Catholic Social Teaching Since Pope John.* Maryknoll, New York: Orbis Books, 1976.

Groethuysen, Bernard. *The Bourgeois: Catholicism and Capitalism in Eighteenth Century France*. Translated by Mary Ilford. New York: Holt Rinehart and Winston, 1968.

Gundlach, Gustav. *Zur Soziologie der katholischen Ideenwelt und des Jesuitenordens*. Inaugural-Dissertation zur Erlangung der Doktorwürde genehmigt von der Philosophischen Fakultät der Friedrich-Wilhelms-Universität zu Berlin. Berlin: Emil Ebering, 1927.

Gunn, Christopher Eaton. *Workers' Self-Management in the United States*. Ithaca, New York: Cornell University Press, 1984.

Habermas, Jürgen. *Knowledge and Human Interests*. Translated by Jeremy J. Shapiro. Boston: Beacon Press, 1971.

________. *Theory and Practice*. Translated by John Viertel. Boston: Beacon Press, 1973.

Halévy, Elie. *The Growth of Philosophic Radicalism*. Translated by Mary Morris. London: Faber and Faber, 1928.

Hanney, Alastair. *Kierkegaard*. London: Routledge and Kegan Paul, 1982

Hasbach, Wilhelm. *Die allgemeinen philosophischen Grundlagen der von François Quesnay und Adam Smith begründeten politischen Ökonomie*. Staats- und sozialwissenschaftlichen Forschungen herausgegeben von Gustav Schmoller. 10. Bd., 2. Heft (Der ganzen Reihe 43. Heft). Leipzig: Duncker und Humblot, 1890.

________. *Untersuchungen über Adam Smith und die Entwicklung der politischen Ökonomie*. Leipzig: Duncker und Humblot, 1891.

Heckscher, Eli F. *Mercantilism*. 2 vols. Translated by Mendel Shapiro. Revised, second edition by E.F. Söderlund. London: George Allen and Unwin, 1962.

Heidegger, Martin. *Basic Writings*. Edited and introduced by David Farrell Krell. New York: Harper and Row, 1977.

________. *Being and Time*. Translated by John Macquarrie and Edward Robinson. New York: Harper and Row, 1962.

Heidelmeyer, Wolfgang, Hrsg. *Die Menschenrechte: Erklärungen, Verfassungsartikel, Internationale Abkommen*. 2. Auflage. Paderborn: Schöningh, 1972.

Heller, Agnes. *The Theory of Need in Marx*. London: Allison and Busby, 1976.

Herbst, Jürgen. *The German Historical School in American Scholarship: A Study in the Transfer of Culture*. Ithaca, New York: Cornell University Press, 1965.

Hertz, Anselm; Korff, Wilhelm; Rendtorff, Trutz; and Ringeling, Hermann, Hrsg. *Handbuch der christlichen Ethik*, 2 Bände. Freiburg and Gütersloh: Herder and Gütersloher Verlagshaus Gerd Mohn, 1978.

Hirschman, Albert O. *The Passions and the Interests: Political Arguments for Capitalism Before its Triumph*. Princeton: Princeton University Press, 1977.

Hla Myint. *Theories of Welfare Economics*. London: Longmans, Green and Co., 1948.

Hollander, Samuel. *The Economics of Adam Smith*. Studies in Classical Political Economy, 1. Toronto: University of Toronto Press, 1973.

________. *The Economics of David Ricardo*. Studies in Classical Political Economy, 2. Toronto: University of Toronto Press, 1979.

Hollis, Martin. *Models of Man: Philosophical Thoughts on Social Action*. Cambridge: Cambridge University Press, 1977.

Hollis, Martin and Noll, Edward J. *Rational Economic Man: A Philosophical Critique of Neo-Classical Economics*. Cambridge: Cambridge University Press, 1975.

Horkheimer, Max. *Eclipse of Reason*. New York: Continuum, 1974.

Horvat, Branko. *The Political Economy of Socialism: A Marxist Social Theory*. Armonk, New York: M.E. Sharpe, Inc., 1982.

Horvat, Branko; Marković Mihailo; and Supek, Rudi. *Self-Governing Socialism: A Reader*. Vol. 1: *Historical Development and Social and Political Philosophy*. White Plains, New York: International Arts and Sciences Press, 1975.

Houck, John W. and Williams, Oliver F. *Catholic Social Teaching and the U.S. Economy: Working Papers for a Bishops' Pastoral*. Washington, D.C.: University Press of America, 1984.

Huber, Joseph. *Die zwei Gesichter der Arbeit: Ungenutzte Möglichkeiten*. Frankfurt am Main: S. Fischer Verlag, 1984.

Huemmer, Karl. *Der ständische Gedanke in der katholisch-sozialen Literatur des 19. Jahrhunderts*. Inaugural-Dissertation, Universität Würzburg. Würzburg: Buchdrückerei Bavaria, 1927.

Hug, James E. *Tracing the Spirit: Communities, Social Action and Theological Reflection*. New York: Paulist Press, 1983.

Hunnius, Gerry; Garson, G. David; and Case, John, eds. *Workers' Control: A Reader on Labor and Social Change*. New York: Random House, 1973.

Hutcheson, Francis. *A System of Moral Philosophy*. London: A. Millar and T. Longman, 1755; reprint ed., 2 vols. in one. New York: Augustus M. Kelley, 1968.

Hyppolite, Jean. *Genesis and Structure of Hegel's "Phenomenology of Spirit."* Translated by Samuel Cherniak and John Hackman. Evanston: Northwestern University Press, 1974.

_______. *Studies on Marx and Hegel.* Translated and edited by John O'Neill. New York: Basic Books, 1969.

Ignatieff, Michael. *The Needs of Strangers: An Essay on Privacy, Solidarity, and the Politics of Being Human.* New York: Elisabeth Sifton and Penguin Books, 1984.

Industrial Democracy in Europe International Research Group. *Industrial Democracy in Europe.* Oxford: Clarendon Press, 1981.

Irwin, Terence. *Plato's Moral Theory: The Early and Middle Dialogues.* Oxford: Clarendon Press, 1977.

Jaspers, Karl. *Von der Logik*, Erster Bd.: *Philosophische Logik.* München: R. Piper und Co. Verlag, 1947.

John Paul II. *Encyclical on Human Work: Laborem Exercens.* Boston: St. Paul Editions, [1981].

Jones, Derek C. and Svejnar, Jan, eds. *Participatory and Self-Managed Firms.* Lexington, Massachusetts: Lexington Books, 1982.

Jünger, Ernst. *Der Arbeiter: Herrschaft und Gestalt.* Hamburg: Hanseatische Verlagsanstalt, 1932.

Kafka, Franz. *The Complete Stories.* Ed. by Nahum N. Glatzer. New York: Schocken Books, 1971.

Kain, Philip J. *Schiller, Hegel and Marx: State, Society and the Aesthetic Ideal of Ancient Greece.* McGill-Queen's Studies in the History of Ideas, 4. Kingston and Montreal: McGill-Queen's University Press, 1982.

Kaiser, Edwin G. *Theology of Work.* Westminster, Maryland: The Newman Press, 1966.

Kamenka, Eugene. *The Ethical Foundations of Marxism.* 2nd edition. London: Routledge and Kegan Paul, 1972.

Katterle, Siegfried. *Sozialwissenschaft und Sozialethik: Logische und theoretische Probleme praktischer Sozialwissenschaften, besonders christlicher Soziallehre.* Wissenschaft und Gesellschaft, 5. Schriften des Forschungsinstituts für Gesellschaftspolitik und beratende Sozialwissenschaft e.V. Hrsg. von Gerhard Weisser. Göttingen: Verlag Otto Schwartz und Co., 1972.

Kempel, Franz. *Göttliches Sittengesetz und neuzeitliches Erwerbsleben: Eine Wirtschaftslehre in sittlich-organischer Auffassung der gesellschaftlichen Erwerbsverhältnisse.* Mainz: Verlag von Franz Kirchheim, 1901.

Keynes, John Neville. *The Scope and Method of Political Economy*, 4th edition. New York: Kelley and Millman, 1955.

Kiker, B.F. *Human Capital: In Retrospect*. Essays in Economics, June, 1968, No. 16. Columbia, South Carolina: University of South Carolina, 1968.

Kirzner, Israel M. *Competition and Entrepreneurship*. Chicago: University of Chicago Press, 1973.

Klages, Helmut. *Technischer Humanismus: Philosophie und Soziologie der Arbeit bei Karl Marx*. Soziologische Gegenwartsfragen, Neue Folge. Stuttgart: Ferdinand Enke Verlag, 1964.

Klüber, Franz. *Katholische Gesellschaftslehre*, 1. Bd.: *Geschichte und System*. Osnabrück: Verlag Fromm, 1968.

Knight, Frank H. *The Economic Organization* with an article, "Notes on Cost and Utility." New York: Augustus M. Kelley Publishers, 1951; reprint ed., New York: Augustus M. Kelley Publishers, 1967.

________. *The Ethics of Competition and Other Essays*. New York: Augustus M. Kelley, Inc., 1935.

________. *On the History and Method of Economics: Selected Essays*. Chicago: The University of Chicago Press, 1956.

Kojève, Alexandre. *Introduction to the Reading of Hegel: Lectures on the "Phenomenology of Spirit."* Assembled by Raymond Queneau; edited by Allan Bloom; translated by James H. Nichols, Jr. New York: Basic Books, Inc., 1967.

Kolakowski, Leszek. *Main Currents of Marxism: Its Origins, Growth and Dissolution*. Vol. I: *The Founders*. Translated from the Polish by P.S. Falla. Oxford: Oxford University Press, 1978.

Krämer-Badoni, Thomas. *Zur Legitimität der bürgerlichen Gesellschaft: Eine Untersuchung des Arbeitsbegriffs in den Theorien von Locke, Smith, Ricardo, Hegel und Marx*. Frankfurt: Campus Verlag, 1978.

Kramer, Rolf. *Arbeit: theologische, wirtschaftliche und soziale Aspekte*. Göttingen: Vandenhoeck und Ruprecht, 1948.

Krieger, Leonard. *The German Idea of Freedom: History of a Political Tradition*. Boston: Beacon Press, 1957.

Kühne, Robert J. *Co-determination in Business: Workers' Representatives in the Boardroom*. New York: Praeger, 1980.

Kuttner, Robert. *The Economic Illusion: False Choices Between Property and Social Justice*. Boston: Houghton Mifflin Company, 1984.

Kwant, Remy C. *Philosophy of Labor*. Duquesne Studies: Philosophical Series, 10. Louvain, Belgium: Editions E. Nauwelaerts, 1960.

Lauer, Quentin. *A Reading of Hegel's "Phenomenology of Spirit."* New York: Fordham University Press, 1976.

Le Goff, Jacques. *Time, Work and Culture in the Middle Ages*. Translated by Arthur Goldhammer. Chicago: University of Chicago Press, 1980.

Letwin, William. *The Origins of Scientific Economics: English Economic Thought—1660-1776*. London: Methuen and Co., Ltd., 1963.

Levinas, Emmanuel. *Totality and Infinity: An Essay in Exteriority*. Translated by Alphonso Lingis. Duquesne Studies, Philosophical Series, 24. Pittsburgh: Duquesne University Press, 1969.

Lim, Sok-Zim. *Der Begriff der Arbeit bei Hegel: Versuch einer Interpretation der "Phänomenologie des Geistes."* Abhandlungen zur Philosophie, Psychologie und Pädagogik, Bd. 27. Bonn: H. Bouvier u. Co. Verlag, 1963.

Lindblom, Charles E. *Politics and Markets: The World's Political-Economic Systems*. New York: Basic Books, 1977.

Lindenfeld, Frank and Rothschild-Whitt, Joyce, eds. *Workplace Democracy and Social Change*. Boston: Porter Sargent Publishers, Inc., 1982.

Lindgren, J. Ralph. *The Social Philosophy of Adam Smith*. International Archives of the History of Ideas, Series Minor, No. 8. The Hague: Martinus Nijhoff, 1973.

Lipschitz, Eleonore. *Die theoretischen Grundlagen David Ricardos im Lichte des Briefwechsels*. Wirtschaftswissenschaftliche Abhandlungen. Berlin: Duncker und Humblot, 1957.

Löwith, Karl. *From Hegel to Nietzsche: The Revolution in Nineteenth-century Thought*. Translated by David E. Greene. New York: Holt, Rinehart and Winston, 1964.

Locke, John. *Two Treatises of Government*. Second edition. Ed. by Peter Laslett. Cambridge: Cambridge University Press, 1970.

Lukács, Georg. *Hegel's False and His Genuine Ontology*. Chap. 3 of Pt. 1 of *Toward the Ontology of Social Being*. Translated by David Fernbach. London: Merlin Press, 1978.

________. *History and Class Consciousness: Studies in Marxist Dialectics*. Cambridge: MIT Press, 1971.

________. *Labour*. Chap. 1 of Pt. 2 of *Toward the Ontology of Social Being*. Translated by David Fernbach. London: Merlin Press, 1980.

________. *Marx's Basic Ontological Principles*. Chap. 4 of Pt. 1 of *Toward the Ontology of Social Being*. Translated by David Fernbach. London: Merlin Press, 1978.

________. *The Young Hegel: Studies in the Relations Between Dialectics and Economics*. Translated by Rodney Livingstone. London: Merlin Press, 1975.

Macfie, A.L. *The Individual in Society: Papers on Adam Smith*. London: George Allen and Unwin, Ltd., 1967.

Macpherson, C.B. *The Political Theory of Possessive Individualism*. Oxford: Oxford University Press, 1962.

Mandel, Ernest. *The Formation of the Economic Thought of Karl Marx: 1843 to "Capital."* Translated by Brian Pearce. New York and London: Monthly Review Press, 1971.

________. *Late Capitalism*. London: NLB, 1975.

________. *Marxist Economic Theory*, 2 vols. New York: Monthly Review Press, 1968.

Mantoux, Paul. *The Industrial Revolution in the Eighteenth Century: An Outline of the Beginnings of the Modern Factory System in England*. New York: Harper and Row, 1961.

Marcuse, Herbert. *Hegels Ontologie und die Grundlegung einer Theorie der Geschichtlichkeit*. Frankfurt am Main: Vittorio Klostermann Verlag, 1932.

________. *Reason and Revolution: Hegel and the Rise of Social Theory*. 2nd ed. New York: Humanities Press, 1954.

Maritain, Jacques. *Freedom in the Modern World*. New York: Scribners, 1936.

Marković, Mihailo. *From Affluence to Praxis: Philosophy and Social Criticism*. Ann Arbor: The University of Michigan Press, 1974.

Mason, Ronald M. *Participatory and Workplace Democracy: A Theoretical Development in Critique of Liberalism*. Carbondale and Edwardsville: Southern Illinois University Press, 1982.

McLellan, David. *The Thought of Karl Marx: An Introduction*. New York: Harper and Row, 1971.

McNulty, Paul J. *The Origins and Development of Labor Economics: A Chapter in the History of Social Thought*. Cambridge: MIT Press, 1980.

Meek, Ronald L. *Economics and Ideology and Other Essays: Studies in the Development of Economic Thought*. London: Chapman and Hall, 1967.

________. *The Economics of Physiocracy: Essays and Translations*. University of Glasgow Social and Economic Studies, New Series, No. 2. London: George Allen and Unwin Ltd., 1962.

________. *Smith, Marx and After: Ten Essays in the Development of Economic Thought*. London: Chapman and Hall, 1977.

_______. *Studies in the Labor Theory of Value*. 2nd ed. New York and London: Monthly Review Press, 1976.

Meszaros, Istvan. *Marx's Theory of Alienation*. London: Merlin Press, 1970.

Meunier, Jean-Guy. *Genèse du Matérialisme dans les Écrits de Jeunesse de Karl Marx*. Ottawa, Canada: Editions de l'Universite d'Ottawa, 1981.

Miller, Richard W. *Analyzing Marx: Morality, Power and History*. Princeton: Princeton University Press, 1984.

Model, Otto and Crefelds, Carl. *Staatsbürgertaschenbuch*. 18. Auflage. München: C.H. Beck'sche Verlagsbuchhandlung, 1979.

Mommsen, Wolfgang J. *Max Weber and German Politics: 1890-1920*. Translated by Michael S. Steinberg. Chicago: University of Chicago Press, 1984.

Montgomery, David. *Workers' Control in America*. Cambridge: Cambridge University Press, 1979.

Müller, Hans Gustav. *Die Nationalökonomie Adam Smiths als Ende der weltanschaulichen Emanzipation des naturalistischen Selbst-Interesses in der europäischen Geistesgeschichte: Versuch einer geistesgeschichtlichen Orientierung*. Inaugural Dissertation zur Erlangung des Doktorgrades der Eberhard-Karls-Universität zu Tübingen. 1969.

Müssiggang, Albert. *Die soziale Frage in der historischen Schule der deutschen Nationalökonomie*. Tübinger wirtschaftswissenschaftliche Abhandlungen, Bd. 2. Tübingen: J.C.B. Mohr (Paul Siebeck), 1968.

Mueller, Franz H. *Economic Aspects of Industrial Decentralization*. Aquin Papers, No. 8. St. Paul, Minnesota: The College of St. Thomas, 1942.

_______. *Heinrich Pesch and His Theory of Christian Solidarism*. Aquin Papers, no. 7. St. Paul, Minnesota: The College of St. Thomas, 1941.

_______. *Heinrich Pesch: Sein Leben und seine Lehre*. Köln: Verlag J.P. Bachem, 1980.

Mulcahy, Richard E. *The Economics of Heinrich Pesch*. New York: Henry Holt and Co., 1952.

Muszynski, Bernhard. *Wirtschaftliche Mitbestimmung zwischen Konflict- und Harmoniekonzeptionen: Theoretische Voraussetzungen, geschichtliche Grundlagen und Hauptprobleme mit der Mitbestimmungsdiskussion der BRD*. Studien zum politischen System der Bundesrepublik Deutschland, hrsg. von Heine Kaack, Bd. 5. Meisenheim am Glan: Verlag Anton Hain, 1975.

Myers, Milton L. *The Soul of Modern Economic Man: Ideas of Self-Interest, Thomas Hobbes to Adam Smith*. Chicago: University of Chicago Press, 1983.

Myrdal, Gunnar. *The Political Element in the Development of Economic Theory*. Translated by Paul Streeten. New York: Simon and Schuster, 1954.

Nahrgang, Alfred. *Die Aufnahme der wirtschaftspolitischen Ideen von Adam Smith in Deutschland zu Beginn des 19. Jahrhunderts*. Inaugural Dissertation. Goethe Universität, Frankfurt am Main, 1933.

Naphtali, Fritz. *Wirtschaftsdemokratie: Ihr Wesen, Weg und Ziel*. Sammlung "Res Novae"—Veröffentlichungen zu Politik, Wirtschaft, Soziologie und Geschichte, Bd. 42. Berlin: Verlagsgesellschaft des allgemeinen deutschen Gewerkschaftsbund, Gmbh., 1928; reprint edition, Frankfurt am Main: Europäische Verlagsanstalt, 1966.

Naville, Pierre. *De l'alienation à la jouissance: la genèse de la sociologie du travail chez Marx et Engels*. Le nouveau Léviathan, 1. 2ème édition. Paris: Editions Anthropos, 1970.

Nelson, Richard R. and Winter, Sidney G. *An Evolutionary Theory of Economic Change*. Cambridge: The Belknap Press of Harvard University Press, 1982.

Nicholson, J. Shield. *A Project of Empire: A Critical Study of the Economics of Imperialism, with Special Reference to the Ideas of Adam Smith*. London: Macmillan and Co., 1909.

Niedenhoff, Horst-Udo. *Mitbestimmung in der Bundesrepublik Deutschland*. 2. ergänzte Auflage. Köln: Deutscher Instituts-Verlag, 1977.

Nove, Alec. *The Economics of Feasible Socialism*. London: George Allen and Unwin, 1983.

O'Driscoll, Gerald P., Jr. *Adam Smith and Modern Political Economy: Bicentennial Essays on the "Wealth of Nations."* Ames, Iowa: The Iowa State University Press, 1979.

Okun, Arthur. *Equality and Efficiency: The Big Trade-Off*. Washington: The Brookings Institution, 1975.

Ollman, Bertell. *Alienation: Marx's Conception of Man in Capitalist Society*. 2nd ed. Cambridge: Cambridge University Press, 1976.

Oncken, August. *Adam Smith und Immanuel Kant: Der Einklang und das Wechselverhältniss ihrer Lehren über Sitte, Staat und Wirthschaft*. Leipzig: Verlag von Duncker und Humblot, 1877.

Orwell, George. *The Road to Wigan Pier*. New York: Harcourt Brace Jovanovich, 1958.

Pannenberg, Wolfhart. *Anthropologie in theologischer Perspektive*. Göttingen: Vandenhoeck und Ruprecht, 1983.

Pateman, Carole. *Participation and Democratic Theory*. Cambridge: Cambridge University Press, 1970.

Pelczynski, Z.A., ed. *Hegel's Political Philosophy—Problems and Perspectives: A Collection of New Essays*. Cambridge: Cambridge University Press, 1971.

_______. *The State and Civil Society: Studies in Hegel's Political Philosophy*. Cambridge: Cambridge University Press, 1984.

Perelman, Michael. *Classical Political Economy: Primitive Accumulation and the Social Division of Labor*. London: F. Pinter; Totoway, New Jersey: Rowman and Allanheld, 1983.

Pieper, Joseph. *Leisure: the Basis of Culture*. Translated by Alexander Dru. New York: New American Library, 1952.

Pius XI, *Quadragesimo Anno*. In *Seven Great Encyclicals*. William J. Gibbons, ed. Glen Rock, New Jersey: Paulist Press, 1963, pp. 125-168.

Plant, Raymond. *Hegel: An Introduction*. 2nd ed. Oxford: Basil Blackwell, 1983.

Raphael, D.D. *Adam Smith*. Oxford: Oxford University Press, 1985.

Reisman, D.A. *Adam Smith's Sociological Economics*. New York: Barnes and Noble Books, 1976.

Rendtorff, Trutz. *Ethik: Grundelemente, Methodologie und Konkretionen einer ethischen Theologie*. 2 Bände. Stuttgart: Verlag W. Kohlhammer, 1980-1981.

Rich, Arthur. *Mitbestimmung in der Industrie: Probleme, Modelle, Kritische Beurteilung—Eine sozialethische Orientierung*. Zürich: Buchclub "Ex Libris," 1973.

Riedel, Manfred. *Between Tradition and Revolution: The Hegelian Transformation of Political Philosophy*. Cambridge: Cambridge University Press, 1984.

Robinson, Joan. *Economic Philosophy*. Harmondsworth: Penguin, 1964.

Roll, Eric. *A History of Economic Thought*. Fourth ed. Homewood, IL: Richard D. Irwin, Inc., 1974.

Rovira, German. *Das Persönlichkeitsrecht auf Arbeit: Trinitarische Konstitution und personale Berufung*. Salzburg: Universitätsverlag Anton Pustet, 1978.

Ruhnau, Clemens. *Der Katholizismus in der sozialen Bewährung: Die Einheit theologischen und sozialethischen Denkens im Werk Heinrich Peschs*. Abhandlungen zur Sozialethik, Bd. 18. Hrsg. von Wilhelm Weber und Anton Rauscher. Paderborn: Ferdinand Schöningh, 1980.

Savary, Louis. *Man: His World and His Work*. New York: Paulist Press, 1967.

Schiller, Friederich. *On the Aesthetic Education of Man: In a Series of Letters*. Translated and introduced by Reginald Small. New York: Frederick Ungar Publishing Co., 1954.

Schmidt, Alfred. *The Concept of Nature in Marx*. Translated by Ben Fowkes. London: NLB, 1971.

Schrecker, Paul. *Work and History: An Essay on the Structure of Civilization*. Princeton: Princeton University Press, 1948.

Schumpeter, Joseph A. *Capitalism, Socialism and Democracy*. New York: Harper and Row, 1976.

________. *History of Economic Analysis*. Ed. by Elizabeth Boody Schumpeter. New York: Oxford University Press, 1954.

Schütz, Alfred. *The Phenomenology of the Social World*. Translated by George Walsh and Frederick Lehnert. Evanston: Northwestern University Press, 1967.

Schwarte, Johannes. *Gustav Gundlach, S.J. (1892-1963): Massgeblicher Repräsentant der katholischen Soziallehre während der Pontifikate Pius' XI und Pius' XII*. Abhandlungen zur Sozialethik, Bd. 9. Hrsg. von Wilhelm Weber und Anton Rauscher. München: Verlag Ferdinand Schöningh, 1975.

Schweickart, David. *Capitalism or Worker Control? An Ethical and Economic Appraisal*. New York: Praeger, 1980.

Shapiro, Ian. *The Evolution of Rights in Liberal Theory: An Essay in Critical Anthropology*. Cambridge: Cambridge University Press, 1986.

Sheridan, Alan. *Michel Foucault: The Will to Truth*. London: Tavistock Publications, 1980.

Simmel, Georg. *The Philosophy of Money*. Translated by Tom Bottomore and David Frisby. Boston and London: Routledge and Kegan Paul, 1978.

Simon, Yves. *Work, Society and Culture*. Ed. by Vukan Kuic. New York: Fordham University Press, 1971.

Skinner, Andrew S. and Wilson, Thomas, eds. *Essays on Adam Smith. The Glasgow Edition of the Works and Correspondence of Adam Smith*. Oxford: Clarendon Press, 1975.

Sohn-Rethel, Alfred. *Intellectual and Manual Labour: A Critique of Epistemology*. London: Macmillan, 1978.

Spiegel, Henry William. *The Growth of Economic Thought*. Revised ed. Durham, N.C.: Duke University Press, 1983.

Spiro, Herbert J. *The Politics of German Codetermination*. Cambridge: Harvard University Press, 1958.

Stegmann, Franz Joseph. *Der soziale Katholizismus und die Mitbestimmung in Deutschland: Von Beginn der Industrialisierung bis zum Jahre 1933*. Beiträge zur Katholizismusforschung. Hrsg. von Anton Rauscher. München: Verlag Ferdinand Schoeningh, 1974.

Stern, J.P. *Ernst Juenger*. New Haven: Yale Unversity Press, 1953.

Steuart, James. *An Inquiry into the Principles of Political Oeconomy*. 2 vols. Ed. by Andrew S. Skinner. Published for the Scottish Economic Society. Edinburgh: Oliver and Boyd, 1966.

Stigler, George J. *Essays in the History of Economics*. Chicago: University of Chicago Press, 1965.

Strauss, Leo. *Natural Right and History*. Chicago: University of Chicago Press, 1953.

Strieder, Jakob und Messner, Johannes, Hrsg. *Sektion für Sozial- und Wirtschaftswissenschaft der Görres-Gesellschaft*. Heft 2. Paderborn: Verlag von Ferdinand Schöningh, 1931.

Sturmthal, Adolf. *Workers Councils: A Study of Workplace Organization on Both Sides of the Iron Curtain*. Harvard: Harvard University Press, 1964.

Tawney, R.H. *Religion and the Rise of Capitalism: A Historical Study*. Holland Memorial Lectures, 1922. New York: Harcourt Brace and Company, 1926.

Taylor, Charles. *Hegel and Modern Society*. Cambridge: Cambridge University Press, 1979.

Taylor, W.L. *Frances Hutcheson and David Hume as Predecessors of Adam Smith*. Durham, North Carolina: Duke University Press, 1965.

Thal, Peter et al., Hrsg. *Adam Smith Gestern und Heute: 200 Jahre "Reichtum der Nationen."* Berlin: Akademie-Verlag, 1976.

Thomas Aquinas, *Summa Theologica*. Complete English Edition in Five Volumes. Translated by Fathers of the English Dominican Province. New York: Benziger Brothers, 1948; reprint ed., Westminster, Maryland: Christian Classics, 1981.

Thomas, Henk and Logan, Chris. *Mondragon: An Economic Analysis*. London: George Allen and Unwin, 1982.

Thompson, John B. and Held, David, eds. *Habermas: Critical Debates*. Cambridge: The MIT Press, 1982.

Tilgher, Adriano. *Work: What It Has Meant to Men Through the Ages*. Translated by Dorothy Canfield Fisher. New York: Harcourt Brace and Company, 1930; reprint ed., Harcourt Brace Jovanovich, 1977.

Todd, John M., ed. *Work—Christian Thought and Practice: A Symposium*. London: Darton, Longman and Todd, 1960.

Trenck, Hedwig v. d. *Der Arbeitslohn in Ricardos "Grundsätzen der Volkswirtschaft und Besteuerung*." Inaugural-Dissertation zur Erlangung der staatswissenschaftlichen Doktorwürde der Albertus Universität zu Königsberg in Preussen. Königsberg in Preussen: Buchdrückerei Richard Lankeit, 1927.

Tucker, Robert C. *Philosophy and Myth in Karl Marx*. Cambridge: Cambridge University Press, 1961.

Unger, Robeto Mangabeira. *Knowledge and Politics*. New York: Free Press, 1975.

Vanek, Jaroslav. *The General Theory of Labor-Managed Market Economies*. Ithaca: Cornell University Press, 1970.

________. *The Labor-Managed Economy: Essays*. Ithaca: Cornell University Press, 1977.

________. *The Participatory Economy: An Evolutionary Hypothesis and a Strategy for Development*. Ithaca: Cornell University Press, 1971.

________, ed. *Self-Management: Economic Liberation of Man*. Harmondsworth: Penguin, 1975.

Venable, Vernon. *Human Nature: The Marxian View*. New York: Alfred A. Knopf, 1945.

Verene, Donald Phillip, ed. *Hegel's Social and Political Thought: The Philosophy of Objective Spirit*. Papers of 1976 Meeting of the Hegel Society. Atlantic Highlands, New Jersey: Humanities Press, 1980.

Vetter, Heinz O. *Mitbestimmung—Idee, Wege, Ziel: Beiträge zur Gesellschaftspolitik, 1969-1979*. Hrsg. von Hans O. Hemmer. Schriftenreihe Hans-Böckler Stiftung, 9. Köln: Bund Verlag, 1979.

Viner, Jacob. *The Role of Providence in the Social Order: An Essay in Intellectual History*. Jayne Lectures for 1966. The American Philosophical Society. Princeton: Princeton University Press, 1972.

Völkerling, Fritz. *Der deutsche Kathedersozialismus*. Berlin: Verlag die Wirtschaft, 1959.

Volkmann, G.; Zachert, U.; and Eitel, L., Hrsg. *Mitbestimmung in Unternehmen und Betrieb*. WSI-Studeien zur Wirtschafts- und Sozialforschung. Köln: Bund-Verlag, 1981.

Walsh, W.H. *Hegelian Ethics*. London: Macmillan, 1969.

Weber, Heinrich und Tischleder, Peter. *Handbuch der Sozialethik*. Vol. 1: *Wirtschaftsethik*. Essen: G.D. Baedecker, 1931.

Weber, Max. *Economy and Society: An Outline of Interpretive Sociology*. 2 vols. Ed. by Guenther Roth and Claus Wittich. Berkeley and Los Angeles: University of California Press, 1978.

________. *The Protestant Ethic and the Spirit of Capitalism*. Translated by Talcott Parsons. New York: Scribners, 1958.

Weisskopf, Walter A. *Alienation and Economics*. New York: E.P. Dutton and Co., 1971.

Welty, Eberhard. *Vom Sinn und Wert der menschlichen Arbeit: Aus der Gedankenwelt des heiligen Thomas von Aquin*. Heidelberg: F.H. Kerle Verlag, 1946.

West, E.G. *Adam Smith*. New Rochelle, New York: Arlington House, 1969.

Wilber, Charles K. and Jameson, Kenneth P. *An Inquiry into the Poverty of Economics*. Notre Dame: University of Notre Dame Press, 1984.

Wolin, Sheldon. *Politics and Vision: Continuity and Change in Western Political Thought*. Boston: Little, Brown and Co., 1960.

Wood, Allen W. *Karl Marx*. London: Routledge and Kegan Paul, 1981.

Wood, John Cunningham, ed. *Adam Smith: Critical Assessments*. 4 vols. London: Croom Helm, 1984.

________, ed. *Alfred Marshall: Critical Assessments*. 4 vols. London: Croom Helm, 1982.

________, ed. *David Ricardo: Critical Assessments*. 4 vols. London: Croom Helm, 1985.

Zwerdling, Daniel. *Workplace Democracy: A Guide to Workplace Ownership, Participation and Self-Management Experiments in the United States and Europe*. New York: Harper and Row, 1980.

Secondary Sources: Articles

Alkon, Paul K. "Johnson and Chronology." In *Greene Centennial Studies: Essays Presented to Donald Greene in the Centennial Year of the University of Southern California*. Ed. by Paul J. Korshin and Robert R. Allen. Charlottesville: University Press of Virginia, 1984.

Anderson, Pauline Relyea. "Gustav von Schmoller." In *Some Historians of Modern Europe*, pp. 415-443. Ed. by Bernadotte E. Schmitt. Chicago: University of Chicago Press, 1942.

Attfield, Robin. "Work and the Human Essence." *Journal of Applied Philosophy* Vol. 1, No. 1 (1984): 141-150.

Aubert, Jean-Marie. "Human Work and Christian Faith." *Theology Digest* Vol. 30, No. 1 (Spring, 1982): 7-12.

Balbaky, E. Mary Lou and Mills, D. Quinn. "Buy Out: Hyatt-Clark Industries, Inc. Parts A and B." Harvard Business School, 1983.

Barta, Russell. "Work: In Search of New Meanings." *Chicago Studies* Vol. 23, No. 2 (August, 1984): 155-168.

Bauer, Clemens. "Wandlungen der sozialpolitischen Ideenwelt in deutschen Katholizismus des 19. Jahrhunderts." In *Sektion für Sozial- und Wirtschaftswissenschaft der Goerres-Gesellschaft*, pp. 11-46. Hrsg. von J. Strieder and J. Messner.

Baumol, William J. "Smith versus Marx on Business Morality and the Social Interest." In *Adam Smith and the Wealth of Nations*, pp. 111-122. Ed. by Fred R. Glahe.

Benhabib, Seyla. "Obligation, Contract and Exchange: On the Significance of Hegel's Abstract Right." In *The State and Civil Society: Studies in Hegel's Political Philosophy*, pp. 159-177. Ed. by Z.A. Pelczynski.

Bernstein, J.M. "From Self-consciousness to Community: Act and Recognition in the Master-Slave Relationship." In *The State and Civil Society: Studies in Hegel's Political Philosophy*, pp. 14-39. Ed. by Z.A. Pelczynski.

Bieler, André. "Work and Religion: A Protestant Outlook." *Labour and Society* Vol. 8, No. 1 (January-March, 1983): 75-102.

Billet, Leonard. "Justice, Liberty, and Economy." In *Adam Smith and the Wealth of Nations*, pp. 83-110. Ed. by F.R. Glahe.

Blandy, R. "Marshall on Human Capital: A Note." In *Adam Smith: Critical Assessments*, Vol. III, pp. 433-435. Ed. by J.C. Wood.

Blaug, Mark. "The Economics of Education in English Classical Political Economy: A Re-examination." In *Essays on Adam Smith*, pp. 568-599. Ed. by Andrew S. Skinner and Thomas Wilson.

Blumberg, P. "Alienation and Participation: Conclusions." In *Self-Management: Economic Liberation of Man*, pp. 324-338. Ed. by Jaroslav Vanek.

Bowley, Marian. "Some Aspects of the Treatment of Capital in *The Wealth of Nations*." In *Essays on Adam Smith*, pp. 361-376. Ed. by Andrew S. Skinner and Thomas Wilson.

________. "Some Seventeenth Century Contributions to the Theory of Value." *Economica* Vol. 30, New Series (May, 1963): 122-139.

Brakelmann, Gunther. "Arbeit." In *Christlicher Glaube in moderner Gesellschaft*, Teilband 8: *Enzyklopädische Bibliothek*, pp. 99-135. Hrsg. von Fr. Böckle, F.X. Kaufmann, K. Rahner, B. Welte. Freiburg: Herder, 1980.

________. "Arbeit: Humanisierung der industriellen Arbeitswelt" In *Theologische Realenzyklopädie*, Bd. 3, pp. 657-669. In Gemeinschaft mit H.R. Balz. Hrsg. von G. Krause und G. Müller. Berlin: Walter de Gruyter, 1978.

Brems, H. "Ricardo's Long-Run Equilibrium." In *David Ricardo: Critical Assessments*, Volume III, pp. 295-311. Ed. by J.C. Wood.

Briefs, Goetz. "Pesch and his Contemporaries: Nationalökonomie vs. Contemporary Economic Theories." *Review of Social Economy* Vol. XLI, No. 3 (Dec., 1983): 235-245.

Brinkmann, Carl. "Schmollers Gerechtigkeit." *Schmollers Jahrbuch* 62. Jahrgang, 2. Halbband (1938): 437-447.

Bruyn, S.T. "A New Direction for Community Development in the United States." *The Social Report* (Boston College Program in Social Economy and Social Policy) Vol. V, No. 1 (Feb., 1985): 1-4.

Burgalassi, Silvano. "Towards a Theology of Man as Worker." In *Work and Religion*, Ed. by Gregory Baum. *Concilium* 131 (1980): 103-116.

Burke, Richard. "'Work' and 'Play'." *Ethics* Vol. 82 (Oct, 1971-1972): 33-47.

Cable, John R., and Fitzroy, Felix R. "Productive Efficiency Incentives and Employee Participation: Some Evidence from West German Experience." *Kyklos* 33, Fasc. 1 (1980): 100-121.

Cahill, Edward, S.J. "The Catholic Social Movement: Historical Aspects." In *Readings in Moral Theology No. 5: Official Catholic Social Teaching*. Ed. by Charles E. Curran and Richard A. McCormick, S.J. New York: Paulist Press, 1986.

Canadian Bishops' Commission. "Alternatives to Present Economic Structures." *Origins: National Catholic Documentary Service* 12, No. 33 (January 27, 1983): 521-527.

Cannan, Edwin, ed. "Editor's Introduction." In *An Inquiry into the Nature and Causes of the Wealth of Nations*, by Adam Smith. New York: The Modern Library, 1965.

Chamley, Paul. "La Doctrine économique de Hegel et la Conception hégélienne du Travail." *Hegel Studien*, Beiheft 4 (1965): 147-159.

________. "Les Origines de la Pensée économique de Hegel." *Hegel Studien* Bd. 3 (1965): 225-261.

Chasse, John Dennis. "Marshall, the Human Agent and Economic Growth: Wants and Activities Revisited." *History of Political Economy* Vol. 16, No. 3 (1984): 381-405.

Checkland, S.G. "Marshall and the Wages-Wealth Paradox." In *Alfred Marshall: Critical Assessments*, Vol III, pp. 338-343. Ed. by J.C. Wood.

Chenu, M.D. "Arbeit." In *Historisches Wörterbuch der Philosophie*, Bd. 1, Sp. 480-482. Hrsg. von J. Ritter. Basel: Schwabe und Co., 1971.

________. "Work." In *Sacramentum Mundi*, Vol. 6, pp. 368-373. Ed. by Karl Rahner. New York: Herder and Herder, 1970.

Christie, Drew. "Recent Calls for Economic Democracy." *Ethics* 95, No. 1 (October, 1984): 112-128.

Coats, A.W. "Adam Smith's Conception of Self-Interest in Economic and Political Affairs." In *Adam Smith: Critical Assessments*, Vol. II, pp. 135-139. Ed. by J.C. Wood.

_______. "Alfred Marshall and Richard T. Ely: Some Unpublished Letters." In *Alfred Marshall: Critical Assessments*, Vol. IV, pp. 114-118. Ed. by J.C. Wood.

_______. "Changing Attitudes to Labor in the Mid-Eighteenth Century." *Economic History Review*, XI (1958): 35-51.

Cohen, G.A. "The Labour Theory of Value and the Concept of Exploitation." *Philosophy and Public Affairs* 8, No. 4 (Summer, 1979): 338-360.

_______. "Marx's Dialectic of Labor." *Philosophy and Public Affairs* 3, No. 3 (Spring, 1974): 235-261.

Cole, G.D.H. "Collectivism, Syndicalism and Guilds." In *Self-Management: Economic Liberation of Man*, pp. 62-68. Ed. by Jaroslav Vanek.

Coleman, D.C. "Labour in the English Economy of the Seventeenth Century." *Economic History Review*, Second Series, Vol. VIII, No. 3 (1956): 280-295.

Coleman, Jules L. "Economics and the Law: A Critical Review of the Foundations of the Economic Approach to Law." *Ethics* 94, No. 4 (July, 1984): 649-679.

Colm, Gerhard. "Adolf Wagner." In *Encyclopaedia of the Social Sciences*, Vol. 15, pp. 320-321. New York: Macmillan, 1934.

Constantin, Charles. "The Puritan Ethic and the Dignity of Labor: Hierarchy vs. Equality." *Journal of the History of Ideas* 40 (1979): 543-561.

Conze, Werner. "Arbeiter." In *Geschichtliche Grundbegriffe: Historisches Lexikon zur politisch-sozialen Sprache in Deutschland*. Bd. A-D, pp. 216-243. Hrsg. von Otto Brunner, Werner Conze, and Reinhart Koselleck. Stuttgart: Ernst Klett Verlag, 1972.

_______. "Arbeiter, Arbeiterfrage." In *Historisches Wörterbuch der Philosophie*, Bd. 1, Sp. 487-489. Hrsg. von J. Ritter. Basel: Schwabe und Co., 1971.

Conze, Werner and Riedel, Manfred. "Arbeit." In *Geschichtliche Grundbegriffe: Historisches Lexikon zur politisch-sozialen Sprache in Deutschland*. Bd. A-D, pp. 154-215. Hrsg. von Otto Brunner, Werner Conze, and Reinhart Koselleck. Stuttgart: Ernst Klett Verlag, 1972.

Cormie, Lee. "Work and Salvation." In *Work and Religion*, Ed. by Gregory Baum. *Concilium* 131 (1980): 129-136.

Curran, Charles E. "The Changing Anthropological Bases of Catholic Social Ethics." In *Moral Theology: A Continuing Journey*, pp. 173-208. University of Notre Dame Press, 1982.

Dautzenberg, Gerhard. "Biblische Perspektiven zum Problemfeld Arbeit." In *Handbuch der Christlichen Ethik*, Bd. 2, pp. 344-352. 2. Auflage. Hrsg. von Anselm Hertz, Wilhelm Korff, Trutz Rendtorff und Hermann Ringeling. Freiburg und Gütersloh: Herder und Verlagshaus Gerd Mohn, 1978.

Davis, J.S. "Adam Smith and the Human Stomach." In *Adam Smith: Critical Assessments*, Vol. II, pp. 38-46. Ed. by J.C. Wood.

Deutscher Gewerkschaftsbund. "Codetermination in the Federal Republic of Germany." In *Workers' Control*, pp. 194-210. Ed. by G. Hunnius, G.O. Gerson and J. Case.

Dobb, Maurice. "Ricardo and Adam Smith." In *Essays on Adam Smith*, pp. 324-335. Ed. by Andrew S. Skinner and Thomas Wilson.

Douglas, Paul H. "Smith's Theory of Value and Distribution." In *Adam Smith: 1776-1926*, pp. 77-115. J.M. Clark, et al., contributors.

Dupré, Louis. "Marx's Idea of Alienation Revisited." *Man and His World* 14 (1981): 387-410.

Ellerman, David. "Capitalism and Workers' Self-Management." In *Self-Management: Economic Liberation of Man*. pp. 145-158. Ed. by J. Vanek.

_______. "The Employment Relation, Property Rights and Organizational Democracy." In *Organizational Democracy and Political Processes*, pp. 265-278. Ed. by C. Crouch and F. Heller. Chichester, New York: John Wiley and Sons, 1983.

_______. "Theory of Legal Structure: Worker Cooperatives." *Journal of Economic Issues* XVIII, No. 3 (Sept, 1984): 861-891.

_______. "Workers' Cooperatives: The Question of Legal Structure." In *Worker Cooperatives in America*. Ed. by Robert Jackall and Henry M. Levin. Berkeley: University of California Press, 1984.

Elwert, Georg; Evers, Hans-Dieter; und Wilkens, Werner. "Die Suche nach Sicherheit: Kombinierte Productionsformen im sogenannten Informellen Sektor," *Zeitschrift für Soziologie* 12, Heft 4 (Okt., 1983), pp. 281-296.

Eucken, Walter. "Die Überwindung des Historismus." *Schmollers Jahrbuch* 62. Jahrgang, 1. Halbband (1938): 191-214.

Fischer, Wolfram. "Gustav Schmoller." In *International Encyclopedia of the Social Sciences*, Vol. 14, pp. 60-63. New York: Macmillan and the Free Press, 1968.

Foucault, Michel. "Discourse on Language." In *The Archaeology of Knowledge*. Translated by A.M. Sheridan Smith. New York: Harper and Row, 1972.

Fraser, Nancy. "Michel Foucault: A Young Conservative?" *Ethics* 96 (October, 1985): 165-184.

Freyer, Hans. "Zur Philosophie der Technik." *Blätter für deutsche Philosophie* 3 (1929-1930): 192-201.

Fry, G.K. "The Marshallian School and the Role of the State." In *Alfred Marshall: Critical Assessments*, Vol. IV, pp. 287-301. Ed. by J.C. Wood.

Fürstenberg, F. "West-German Experience with Industrial Democracy." *Annals of the American Academy of Political and Social Science* 431, *Industrial Democracy in International Perspective* (May, 1977): 44-53.

Gee, J.M.A. "Adam Smith's Social Welfare Function." In *Adam Smith: Critical Assessments*, Vol. IV, pp. 84-97. Ed. by J.C. Wood.

Giddens, Anthony. "Labour and Interaction." In *Habermas: Critical Debates*. Ed. by John B. Thompson and David Held. Cambridge: The MIT Press, 1982.

Gill, Emily R. "Civic Virtue, Adam Smith and the Division of Labor." Address at the Annual Meeting of the American Political Science Association, 1984.

Gordon, Colin. "Afterword." In *Power/Knowledge: Selected Interviews and Other Writings, 1972-1977—Michel Foucault*, pp. 229-259. Brighton, England: Harvester Press, 1980.

Gorz, André. "Strategy for Labor." In *Self-Governing Socialism*, Vol. 2, pp. 112-122. Ed. by B. Horvat et al. White Plains, New York: International Arts and Sciences Press, 1975.

Gough, Ian. "Marx's Theory of Productive and Unproductive Labour." *New Left Review* 76 (Nov.-Dec., 1972): 47-72.

Gramp, W.D. "Adam Smith and the Economic Man." In *Adam Smith: Critical Assessments*, Vol. I, pp. 250-272. Ed. by J.C. Wood.

Gremmels, Christian. "Fragen der Arbeit und der Arbeitswelt." In *Handbuch der Christlichen Ethik*, Bd. 2, pp. 362-370. 2. Auflage. Hrsg. von Anselm Hertz, Wilhelm Korff, Trutz Rendtorff und Hermann Ringeling. Freiburg und Gütersloh: Herder und Verlagshaus Gerd Mohn, 1978.

Groenewegen, Peter D. "History and Political Economy: Smith, Marx and Marshall." *Australian Economic Papers* 21, No. 38 (June, 1982): 1-17.

Groethuysen, Bernhard. "Die Idee der Arbeit in der bürgerlichen Welt- und Lebensanschauung." *Blätter für deutsche Philosophie* Bd. 3, Heft 2 (1929-1930): 216-220.

Gundlach, Gustav. "Solidarist Economics." *Social Order* Vol. 1, No. 4 (April, 1951): 181-185.

Habermas, Jürgen. "Hannah Arendt's Communications Concept of Power." *Social Research* 44 (1977): 3-24.

_______. "A Reply to My Critics." In *Habermas: Critical Debates*. John B. Thompson and David Held, eds. Cambridge: MIT Press, 1982.

Haines, Walter W. "The Psychoeconomics of Human Needs: Maslow's Hierarchy and Marshall's Organic Growth." *Journal of Behavioral Economics* Vol. 11, No. 2 (Winter, 1982): 97-121.

Harpham, E.J. "Liberalism, Civic Humanism and the Case of Adam Smith." *American Political Science Review* 78 (1984): 764-774.

Harrison, Royden. "Two Early Articles by Alfred Marshall." *The Economic Journal* LXXIII (Sept., 1963): 422-430.

Hartmann, Klaus. "Towards a New Systematic Reading of Hegel's *Philosophy of Right*." In *The State and Civil Society: Studies in Hegel's Political Philosophy*, pp. 114-135. Ed. by Z.A. Pelczynski.

Heilbroner, Robert L. "The Paradox of Progress: Decline and Decay in the *Wealth of Nations*." In *Essays on Adam Smith*, pp. 524-539. Ed. by Andrew S. Skinner and Thomas Wilson.

Heiman, G. "The Sources and Significance of Hegel's Corporate Doctrine." In *Hegel's Political Philosophy—Problems and Perspectives: A Collection of New Essays*, pp. 111-135. Ed. by Z.A. Pelczynski.

Hengsbach, Friedhelm. "The Church and the Right to Work." *Concilium* 160 (1982): 40-49.

Henry, J.F. "Comment: Ideology in the Ricardo Debate." In *David Ricardo: Critical Assessments*, Vol. IV, pp. 203-205. Ed. by J.C. Wood.

Hicks, J.R. "Edgeworth, Marshall and the Indeterminateness of Wages." In *Alfred Marshall: Critical Assessments*, Vol. III, pp. 96-110. Ed. by J.C. Wood.

Hodgson, G. "Economic Pluralism and Self-Management." From the Fourth International Conference on the Economics of Self-Management. Liege, Belgium; July 15-17, 1985.

Hollander, J.H. "David Ricardo: A Centenary Estimate." *Johns Hopkins University Studies in Historical and Political Science* Vol. XXVIII, No. 4 (1910): 453-589.

Hollenbach, David. "Unemployment and Jobs: A Theological and Ethical Perspective." In *Catholic Social Teaching and the United States Economy: Working Papers for a Bishops' Pastoral*, pp. 110-138. Ed. by John W. Houck and Oliver F. Williams. Washington, D.C.: University Press of America, 1984.

Honecker, Martin. "Arbeit: 18.-20. Jahrhundert." In *Theologische Realenzykopädie*, Bd. 3, pp. 639-657. In Gemeinschaft mit H.R. Balz hrsg. von Gerhard Krause und Gerhard Müller. Berlin: Walter de Gruyter, 1978.

Horvat, Branko. "The Theory of [the] Worker-Managed Firm Revisited: A Short Essay in the Sociology of Knowledge." Yale University, Institute of Social and Policy Studies. 22 pp. 7. December, 1984.

Ilting, K.-H. "Hegel's Concept of the State and Marx's Early Critique." In *The State and Civil Society: Studies in Hegel's Political Philosophy*, pp. 93-113. Ed. by Z.A. Pelczynski.

________. "The Structure of Hegel's *Philosophy of Right*." In *Hegel's Political Philosophy—Problems and Perspectives: A Collection of New Essays*. pp. 90-110. Ed. by Z.A. Pelczynski.

Jacobs, Jane. "Cities and the Wealth of Nations." *Atlantic Monthly* 253 (March, 1984): 41-66.

________. "The Dynamic of Decline." *Atlantic Monthly* 253 (April, 1984): 98-114.

Jensen, M.C. and Meckling, W.H. "Rights and Production Functions: An Application to Labor-Managed Firms and Codetermination." *The Journal of Business* 52 (1979): 469-506.

John XXIII. "Mater et Magistra: On Christianity and Social Progress." In *Seven Great Encyclicals*. Glen Rock, New Jersey: Paulist Press, 1963.

________. "Pacem in Terris: On Peace on Earth." In *Seven Great Encyclicals*. Glen Rock, New Jersey: Paulist Press, 1963.

Kantor, Rosabeth Moss. "Work in a New America." *Daedalus* 107 (Winter, 1978): 47-78.

Kaushil, S. "The Case of Adam Smith's Value Analysis." In *Adam Smith: Critical Assessments*, Vol. III, pp. 275-286. Ed. by J.C. Wood.

________. "The Case of David Ricardo's Value Analysis." In *David Ricardo: Critical Assessments*, Vol. III, pp. 335-348. Ed. by J.C. Wood.

Kelly, George Armstrong. "Notes on Hegel's 'Lordship and Bondage'." In *Hegel: A Collection of Critical Essays*. Ed. by Alasdair MacIntyre. Garden City, New York: Doubleday and Company, 1972.

Keynes, J.M. "Alfred Marshall, 1842-1924." In *Alfred Marshall: Critical Assessments*, Vol. I, pp. 7-65. Ed. by J.C. Wood.

Kiker, B.F. "Marshall on Human Capital: Comment." In *Alfred Marshall: Critical Assessments*, Vol. III, pp. 441-443. Ed. by J.C. Wood.

Klüber, Franz. "Das Naturrecht der Mitbestimmung." In *Normen der Gesellschaft*, pp. 92-112. Hrsg. von H. Achinger et al.

Knauer, James T. "Rethinking Arendt's *Vita Activa:* Toward a Theory of Democratic Praxis." *Praxis International* 5 (July, 1985): 185-194.

Knoll, August M. "Karl von Vogelsang und der Standegedanke." In *Sektion für Sozial- und Wirtschaftswissenschaft der Görres-Gesellschaft*, pp. 64-85. Hrsg. von J. Strieder und J. Messner.

Kofler, Leo. "Das Prinzip der Arbeit in der Marxschen und in der Gehlenschen Anthropologie." *Schmollers Jahrbuch* 78. Jahrgang, 1. Halbband (1958): 71-86.

Kolm, Serge-Christophe. "Altruism and Efficiency." *Ethics* 94, No. 1 (Oct., 1983): 18-65.

Kromphardt, Wilhelm. "Die Überwindung der Klassenkämpfe nach Gustav von Schmoller." *Schmollers Jahrbuch* 62. Jahrgang, 2. Halbband (1938): 717-732.

Krüger, F. "Die Arbeit als philosophisches Problem." *Blätter für deutsche Philosophie* Bd. 3, Heft 2 (1929-1930): 159-192.

Krüger, H.J. "Arbeit." In *Historisches Wörterbuch der Philosophie*, Bd. 1, Sp. 482-487. Hrsg. von J. Ritter.

Kuczynski, Jürgen. "Werner Sombart." In *International Encyclopedia of the Social Sciences*, Vol. 15, pp. 57-59. New York: Macmillan and the Free Press, 1968.

Kurth, E.A. "Codetermination." In *New Catholic Encyclopedia*, Vol. 3, pp. 976-977.

Lakebrink, Bernhard. "Geist und Arbeit im Denken Hegels." *Philosophisches Jahrbuch* 70 (1962-1963): 98-108.

Lamb, R. "Adam Smith's Concept of Alienation." In *Adam Smith: Critical Assessments*, Vol. I, pp. 478-488. Ed. by J.C. Wood.

Lane, Frederic C. "Some Heirs of Gustav Schmoller." In *Architects and Craftsmen in History*, pp. 9-39. Festschrift für Abbott Payson Usher. Ed. by Joseph T. Lambio. Veröffentlichungen der List Gesellschaft E.V., Bd. 2. Tübingen: J.C.B. Mohr (Paul Siebeck), 1956.

Lane, Robert. "From Political to Industrial Democracy." 3. December, 1984 revision of Yale University, Institute for Social and Policy Studies, Working Paper No. 1020, July, 1984.

Lauer, Quentin. "Co-Management in Germany: Toward Industrial Democracy." *Social Order* 1 (January, 1951): 11-22.

Laum, Bernhard. "Entgegnung zu Euckens Aufsatz." *Schmollers Jahrbuch* 62. Jahrgang, 1. Halbband (1938): 215-220.

Leo XIII, "Rerum Novarum: On the Condition of Labor." in *Seven Great Encyclicals*. Glen Rock, New Jersey: Paulist Press, 1963.

Levinas, Emmanuel. "Ethics and Infinity." *Cross Currents* (Summer, 1984): 191-203.

Lewis, T.J. "Adam Smith: The Labor Market as the Basis of Natural Right." In *Adam Smith: Critical Assessments*, Vol. I, pp. 656-679. Ed. by J.C. Wood.

Lonergan, Bernard. "Healing and Creating in History." In *Three Lectures*, pp. 55-68. Edited and introduced by R. Eric O'Connor. Thomas More Institute Papers, 75. Montreal: Thomas More Institute for Adult Education, 1975.

Lukens, Nancy. "Adam von Trott: Resistance and Contemplation—A Work Ethic in Nazi Germany." In *Human Responses to the Holocaust: Perpetrators and Victims, Bystanders and Resisters*. Ed. by Michael Ryan. New York: Edwin Mellen Press, 1981.

MacIntyre, Alasdair. "The Magic of the Pronoun 'My'." (A review of B. Williams's *Moral Luck*.) *Ethics* 94 (Oct., 1983): 113-125.

Marcuse, Herbert. "Über die philosophischen Grundlagen des wirtschaftswissenschaftlichen Arbeitsbegriff." *Archiv für Sozialwissenschaft und Sozialpolitik* 69 (1933): 257-292.

Marković, Mihailo. "Philosophical Foundations of the Idea of Self-Management." In *Self-Governing Socialism: A Reader*, Vol. 1, pp. 327-350. Ed. by Branko Horvat, Mihailo Marković and Rudi Supek.

Markovits, Andrei S. "The Legacy of Liberalism and Collectivism in the Labor Movement: A Tense but Fruitful Compromise for Model Germany." In *The Political Economy of West Germany: Modell Deutschland*, pp. 141-181. Ed. by Andrei S. Markovits. New York: Praeger, 1982.

McCain, Roger A. "The Economics of a Labor-Managed Enterprise in the Short Run: An 'Implicit Contracts' Approach." In *Advances in the Economic Analysis of Participatory an Labor-Managed Firms*, Vol. 1, pp. 41-53. Greenwich, Connecticut: JAI Press, Inc., 1985.

________. "A Theory of Codetermination." *Zeitschrift für Nationalökonomie* 40 (1980): 65-90.

McNulty, Paul J. "Adam Smith's Concept of Labor." *Journal of the History of Ideas* XXXIV (1973): 345-366.

________. "A Note on the Division of Labor in Plato and Smith." In *Adam Smith: Critical Assessments*, Vol. III, pp. 348-353. Ed. by J.C. Wood.

McNulty, T. Michael. "Economics and Explanation." *Proceedings of the Jesuit Philosophical Association* (1985): 20-36.

McPherson, Michael. "Efficiency and Liberty in the Productive Enterprise: Recent Works in the Economics of Work Organization." *Philosophy and Public Affairs* 12, No. 4 (Fall, 1983): 354-368.

McWilliams-Tullberg, R. "Marshall's Tendency to Socialism.: In *Alfred Marshall: Critical Assessments*, Vol. I, pp. 374-408.

Messner, Johannes. "Freedom as a Principle of Social Order." *Modern Schoolman* 28, No. 2 (Jan., 1951): 97-110.

_______. "Zum Begriff der sozialen Gerechtigkeit." In *Sektion für Sozial- und Wirtschaftswissenschaft der Goerres-Gesellschaft*, pp. 416-435. Hrsg. von J. Strieder and J. Messner.

_______. "Zur philosophischen und theologischen Begründung des Solidarismus." In *Normen der Gesellschaft*, pp. 72-91. Hrsg. von Hans Achinger, Ludwig Preller und Hermann Josef Wallraff.

Meyer-Faje, Arnold. "Versuch zur Strukturierung einer Anthropologie wirtschaftlicher Tätigkeit." *Die Mitarbeit: Zeitschrift zur Gesellschafts- und Kulturpolitik* 32. Jahrgang, Heft 2 (Juli, 1983): 83-95.

Meyer, Gerhard. "Adolf Wagner." *International Encyclopedia of the Social Sciences*, Vol. 16, pp. 429-432. New York: Macmillan and the Free Press, 1968.

Mieth, Dietmar. "Solidarity and the Right to Work." *Concilium* 160 (1982): 58-65.

Mill, John Stuart. "The Classes of Labour." In *Dissertations and Discussions: Political, Philosophical and Historical*, Vol. 2, pp. 181-217. London: John W. Parker and Son, 1859.

Mönch, Hermann. "Über einige Ordnungsgedanken im Werke Gustav von Schmollers." *Schmollers Jahrbuch* 62. Jahrgang, 2. Halbband (1938): 474-496.

Moltmann, Jürgen. "The Right to Meaningful Work." In *On Human Dignity: Political Theology and Ethics*, pp. 37-58. Translated and introduced by M. Douglas Meeks, 1984.

Moody, J.N. and Jostock, P. "Social Movements, Catholic." *New Catholic Encyclopedia*, Vol. 13, pp. 321-326.

Morrow, Glenn R. "Adam Smith: Moralist and Philosopher." In *Adam Smith, 1776-1926*. J.M. Clark et al., contributors.

Moser, Simon. "Zum philosophischen und sozialwissenschaftlichen Begriff der Arbeit." *Archiv für Rechts- und Sozialphilosophie* 50 (1964): 87-103.

Mueller, Franz H. "Social Economics: The Perspective of Pesch and Solidarism." *Review of Social Economy* XXXV, No. 3 (Dec., 1977): 293-297.

Mueller, Max. "Freedom—Philosophical." *Sacramentum Mundi*, Vol. 2, pp. 352-361. New York: Herder and Herder, 1968.

Mulcahy, Richard E. "Economic Freedom in Pesch." *Social Order* 1, No. 4 (April, 1951): 161-168.

Murphy, James. "Adam Smith, Karl Marx and the Division of Labor." Department of Political Science, Yale University, 15. January, 1986.

________. "Automation and Labor: Neoclassical and Marxian Perspectives." Department of Political Science, Yale University, 1. July, 1985.

O'Connor, J. "Smith and Marshall on the Individual's Supply of Labor: A Note." In *Alfred Marshall: Critical Assessments*, Vol. III, pp. 399-404. Ed. by J.C. Wood.

Onyewuenyi, Innocent. "The Realization of Ethicality Through Labor by Karl Marx." *Indian Philosophical Quarterly* 9, No. 2 (Jan., 1982): 139-145.

Parsons, Talcott. "Economics and Sociology: Marshall in Relation to the Thought of his Time." In *Alfred Marshall: Critical Assessments*, Vol. I, pp. 209-231. Ed. by J.C. Wood.

________. "Service." In *Encyclopaedia of the Social Sciences*. Vol. 13, pp. 673-674. New York: Macmillan, 1934.

________. "Wants and Needs in Marshall." *Alfred Marshall: Critical Assessments*, Vol. I, pp. 179-208. Ed. by J.C. Wood.

Pawlas, Andreas. "Gewissen und ökonomisches Handeln." *Die Mitarbeit: Zeitschrift zur Gesellschafts- und Kulturpolitik* 32. Jahrgang, Heft 2 (Juli, 1983): 96-109.

Petridis, A. "Alfred Marshall's Attitudes to and Economic Analysis of Trade Unions: A Case of Anomalies in a Competitive System." In *Alfred Marshall: Critical Assessments*, Vol. III, pp. 480-507. Ed. by J.C. Wood.

Petrovic, Gajo. "Praxis." In *A Dictionary of Marxist Thought*, pp. 384-389. Ed. by Thomas Bottomore. Oxford: Blackwell, 1983.

Pfister, Bernhard. "Schmoller." In *Staatslexikon*, 1932 ed., Bd. 4, Sp. 1282-1287.

Philippovich, Eugen v. "Das Eindringen der sozialpolitischen Ideen in die Literatur." In *Die Entwicklung der deutschen Volkswirtschaftslehre im neunzehnten Jahrhundert*, Zweiter Teil, pp. 1-51. S.P. Altmann et al., Mitarbeiter. Leipzig: Verlag von Duncker und Humblot, 1908.

Piana, Giannino. "Human Work: Blessing and/or Curse?" *Concilium* 160 (1982): 66-71.

Pieper, August. "Franz Hitze und die Korporationsidee." In *Sektion für Sozial- und Wirtschaftswissenschaft der Goerres-Gesellschaft*, pp. 86-98. Hrsg. von J. Strieder and J. Messner.

Pigou, A.C. "Review of the Fifth Edition of Marshall's Principles of Economics." In *Alfred Marshall: Critical Assessments*, Vol. II, pp. 54-57. Ed. by J.C. Wood.

Plant, Raymond. "Hegel and Political Economy—I." *New Left Review* No. 103 (1977): 79-92.

_______. "Hegel and Political Economy—II." *New Left Review* No. 104 (1977): 103-113.

_______. "Hegel on Identity and Legitimation." In *The State and Civil Society: Studies in Hegel's Political Philosophy*, pp. 227-243. Ed. by Z.A. Pelczynski.

Rahner, Karl. "Theological Remarks on the Problem of Leisure." In *Theological Investigations*, Vol. 4, pp. 368-390. Translated by Kevin Smith. Baltimore: Helicon Press, 1966.

_______. "The Theology of Power." In *Theological Investigations*, Vol. 4, pp. 391-409. Translated by Kevin Smith. Baltimore: Helicon Press, 1966.

_______. "The Theology of the Symbol." In *Theological Investigations*, Vol. 4, pp. 221-252. Translated by Kevin Smith. Baltimore: Helicon Press, 1966.

Rajchman, John. "Ethics after Foucault." *Social Text* 13/14 (Winter/Spring, 1986): 165-183.

Ranadive, K.R. "The Relevance of the Ricardian System." In *David Ricardo: Critical Assessments*, Vol. III, pp. 554-575. Ed. by J.C. Wood.

_______. "The *Wealth of Nations:* The Vision and the Conceptualization." In *Adam Smith: Critical Assessments*, Vol. II, pp. 244-270. Ed. by J.C. Wood.

Régnier, Jérome. "'Right to Work': Evolving Christian Thought." *Theology Digest* 30, No. 1 (Spring, 1982): 12-14.

Reder, Fritz. "Die Grundlagen der Wirtschaftstheorie Heinrich Pesch S.J.: Das Verhältnis von Wirtschaft und Moral." *Jahrbücher für Nationalökonomie* 129 (1928): 747-760.

Rees, Albert. "Compensating Wage Differentials." In *Essays on Adam Smith*, pp. 336-349. Ed. by Andrew S. Skinner and Thomas Wilson.

Reibel, R. "The Workingman's Production Association or the Republic of the Workshop." In *Self-Management: The Economic Liberation of Man*, pp. 39-46. Ed. by Jaroslav Vanek.

Reich, M. "Empirical and Ideological Elements in the Decline of Ricardian Economics." In *David Ricardo: Critical Assessments*, Vol. III, pp. 650-671. Ed. by J.C. Wood.

Richards, Howard. "A Normative Economics." *The Center Magazine* 9 (May/June, 1976): 36-51.

Riedel, Manfred. "Arbeit." In *Handbuch philosophischer Grundbegriffe*, Bd. 1, pp. 125-141. Hrsg. von H. Krings, H.M. Baumgartner und C. Wild. München: Kösel, 1973.

Rimlinger, G.V. "Smith and the Merits of the Poor." In *Adam Smith: Critical Assessments*, Vol. IV, pp. 224-232. Ed. by J.C. Wood.

Robertson, H.M. "Alfred Marshall." In *Alfred Marshall: Critical Assessments*, Vol. I, pp. 442-452. Ed. by J.C. Wood.

Rombach, Heinrich. "Leistung und Musse." In *Christlicher Glaube in moderner Gesellschaft. Enzyklopädische Bibliothek.* Teilband 8, pp. 39-69. Hrsg. von Fr. Böckle, F.X. Kaufmann, K. Rahner, B. Welte. Freiburg: Herder, 1980.

Rosenberg, N. "Adam Smith on the Division of Labour: Two Views or One?" In *Adam Smith: Critical Assessments*, Vol. III, pp. 171-183. Ed. by J.C. Wood.

________. "Some Institutional Aspects of the *Wealth of Nations*." *In Adam Smith: Critical Assessments*, Vol. II, pp. 105-120.

Rothschild-Whitt, Joyce. "Who Will Benefit from ESOPs?" *Labor Research Review* 1, no. 6 (Spring, 1985): 71-80.

Ryan, Alan. "Hegel on Work, Ownership and Citizenship." In *The State and Civil Society: Studies in Hegel's Political Philosophy*, pp. 178-196. Ed. by Z.A. Pelczynski.

Samuelson, Paul, "A Modern Treatment of the Ricardian Economy I and II." In *David Ricardo: Critical Assessments*, Vol. III, pp. 179-219. Ed. by J.C. Wood.

Schauer, Helmut. "Critique of Co-Determination." In *Workers' Control*, pp. 210-224. Ed. by G. Hunnius, G.O. Garson and J. Case. New York: Random House, 1973.

Scheler, Max. "Arbeit und Ethik." In *Gesammelte Werke*, Bd. 1: *Frühe Schriften*, pp. 163-195. Hrsg. von Maria Scheler und Manfred S. Frings. Bern und München: Francke Verlag, 1971.

Schmidt, Alfred. "Praxis." In *Handbuch Philosophischer Begriffe*, Bd. II, pp. 1107-1138. Hrsg. von Hermann Frings, Hans Michael Baumgartner und Christoph Wild. München: Kösel-Verlag, 1973.

Schröter, Manfred. "Buchbesprechungen: Zur neueren Literatur über Philosophie der Technik." *Blätter für deutsche Philosophie*, Bd. 3, Heft 2 (1929-1930): 221-225.

Schüssler Fiorenza, Francis. "Religious Beliefs and Praxis: Reflections on Catholic Theological Views of Work." In *Work and Religion*, Ed. by Gregory Baum. *Concilium* 131 (1980): 92-102.

________. "Work and Critical Theology." In *A Matter of Dignity: Inquiries into the Humanization of Work*. Ed. by W.J. Heisler and John W. Houck. Notre Dame: University of Notre Dame Press, 1977.

Schumpeter, Joseph A. "Alfred Marshall's *Principles:* A Semi-Centennial Appraisal." In *Alfred Marshall: Critical Assessments*, Vol. II, pp. 100-113. Ed. by J.C. Wood.

________. "Gustav v. Schmoller und die Probleme von heute." *Schmollers* Jahrbuch für Gesetzgebung, Verwaltung und Volkswirtschaft im *Deutschen Reiche* 50 (1926): 337-388.

Schütz, Alfred. "On Multiple Realities." In *Collected Papers*, Vol. 1: *The Problem of Social Reality*. Ed. by Maurice Natanson. The Hague: Martinus Nijhoff, 1967.

Selucky, R. "Marxism and Self-Management." In *Self-Management: Economic Liberation of Man*. pp. 47-61. Ed. by Jaroslav Vanek.

Shove, G.F. "The Place of Marshall's *Principles* in the Development of Economic Theory." *The Economic Journal* 52 (Dec., 1942): 294-329. Ed. by J.C. Wood.

Siebert, Rudolf. "Work and Religion in Hegel's Thought." In *Work and Religion*. Ed. by Gregory Baum. *Concilium* 131 (1980): 117-128.

Simitis, S. "Workers' Participation in the Enterprise—Transcending Company Law." *Modern Law Review* 38, No. 1 (Jan., 1975): 1-23.

Simonds, A.P. "Marxism and Morals: Review Essay." *Ethics* 93, No. 4 (July, 1983): 792-799.

Sirianni, Carmen. "Production and Power in a Classless Society: A Critical Analysis of the Utopian Dimensions of Marxist Theory." *Socialist Review*, Issue 59, 11, No. 5 (Sept.-Oct., 1981): 33-82.

Slott, Mike. "The Case Against Worker Ownership." *Labor Research Review*, 1, No. 6 (Spring, 1985): 83-97.

Spengler, J.J. "Adam Smith on Human Capital." In *Adam Smith: Critical Assessments*, Vol. III, pp. 491-497. Ed. by J.C. Wood.

Spiethoff, Arthur, Hrsg. *Gustav von Schmoller und die deutsche geschichtliche Volkswirtschaftslehre*. *Schmollers Jahrbuch*. 62. Jahrgang, 2. Halbband (1938).

Stigler, G.J. "Alfred Marshall's Lectures on Progress and Poverty." In *Alfred Marshall: Critical Assessments*, Vol. IV, pp. 146-192. Ed. by J.C. Wood.

Summers, Clyde W. "Co-determination in the United States: A Projection of Problems and Potentials." *Journal of Comparative Corporate Law and Securities Regulation* 4 (1982): 155-191.

________. "Worker Participation in the U.S. and West Germany: A Comparative Study from an American Perspective." *American Journal of Comparative Law* 28, No. 3 (Summer, 1980): 367-392.

Sunoris, Helmut. "Remarks [Concerning the 1976 Decision of the West German Constitutional Court in the Matter of Codetermination]." Yale Law School. 18. Feb., 1983.

Surpin, Rick. "Enterprise Development and Worker Ownership: A Strategy for Community Economic Development." *Working Papers in Community Development*. (Mar. 1984). New York: Community Service Society of New York.

Suter, J.-F. "Burke, Hegel and the French Revolution." In *Hegel's Political Philosophy—Problems and Perspectives: A Collection of New Essays*, pp. 52-72. Ed. by Z.A. Pelczynski.

Svejnar, Jan. "Codetermination and Productivity: Empirical Evidence from the Federal Republic of Germany." In *Participatory and Self-Managed Firms*, pp. 199-212. Ed. by Derek C. Jones and Jan Svejnar. Lexington, Massachusetts: Lexington Books, 1982.

Swinney, Dan. "Worker Ownership: A Tactic for Labor." *Labor Research Review* 6 (Spring, 1985): 99-112.

Tabb, J.Y. and Goldfarb, A. "The Typology of Systems of 'Participation'." In *Self-Governing Socialism*, Vol. 2, pp. 61-63. Ed. by B. Horvat, M. Marković and R. Supek.

Thomas, Keith. "Work and Leisure in Pre-Industrial Society." *Past and Present* 29 (Dec., 1964): 50-62.

Thompson, E.P. "Time, Work-Discipline and Industrial Capitalism." *Past and Present* 38 (Dec., 1967): 56-97.

Titmuss, R.M. "The Social Division of Welfare." Some Reflections on the Search for Equality." In *Essays on the "Welfare State,"* pp. 34-55. 3rd ed. By R.M. Titmuss. London: Allen and Unwin, 1976.

United States Congress. Senate. *Administrative Committee, National Catholic War Council: Bishops' Program of Social Reconstruction.* S. Doc. 79, 76th Cong., 1st sess, 1939.

Vahanian, Gregory. "Utopia as Ethic of Leisure." *Humanities* 8 (Nov., 1972): 347-365.

Vanek, Jan. "The Worker-Managed Enterprise as an Institution." In *Self-Management: Economic Liberation of Man*, pp. 257-270. Ed. by Jaroslav Vanek.

Vanek, Jaroslav. "Decentralization Under Workers' Management: A Theoretical Approach." In *Self-Management: Economic Liberation of Man*, pp. 352-365.

________. Interview. Cornell University. Ithaca, New York. 8. January, 1986.

________. "The Kingdom of God is Among You." 45 pp. Jan., 1985.

________. "Lifestyle Toward Eternity—Result of Round the World Mission: Kenya/Sudan." 12 pp. n.d.

________. "Technology, Self-Management and the Social Order: An Inquiry with a Concrete Dimension." *Socialism in the World* 6, No. 33 (1982): 59-65.

Vanek, Jaroslav and Skalicky, Karel. "The Principle of Participation and the Renewal of Society." *Didaskalia* 5 (1975): 283-297.

Vatican II, Ecumenical Council, "Gaudium et Spes." In *The Documents of Vatican II*, pp. 199-308. Ed. by Walter M. Abbot and Joseph Gallagher. London: Geoffrey Chapman, 1967.

Veblen, Thorstein. "Gustav Schmoller's Economics." In *The Place of Science in Modern Civilization and Other Essays*, pp. 252-278. New York: B.W. Huebsch, 1919.

ver Eecke, W. "Ethics in Economics: From Classical Economics to Neo-liberalism." *Philosophy and Social Criticism* (1983): 147-168.

________. "Relation Between Economics and Politics in Hegel." In *Hegel's Social and Political Thought*, pp. 91-101. Ed. by D.P. Verene.

Viner, Jacob. "Adam Smith and Laissez Faire." In *Adam Smith: 1776-1926*, pp. 116-155. J.M. Clark et al., contributors.

________. "Marshall's Economics in Relation to the Man and his Times." In *Alfred Marshall: Critical Assessments*, Vol. I, pp. 241-255. Ed. by J.C. Wood.

Vleugels, Wilhelm. "Gustav von Schmoller und die ethisch-politische Theorie der Volkswirtschaft." *Schmollers Jahrbuch* 62. Jahrgang, 2. Halbband (1938): 420-436.

Volf, Miroslav. "On Human Work: An Evaluation of the Key Ideas of the Encyclical *Laborem Exercens*." *Scottish Journal of Theology* 37 (1984): 65-79.

Wagner, A. "Marshall's Principles of Economics." In *Alfred Marshall: Critical Assessments*, Vol. II, pp. 10-23. Ed. by J.C. Wood.

Walker, D.A. "Marshall on the Long-Run Supply of Labor." In *Alfred Marshall: Critical Assessments*, Vol. III, pp. 527-542. Ed. by J.C. Wood.

________. "Marshall on the Short-run Supply of Labor." In *Alfred Marshall: Critical Assessments*, Vol. III, pp. 543-561. Ed. by J.C. Wood.

Walton, A.S. "Economy, Utility and Community in Hegel's Theory of Civil Society." In *The State and Civil Society: Studies in Hegel's Political Philosophy*, pp. 244-261. Ed. by Z.A. Pelczynski.

________. "Hegel, Utilitarianism and the Common Good." *Ethics* 93, No. 4 (July, 1983): 753-771.

Weakland, Rembert. "Where Does the Economic Pastoral Stand?" *Origins—NC Documentary Service* 13, No. 46 (Ap. 26, 1984): 753-759.

Wearne, Bruce C. "Talcott Parson's Appraisal and Critique of Alfred Marshall." *Social Research* 48, No. 4 (Winter, 1981): 816-851.

Wehrle, Emil. "Genossenschaftliche Selbsthilfe." In *Sektion für Sozial- und Wirtschaftswissenschaft der Goerres-Gesellschaft*, pp. 350-360. Hrsg. von J. Strieder and J. Messner.

Weinberger, Otto. "Heinrich Pesch, S.J.: 1854-1926." *Zeitschrift für die gesamte Staatswissenschaft* 82 (1927): 512-535.

Weizsäcker, Viktor v. "Zum Begriffe 'Arbeit': Eine Habeas Corpus-Akte der Medizin." In *Synopsis: Festgabe für Alfred Weber*. Hrsg. von Edgar Salin. Heidelberg: Verlag Lambert Schneider, 1948.

West, E.G. "Adam Smith and Alienation: 'Wealth Increases, Men Decay'?" In *Essays on Adam Smith*, pp. 540-552. Ed. by Andrew S. Skinner and Thomas Wilson.

________. "Scotland's Resurgent Economist: A Survey of the New Literature on Adam Smith." In *Adam Smith: Critical Assessments*, Vol. IV, pp. 278-301. Ed. by J.C. Wood.

Westermann, Claus. "Work, Civilization and Culture in the Bible." In *Work and Religion*. Ed. by Gregory Baum. *Concilium* 131 (1980): 81-91.

Whitaker, J.K. "Some Neglected Aspects of Alfred Marshall's Economic and Social Thought." In *Alfred Marshall: Critical Assessments*, Vol. I, pp. 453-486. Ed. by J.C. Wood.

Wiles, Richard C. "The Theory of Wages in Later English Mercantilism." *The Economic History Review* 2nd Series, No. 21 (1968): 113-126.

Yenni, Jacques. "Pesch's Goal of the Economy." *Social Order* 1, No. 4 (Apr., 1951): 169-176.

Youngson, A.J. "Marshall on Economic Growth." In *Alfred Marshall: Critical Assessments*, Vol IV, pp. 95-111. Ed. by J.C. Wood.

Zemelka, Stanley A. "Towards a Self-Regulating Economy: Heinrich Pesch and the Changing American Economic System." *Catholicism in Crisis* 2, No. 3 (February, 1984): 25-28.

Zwiedineck-Südenhorst, Otto v. "Gustav von Schmollers Gewerbepolitik." *Schmollers Jahrbuch* 62. Jahrgang, 2. Halbband (1938): 513-535.

Index